Millimeter-Wave/Sub-Terahertz Ultra-Massive MIMO Transmission Technology

Zhen Gao · Ziwei Wan · Yikun Mei · Keke Ying · Kuiyu Wang

Millimeter-Wave/ Sub-Terahertz Ultra-Massive MIMO Transmission Technology

Springer

Zhen Gao
Advanced Research Institute
of Multidisciplinary Science
Beijing Institute of Technology
Beijing, China

Ziwei Wan
Advanced Research Institute
of Multidisciplinary Science
Beijing Institute of Technology
Beijing, China

Yikun Mei
Beijing Institute of Technology
Beijing, China

Keke Ying
Beijing Institute of Technology
Beijing, China

Kuiyu Wang
Beijing Institute of Technology
Beijing, China

ISBN 978-981-99-2387-8 ISBN 978-981-99-2388-5 (eBook)
https://doi.org/10.1007/978-981-99-2388-5

This Springer imprint is published by the registered company Springer Nature Singapore Pte Ltd.
The registered company address is: 152 Beach Road, #21-01/04 Gateway East, Singapore 189721, Singapore

Contents

Acronyms

5G	Fifth-Generation
6G	Sixth-Generation
AI	Artificial Intelligence
AMP	Approximate Message Passing
AoA	Angle of Arrival
AoD	Angle of Departure
ASE	Average Spectrum Efficiency
ASN	Antenna Switching Network
AUD	Active User Detection
AWGN	Additive White Gaussian Noise
B5G	Beyond Fifth-Generation
BD	Block Diagonalization
BER	Bit Error Rate
BS	Base Station
CE	Channel Estimation
CRLB	Cramér-Rao Lower Bound
CS	Compressive Sensing
CSI	Channel State Information
DADD	Data-Aided Decision-Directed
DC	Dual Crossing
DCS	Distributed Compressive Sensing
DFT	Discrete Fourier Transform
DoF	Degree-Of-Freedom
EM	Electromagnetic
ESPRIT	Estimating Signal Parameters via Rotational Invariance Techniques
FD	Full-Dimensional
FFT	Fast Fourier Transformation
GHz	Gigahertz
GMMV	Generalized Multiple-Measurement-Vector
GTTDU	Grouping True-Time Delay Unit
IEEE	Institute of Electrical and Electronics Engineers

IFFT	Inverse Fast Fourier Transformation
IoT	Internet of Things
IUI	Inter-User Interference
KPI	Key Performance Indicator
LoS	Line-of-Sight
LS	Least Square
MIMO	Multiple-Input Multiple-Output
ML	Maximum Likelihood
MMSE	Minimum Mean Squared Error
MMV	Multiple-Measurement-Vector
mmWave	Millimeter-Wave
MPC	Multipath Components
NLoS	Non-Line-of-Sight
NMSE	Normalized Mean Square Error
NOMA	Non-Orthogonal Massive Access
OFDM	Orthogonal Frequency Division Multiplexing
OMP	Orthogonal Matching Pursuit
PSN	Phase Shift Network
QAM	Quadrature Amplitude Modulation
QPSK	Quadrature Phase Shift Keying
RF	Radio Frequency
RIS	Reconfigurable Intelligent Surfaces
SAGIN	Space-Air-Ground Integrated Network
SISO	Single-Input Single-Output
SNR	Signal-to-Noise Ratio
SNS	Spatial Non-Stationary
SSD	Simultaneous Schur Decomposition
SVD	Singular Values Decomposition
SW-OMP	Simultaneous Weighted-Orthogonal Matching Pursuit
Tbps	Terabits per second
TDD	Time Division Duplexing
THz	Terahertz
TLS	Total Least Square
TTDU	True-Time Delay Unit
UAV	Unmanned Aerial Vehicle
UE	User Equipment
ULA	Uniform Linear Array
UM	Ultra-Massive
UPA	Uniform Planar Array
XL	Extra-Large Scale
ZF	Zero-Forcing

Chapter 1
Introduction

After about forty years of leap forward, the mobile communications have greatly reshaped people's life and promoted the development of our society. With the official commercialization of the fifth-generation (5G) technologies in the 2020s, the current mobile networks are basically able to keep up with the versatile demand of users. However, the mobile communication paradigm have gradually expanded from the simple Internet to the Internet of Things (IoT), which requires a plethora of devices to connect, interact, and exchange data anywhere and anytime. Also, as The Times require, various novel concepts and applications, such as ultra-high-definition video, extended reality, holography, digital twin, Metaverse, and so on, have emerged. In order to meet these challenges, the global industry and government have started the research and exploration of the sixth-generation (6G) mobile communication technologies [1, 2]. As shown in Fig. 1.1, compared to the current 5G networks, the key performance indicators (KPIs) of the future 6G networks are expected to significantly increase. For instance, in 6G, the peak data rate, user-experienced data rate, spectral/energy efficiency, supportable mobility, connection density have been improved by about 10–100 times compared to 5G. Fortunately, there are many advanced techniques which support the realization of such KPIs. Specifically, high-frequency communications, such as millimeter-wave (mmWave) or terahertz (THz) communications, can provide adequate spectrum resources to meet the KPI requirements like peak data rate and user-experienced data rate [3, 4]; massive multiple-input multi-output (MIMO) and even ultra-massive MIMO can further improve the system spectral efficiency [3, 5]; hybrid beamforming technique is expected to trade off the system performance, hardware complexity and energy consumption [6]; the novel network architectures, such as space-air-ground integrated network (SAGIN), are able to enhance the coverage of communication service, even in the remote areas or high-mobility scenarios [7]; and massive IoT, non-orthogonal massive access (NOMA), and reconfigurable intelligent surface (RIS) techniques are capable of improving both connection density and coverage [8–10]. The above emerging technologies and applications provide unlimited possibilities for the future

Z. Gao et al., *Millimeter-Wave/Sub-Terahertz Ultra-Massive MIMO Transmission Technology*, https://doi.org/10.1007/978-981-99-2388-5_1

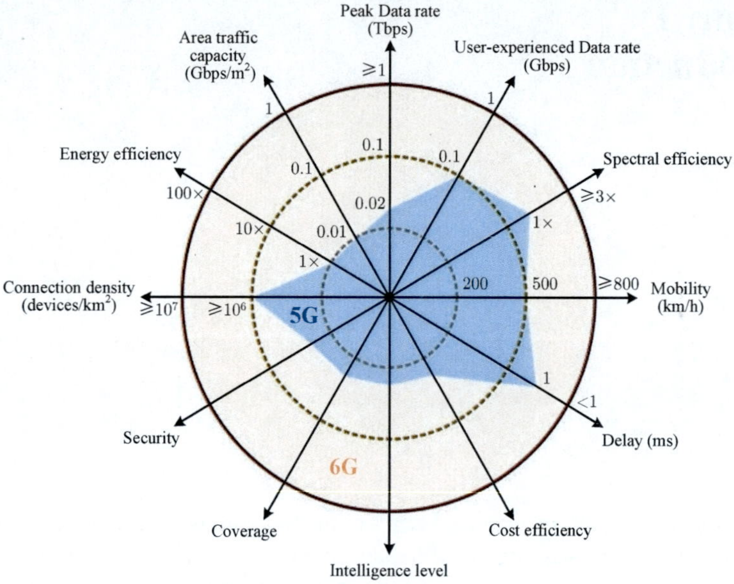

Fig. 1.1 Key performance indicators of 6G communication networks [1].

6G networks, while they also introduce difficult challenges that need to be well studied in the further research.

1.1 MmWave and THz Communications for 6G

The spectrum crunch in the existing sub-6 gigahertz (GHz) frequency bands motivates the future 6G communications to develop available spectrum resources in the higher-frequency bands, such as mmWave and THz bands. As shown in Fig. 1.2, the frequency of mmWave band (which is also called extremely high frequency (EHF) band) ranges from 30 to 300 GHz, and the wavelength of corresponding electromagnetic (EM) signals ranges from 1 to 10 mm. The abundant spectrum resources that have yet been fully utilized in the mmWave band can provide the transmission bandwidth of the order of GHz, which is tens of times of that in current commercial 5G networks [11] with the bandwidth 200 megahertz (MHz) at most. Such a huge advantage on bandwidth resources ensures that the data transmission rate of the mmWave communication systems can reach the order of gigabits per second (Gbps). Furthermore, THz band with a carrier frequency ranging from 0.1 to 10 THz can provide the bandwidth resources far more than the mmWave band. In fact, THz band is the last piece of radio frequency (RF) spectrum, as indicated in Fig. 1.2, and it can support ultra-large bandwidth communications with the peak data rate up to the order of terabits per second (Tbps) [4]. Given its superiority, THz communication

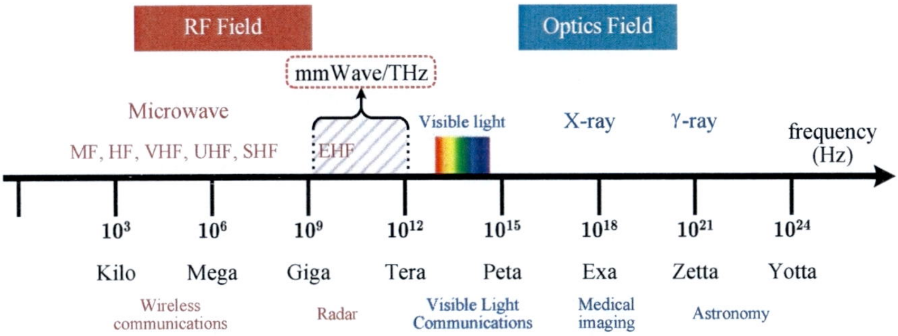

Fig. 1.2 Schematic diagram of radio spectrum and corresponding application scenarios

is considered to be one of the key enabling techniques in 6G networks, and it is expected to be applied in many emerging communication scenarios.

The utilization of mmWave/THz bands also facilitates other advanced physical-layer techniques for wireless communications. Thanks to the extremely short wavelengths of EM signals in the mmWave/THz bands, the antennas operating in the mmWave/THz bands will have very small form factors, and thus a large number of antenna elements can be compactly packaged in a limited physical size, readily realizing the deployment of massive or even ultra-massive MIMO systems [3, 12]. Taking a mmWave uniform linear array (ULA) operating at 60 GHz (corresponding to a wavelength 5 mm) as an example, if the antenna elements are critically spaced, that is to say, the spacing of any two adjacent elements is half wavelength, the size of such ULA is only about 25 cm with 100 antenna elements. On the other hand, the large-scale antenna array is able to provide sufficient array gain for mmWave/THz communications to compensate the low signal-to-noise ratio (SNR) caused by severe path loss in the high-frequency bands [13]. In a nutshell, mmWave/THz communications and massive/ultra-massive MIMO can cooperate well to enhance the performance of communication networks, which lays a solid foundation for our research in this book.

The propagation properties of mmWave/THz bands have also opened up many game-changing opportunities for wireless communications. The EM signals in the mmWave/THz bands will suffer from severe path loss and blockage effect [14]. Particularly, compared to mmWave signals, THz signals may be more vulnerable to the molecule absorption attenuation during propagation. For example, for the THz signals @ 1 THz, the attenuation caused by the absorption of water molecules and oxygen molecules increases by 10–35 dB compared to the mmWave signals @ 26 GHz [15]. Therefore, by taking advantage of the natural anti-interference advantage brought by the propagation properties, the mmWave/THz base station (BS) can reuse the same frequency (within a short distance) to improve the multiplexing capability. This guarantees the feasibility of ultra-dense network (UDN) technique [16]. In addition, the extremely narrow directional beams (i.e., pencil beams) provided by the massive/ultra-massive MIMO can further ensure high security and privacy during mmWave/THz communications.

Due to the aforementioned advantages, the global industry is actively carrying out research work on mmWave/THz communications in order for their upcoming standardization and practical applications. For example, in 2017, the Institute of Electrical and Electronics Engineers (IEEE) released the IEEE 802.15.3d-2017 to standardize consumer wireless communications in the THz frequency band (300 GHz) [17], which is considered to be the first standardization effort for THz communications towards 6G. In 2018, a project called "Spectrum Horizons" was approved by the Federal Communications Commission (FCC), opening the spectrum from 0.95 to 3 THz to encourage the development of new communication technologies and expedite the deployment of new services. At the same time, the academic community is also conducting research work on the specific problems of mmWave/THz massive/ultra-massive MIMO systems, including application scenario planning, hardware design, channel modeling, and signal processing algorithm design [18, 19].

1.2 From Massive MIMO to Ultra-massive MIMO

As the carrier frequency in the mobile communications gradually increases, the form of the antenna array at the transceiver also changes accordingly. For the small-scale MIMO system, the transceiver uses multiple antennas, each equipped with a dedicated RF chain, to send or receive signals at the same frequency resource block. By ensuring a sufficiently low correlation between different antennas, the capacity of the MIMO system can be exponentially increased compared to the traditional single-antenna communication system. Given these advantages, the Long Term Evolution-Advanced (LTE-A) standard in the fourth-generation (4G) considers small-scale MIMO system, supporting transmission with 4–8 antennas. However, it can only provide the spectral efficiency of about 10 bits per second per hertz (bit/s/Hz), which is difficult to keep up with the explosive growth of mobile services [20]. In the 2010s, the concept of massive MIMO came in [13, 21]. By deploying hundreds of or even thousands of antennas at the BS to serve tens of users, massive MIMO is able to counteract the severe path loss of channels [3, 22], achieve finer-grain spatial multiplexing, and thus significantly improve the spectral efficiency. Since then, a series of studies on the theory and application of massive MIMO have been carried out [21, 23, 24]. We present some principle prototype of massive MIMO built by the prestigious research institutions in Fig. 1.3.

As mentioned in previous text, the traditional small-scale MIMO is like to deploy powerful fully-digital architecture, where each antenna requires a dedicated RF chain. However, such complex and power-hungry architecture is not suitable for massive MIMO with hundreds of or even thousands of antennas [28]. In contrast, the hybrid beamforming architecture [3, 6] utilizes the phase shifter network (PSN) and very few RF chains to decompose the fully-digital MIMO processing into a high-dimensional analog beamforming problem and a low-dimensional baseband (digital) processing

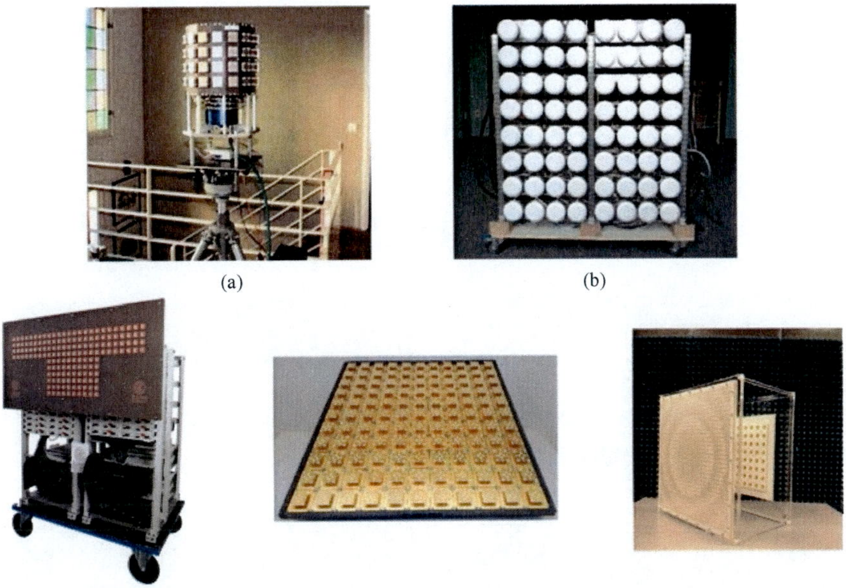

(a) (b) (c) (d) (e)

Fig. 1.3 The prototype of massive MIMO: **a** cylindrical antenna array with 128 antennas built by the Bell Labs and the Linköping University; **b** Argos system built by the Yale University and the Bell Labs [25]; **c** Lund University Massive MIMO (LuMaMi) system built by the Lund University; **d** an 11×11 massive MIMO for indoor applications built by the Leibniz University Hannover (Germany) [26]; and **e** an 8×8 lens antenna array built by the National University of Singapore [27]

problem. In this way, the hybrid beamforming architecture is able to reduce the hardware complexity and power consumption of massive MIMO while reserving its advantages, and thus it is regarded as a promising architecture for mmWave/THz massive MIMO systems. On the other hand, the lens antenna array is also a promising candidate to realize energy-efficient mmWave/THz massive MIMO. The principle of the lens antenna array is to transform the spatial-domain channel into the angular-domain (beamspace) channel via the spatial Fourier transform induced by the EM lens, and then select only a few elements in the beamspace channel for further signal processing, which is, in a way, similar to the hybrid beamforming architecture. Note that the number of effective scatterers in mmWave/THz channels is very limited, so the dimension-reduced beamspace channel may be sufficient to concentrate most of the power of the original channel via appropriate beam selection. Moreover, compared to the hybrid beamforming architecture, the lens antenna array can further improve the energy efficiency of massive MIMO, since the EM lens is passive and it does not require the active PSN. Figure 1.3e shows a prototype of lens antenna array designed and fabricated by the National University of Singapore in 2017 [27].

In recent years, with the development of material and manufacturing technologies, both the physical aperture and the number of antenna elements of massive MIMO

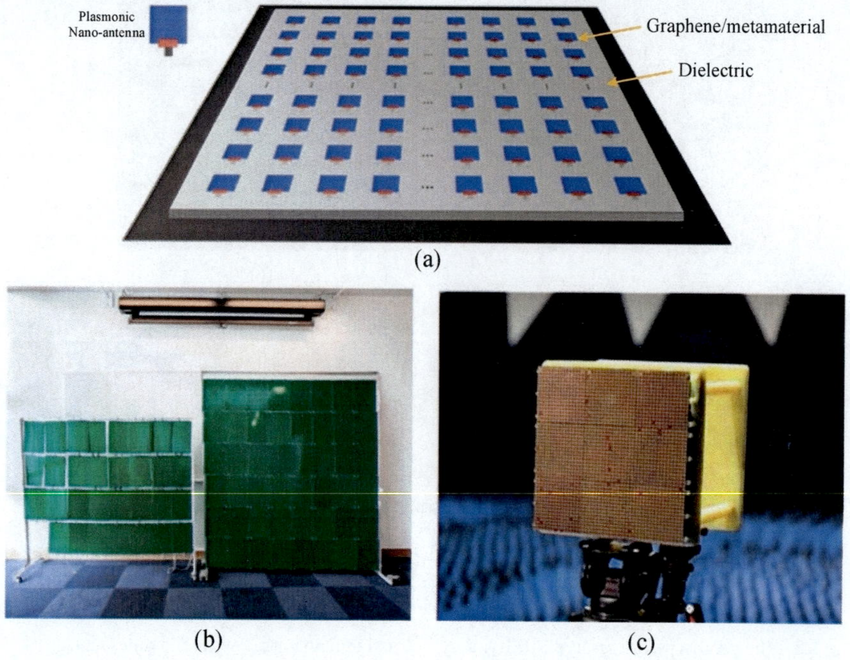

Fig. 1.4 The ultra-massive MIMO: **a** a sketch map of a 1024 × 1024 ultra-massive MIMO @ 0.06–10 THz based on plasmonic nano-antennas made from graphene/metamaterial [5]; **b** RFocus system with 3200 antenna elements built by the Massachusetts Institute of Technology [29]; and **c** an ultra-massive MIMO @ 28 GHz with 2304 antenna elements built by the Tsinghua University [30]

has increased, which lead to the concept of ultra-massive MIMO including both extra-large scale MIMO (XL-MIMO, see Chap. 5) and UM-MIMO (see Chap. 6). It is necessary to clarify that we use the acronym *UM*-MIMO to narrowly refer to a kind of massive MIMO system which suffers from the *squint effect* due to the large signal bandwidth, as detailed in Chap. 6, while we use the term *ultra-massive* MIMO (e.g., in the title of this book) to broadly refer to the evolution of massive MIMO. In Fig. 1.4 we show the sketch map and some existing prototype of ultra-massive MIMO.

The technical evolution from traditional small-scale MIMO to massive MIMO to ultra-massive MIMO is very inspiring and attracts great interest from both industry and academia [31, 32]. However, mmWave/THz massive MIMO and ultra-massive MIMO face serious challenges such as channel modeling, hardware design, physical-layer algorithm design, and so on. There are still many unsolved problems, as will be detailed in the next section.

1.3 Prior Work

The current research work related to this book can be roughly divided into five aspects, namely the channel estimation (CE) for mmWave massive MIMO, the beamforming design for mmWave massive MIMO, the massive MIMO-aided massive IoT access, the near-field CE for mmWave/THz ultra-massive MIMO, and the CE and tracking for high-dynamic time-varying mmWave/THz channels. This section will explain each of them one by one.

1.3.1 Channel Estimation for MmWave Massive MIMO

The existing literature has proposed the codebook-based beam training methods [33–36] and compressive sensing (CS)-based CE schemes [37, 38], to obtain narrowband mmWave channel state information (CSI). Specifically, the beam training scheme was originally used for analog beamforming in the standards such as IEEE 802.11ad [33] and IEEE 802.15.3c [16], where the optimal beam pair was obtained by exhausted search among a predefined codebook to maximize the received SNR. In order to sidestep the high training overhead in tradition beam training methods, the authors of [34] designed a multi-level overlapping beam pattern for low-overhead beam training. These overlapping beam patterns were narrowed step-by-step during the training stage, and finally the optimal finer-grain beam pair would be obtained. Unfortunately, the above beam training schemes [33, 34] only considered analog beamforming under single-stream transmission. For the hybrid beamforming system with multiple RF chains, the authors of [35, 36] proposed beam training solutions with hierarchical multi-beam codebooks, which are the direct extension of the work [34]. On the other hand, by exploiting the inherent angular domain sparsity in mmWave MIMO channels, CS-based low-overhead CE scheme has attracted wide attention [37, 38]. In particular, the authors of [37] modeled the CE problem as a sparse signal recovery problem, and used the orthogonal matching pursuit (OMP) algorithm to estimate sparse mmWave MIMO channels, where a redundant dictionary was also designed to facilitate CS-based CE. The authors of [38] proposed a Bayesian CS-based CE scheme under the impact of hardware impairment. Furthermore, by exploiting the low-rank property of mmWave channels, the authors of [39] proposed a CE scheme based on CANDECOMP/PARAFAC decomposition to further improve the estimation performance.

It is noteworthy that the above literature [34–39] only considered narrowband mmWave MIMO channels with frequency flat fading. In practice, however, the large system bandwidth and multi-path components (MPCs) in mmWave channels will lead to significant delay spread, resulting in frequency selective fading property of mmWave channels [3]. To solve the CE problem under frequency selective fading, the authors of [40] proposed to utilize a distributed grid matching pursuit (DGMP) algorithm to estimate the channels in the MIMO-orthogonal frequency division multiplexing (OFDM) system. Furthermore, the adaptive grid matching pursuit (AGMP) algorithm, as an extension of the DGMP algorithm, was proposed in [41]. The AGMP algorithm uses the adaptive grid matching to reduce the power leakage in the angle domain. The authors of [42] used the OMP algorithm to estimate the sparse mmWave MIMO channels in either the time domain or the frequency domain. The authors of [43] proposed a CE scheme based on the simultaneous weighted-OMP (SW-OMP) algorithm, where the common angular-domain sparsity of channels on different subcarriers was considered. The authors of [44] proposed a block CS-based CE solution, and designed a dedicated training sequences. In [45], the received training signals are represented as high-order tensors using CANDECOMP/PARAFAC decomposition. Based on the low-rank characteristic of mmWave MIMO channels, main channel parameters including delay, angles of departure (AoDs) and angles of arrival (AoAs) are estimated to re-construct the channels. However, most of the CS-based CE schemes use the dictionary composed of discrete AoAs/AoDs grids. This causes the severe mismatch between the dictionary codewords and the AoAs/AoDs in actual mmWave MIMO channels which are continuously distributed, and thus degrades the performance of CS-based CE.

For the mmWave massive MIMO with lens antenna arrays, the authors of [46] decomposed the large-dimensional MIMO channel into several independent single-input single-output (SISO) channels based on the energy focusing property of the lens antenna array, and the cross interference between different SISO channels is ignored or treated as negligible noise. However, this decomposition may suffer from the performance loss due to the severe power leakage caused by the limited angular resolution of the lens antenna array and the continuous distribution of channel AoAs/AoDs. In [47], the CE problem for lens antenna array was formulated as a sparse signal recovery problem, and the OMP algorithm with support detection (SD) was proposed to solve the problem. This idea is further developed in [48] as an OMP algorithm based on dual cross (DC). The literature [47, 48] showed that the CE performance based on the improved OMP algorithms (OMP with SD and OMP with DC) is significantly better than that based on the original OMP algorithm. More sophisticated CE schemes for lens antenna arrays have been proposed [49, 50]. The authors of [49] proposed a model-driven machine learning-based method, where the neural network was applied to unfold the CS-based approximate message passing (AMP) algorithm. Compared to the original AMP algorithm, the deep unfolding of AMP algorithm is able to further de-noise for the coarse estimate of channels and thus the better performance can be achieved. In [50], the channel associated with lens antenna array was treated as the picture information, and an image reconstruction algorithm scheme was applied to accurately estimate the channel. However, the above works [47–50] adopted the large-scale PSN in the lens antenna array to assist CE, which contradicts the idea of using EM lens to replace the PSN in the lens antenna array. Therefore, these works cannot take full advantage of the high energy efficiency of the lens antenna array.

1.3.2 Beamforming Design for MmWave Massive MIMO

As mentioned before, the hybrid beamforming architecture is a mainstream solution to realize mmWave massive MIMO, and the corresponding hybrid beamforming design has become a popular research problem. Many hybrid beamforming schemes have been proposed. In a pioneering paper [6], the authors focused on the sparse property of the mmWave channels and pointed out that the optimal beamforming weight vector can be represented by only a very small number of spatial based vectors. Furthermore, the OMP algorithm was used to respectively design the analog beamformer and the digital beamformer so that their product can mimic the optimal beamforming weight vector. However, [6] does not consider the inter-user interference (IUI) in the multi-user mmWave MIMO, and the applied algorithm has a high computational complexity. The authors of [51] proposed a low-complexity hybrid beamforming design with multiple single-antenna users, which extracted the phase of the conjugate transpose of the channel matrix as the analog beamformer to maximize the array gain, and then obtained the digital beamformer via the zero-forcing (ZF) algorithm. To support multiple multi-antenna users, the authors of [52, 53] proposed a codebook-based low-complexity two-stage hybrid beamforming scheme and a heuristic hybrid beamforming scheme, respectively. The traditional block diagonalization (BD) beamforming scheme was improved in [54] as a hybrid BD beamforming scheme that supports multi-user and multi-stream communications.

The broadband hybrid beamforming design under frequency selective fading channels has also been proposed [55–59]. With the extra dimension in the frequency domain, the broadband hybrid beamforming scheme is more difficult than the narrowband case. Worse still, the PSN in the hybrid architecture is generally the frequency-flat device, and it is impossible to assign different analog beamformers to different subcarrier channels. This introduces new constraints for the broadband hybrid beamforming design. Fortunately, the authors of [55] theoretically proved the optimality of a frequency-flat analog beamformer in the case of the ideally sparse mmWave channels, providing a solid basis for the use of the frequency-flat PSN in the broadband hybrid beamforming design. Given the large dimension of the channels associated with the mmWave massive MIMO, how to reduce the feedback overhead of CSI for beamforming design is also an essential problem. The authors of [56] proposed a broadband hybrid beamforming scheme based on a limited feedback codebook to reduce the CSI feedback overhead. In [57], a broadband beamforming scheme under imperfect CSI feedback was proposed. In addition, in order to support multi-users and multi-stream communications, the authors of [58] proposed a broadband beamforming scheme based on the reciprocity between uplink and downlink channels, which can strike a balance between the algorithm complexity and communication performance. The authors of [59] proposed a low-complexity simultaneous greedy hybrid precoding (S-GHP) scheme, where the beamforming weight matrices at the BS and at the users were designed simultaneously so that they are respectively equivalent to the conjugate transposes of the left and right singular matrices of the MIMO channel matrices.

At the same time, scholars have improved the hardware architecture for the hybrid beamforming [60–62]. The traditional PSN adopts a fully-connected architecture where each RF chain will be connected to all antennas through phase shifters as many as the number of antennas. As a result, the total number of phase shifters required in the fully-connected architecture is the number of RF chains multiplied by the number of antennas, which may be unaffordable in practice. To tackle this problem, the literature [60, 61] considered the hybrid architecture based on the partially-connected structure, namely, each RF chain will only be connected to a part of the antennas (i.e., a sub-array). In this way the number of phase shifters in PSN will be greatly reduced, thereby further reducing hardware cost and power consumption. However, in [60, 61], the connection scheme between the RF chains and antennas is fixed, which limits the degree-of-freedom (DoF) for beamforming design. The authors of [62] further proposed a dynamic partially-connected structure. By adding a dynamic antenna switching network (ASN) between the RF chains and antennas, the system can dynamically adjust the connection scheme according to the real-time CSI. Simulation results showed that the partially-connected hybrid architecture significantly outperforms the fully-connected hybrid architecture in terms of energy efficiency, and the spectral efficiency of such dynamic architecture is better than that of the fixed partially-connected architecture.

1.3.3 Massive MIMO-Aided Massive IoT Access

The joint active user detection (AUD) and CE problem is essential in massive IoT access. Currently, the sporadic traffic property (i.e., the sparsity of active users) has been exploited to solve the problem. Based on such sparsity, the CS-based grant-free random access schemes have been proposed. In [63], the joint AUD and CE problem was formulated as a dynamic CS problem, and a sequential AMP algorithm was designed to recover time-dependent sparse signals. The authors of [64, 65] proposed the improved AMP and orthogonal AMP (OAMP) algorithms, respectively, to realize joint AUD, CE, and data detection. However, the above massive IoT access solutions [63–65] only considered single-antenna BS, which cannot make full use of the spatial DoF to improve the performance of AUD and CE. Taking the massive MIMO into account at the BS, the massive MIMO-aided massive access has been studied [66–68]. Specifically, the authors of [66] designed a vector AMP algorithm for the joint AUD and CE by using statistical CSI. The belief propagation (BP) theory was applied in [67], where an iterative message passing algorithm was proposed to facilitate unsourced random access systems. While only simple Rayleigh fading channels were considered in the literature [66, 67], the authors of [68] adopted the geometric channel model with spatial sparsity, and propose a hybrid message passing algorithm which combines the variational message passing algorithm and AMP algorithm. In this way, the channel sparsity in both the angle domain and the user domain can be jointly exploited. Unlike the narrowband scenarios in [66–68], the Enhanced Mobile Broadband (eMBB) scenario should be considered in massive

IoT access systems. To this end, the authors of [69] applied OFDM technology to deal with frequency selective fading channels in broadband IoT access. Based on the structured sparsity of massive MIMO channels and the distributed CS (DCS) theory, a generalized multiple measurement vector AMP (GMMV-AMP) algorithm was proposed to perform AUD and CE. Also, several joint AUD and CE schemes based on sparse Bayesian learning (SBL) were investigated [70–72], where the the common row sparsity and the clustered sparse structure of massive MIMO channel matrices were exploited to enhance the performance of AUD and CE.

In addition to the aforementioned Bayesian CS algorithms, the greedy CS algorithms [71–74] and CS algorithms based on optimization [75, 76] are also effective solutions for joint AUD and CE in massive IoT access. In particular, the block OMP (BOMP) algorithm was utilized in [71, 72] to solve the block sparse signal recovery problem in joint AUD and CE. The authors of [73, 74] proposed several algorithms based on subspace pursuit (SP), where the spatio-temporal structure of user activity was used to construct Kronecker-CS problem [74]. In [75], the authors proposed an optimization algorithm based on the Riemann trust region to shrink the search space in the multiple measurement vector (MMV)-CS problem. On this basis, the joint AUD and CE problem was transformed into a low-dimensional optimization problem with the constraint of full-column rank. In [76], the famous alternating direction method of multipliers (ADMM) was applied to the joint AUD and CE in massive IoT access, where the second-order statistics of spatially correlated channels were exploited to solve the sparse MMV recovery problem.

On the other hand, with the development of artificial intelligence (AI), deep learning methods can be applied to massive IoT access, realizing intelligent wireless access [8]. For instance, the authors of [77, 78] proposed model-driven deep learning frameworks based on the AMP algorithm and the ADMM algorithm, respectively, to solve joint AUD and CE for grant-free massive access. Using the long short-term memory (LSTM) framework, a deep learning-based AUD and CE scheme was proposed in [79], where a deep neural network mapped the received signals into the indices of the active users' channels.

1.3.4 Near-Field Channel Estimation for MmWave/THz Ultra-massive MIMO

Basically, the ultra-massive MIMO has a very large physical aperture, which will significantly increase the so-called *Rayleigh distance*. As a consequence, the commonly-adopted far-field assumption in traditional channel models (i.e., the distance between the transceiver is much larger than the Rayleigh distance) is no longer applicable, and thus it should be replaced by the practical near-field assumption (i.e., the distance between the transceiver is within the Rayleigh distance). The research on the ultra-massive MIMO communications under near-field scenarios is still at its infancy. In [80, 81], the authors studied the multi-user scheduling problem for the uplink transmission under near-field spatial non-stationary (SNS) channels, with the assumption of the knowledge of perfect CSI. In fact, the acquisition of CSI in the ultra-massive MIMO under near-field scenarios is a very difficult task. In order to obtain the ultra-massive MIMO channels with SNS property, the dedicated CE schemes have been proposed [82–86]. To be specific, in [82], the authors proposed a two-stage sparse CE scheme exploiting the sparsity of SNS channels in the delay domain. In [84], the delay-domain sparsity of ultra-massive MIMO channels was further characterized via the hierarchically sparse Markov prior probability distribution, and then an iterative SBL scheme was proposed to infer channel vectors and unknown hyperparameters. The authors of [83] proposed to treat the near-field SNS channel as an image, and accordingly designed a model-driven downlink CE scheme by introducing the YOLO (You Only Look Once) neural network that is popular in image processing. The ultra-massive MIMO channel parameters, including the angle, delay, and visible region of each scatterer, can be quickly and accurately estimated with the aid of the advanced YOLO neural network. An interesting work on the near-field CE for the ultra-massive MIMO has been demonstrated in [85, 86], where the authors first constructed a new representation of ultra-massive MIMO channels in the *polar domain*, and then proposed an improved OMP algorithm to estimate near-field channels by exploiting the sparsity in the polar domain. In addition, the authors also analysed the conditional transition between the near-field region and the far-field region, based on which the concept of *hybrid-field* communications is initially proposed [85].

1.3.5 Fast Time-Varying Channel Estimation and Tracking

For the estimation for fast time-varying channels in high dynamic scenarios such as the space-air-ground integrated network (SAGIN) and the vehicle-to-everything (V2X) network, The authors of [87–89] proposed some CE and channel tracking schemes to avoid the unaffordable training overhead caused by the frequent CE operations. Specifically, the authors of [87] proposed a data-aided channel tracking scheme with the utilization of lens antenna array. According to the time-varying angular sparsity of channels, the authors of [88, 89] modeled the change of channel parameters as the first-order auto-regressive process and established the expectation-maximization-based SBL frameworks to estimate and track the channel parameters. Furthermore, in [90, 91], some multi-stage channel parameter estimation schemes (rather than directly estimating the whole MIMO channel matrix) were proposed, which also supported the fast channel tracking in narrowband mmWave MIMO systems. Note that these schemes [87–91] only considered the CE and channel tracking in mmWave communications. As for THz communication systems, a position-assisted channel tracking scheme was proposed in [92], which can predict and track the direction of line-of-sight (LoS) component of the THz massive MIMO channel in real time with less pilot overhead. For the indoor short-range THz communications, the authors of [93] proposed an angle estimation method based on Markov process and

Bayesian inference, where a forward-backward algorithm was used. However, the existing literature [87–93] does not address the CE and tracking associated with the ultra-massive MIMO. The CE and tracking scheme for THz ultra-massive MIMO in high-dynamic scenarios needs to be explored further.

1.4 Book Organization

In this book, we provide a comprehensive overview of the recent advances in mmWave/sub-THz[1] massive/ultra-massive MIMO. We originally propose several promising application scenarios and technical solutions to empower the 6G communications.

This book is organized as follows. Chapter 2 presents a closed-loop CE method for mmWave full-dimension MIMO systems. Chapter 3 investigates the application of CS theory to the CE for the promising mmWave lens antenna array. Chapter 4 presents the hybrid beamforming design for mmWave massive MIMO systems. Chapter 5 studies the joint active user detection and CE in mmWave XL-MIMO systems using CS. Chapter 6 investigates THz UM-MIMO in the space-air-ground integrated network. Chapter 7 further discusses some key challenges and open issues for mmWave/THz massive/ultra-massive MIMO.

[1] Sub-THz generally refers to the frequency band ranging from the higher frequency of mmWave band to the lower frequency of THz band.

Chapter 2
Closed-Loop Sparse Channel Estimation for MmWave Full-Dimensional MIMO Systems

Abstract This chapter considers the channel estimation problem in wideband millimeter-wave full-dimensional MIMO systems and proposes a closed-loop channel estimation method, which exploits the channel sparsity in both angular and delay domains. By carefully designing the transmit precoder and receive combiner during the training stage, the proposed scheme is capable of acquiring the super-resolution estimates of angle-of-arrivals, angle-of-departures, and delays based on the multi-dimensional unitary-estimating signal parameters via rotational invariance techniques algorithm. Simulation results have demonstrated that the proposed closed-loop sparse channel estimation scheme offers more accurate channel estimates than state-of-the-art compressive sensing-based schemes while imposing lower computational complexity and storage requirements.

2.1 Introduction

The millimeter-wave (mmWave) frequency band has abundant spectral resources that can increase system throughput by several orders of magnitude, and its combination with the massive multiple-input multiple-output (MIMO) technology is considered to be a key enabling technology for beyond fifth-generation (B5G) and sixth-generation (6G) mobile communications [16, 94]. To cope with the severe path loss in mmWave communications, massive MIMO with beamforming techniques are usually employed to form high-gain directional beams [22]. Compared with unaffordable fully-digital MIMO [28] employing a dedicated radio frequency (RF) chain for each antenna, hybrid MIMO with a much smaller number of RF chains than that of antennas offers a practical solution by using hybrid analog/digital beamforming [3]. Nonetheless, for hybrid MIMO systems, the limited number of RF chains will lead to higher training overhead to estimate the high-dimensional mmWave MIMO channels. Moreover, the low signal-to-noise ratio (SNR) before beamforming can further degrade the accuracy of channel state information (CSI) acquisition [95]. Therefore, channel estimation (CE) is a challenging problem for hybrid MIMO systems [16].

Existing mmWave massive MIMO CE schemes [34–39] only consider narrowband channels with flat fading, while most CE schemes for wideband frequency-selective

Z. Gao et al., *Millimeter-Wave/Sub-Terahertz Ultra-Massive MIMO Transmission Technology*, https://doi.org/10.1007/978-981-99-2388-5_2

fading channels are based on compressive sensing (CS) framework [40–43]. Considering the continuous distribution of the practical angle of arrivals/angle of departures (AoAs/AoDs), the CS-based CE schemes usually adopt discretized AoAs/AoDs grids for the CS dictionary, therefore, the dismatched model may cause severe degradation of CE performance. In addition, most existing CE schemes only consider uniform linear arrays (ULA), while the practical uniform planar arrays (UPA) with full-dimensional (FD) beamforming capability are seldom investigated. Although [96, 97] have studied the CE problem in mmWave FD-MIMO, they only considered fully-digital MIMO or frequency-flat fading channels. Therefore, these schemes cannot adapt to the wideband mmWave FD-MIMO systems.

This chapter proposes a closed-loop sparse CE scheme by exploiting the angular domain and delay domain sparsity of the multipath components (MPCs) in the mmWave MIMO channel. To exhibit the double sparsity in the angular-delay domain, this chapter considers a mmWave FD-MIMO unmanned aerial vehicle (UAV)-based base station (BS). By leveraging its flexible deployment capability, the UAV-based BS can provide high data-rate enhanced mobile broadband services for user equipment (UE) in hotspot areas [98]. Unlike the widely used terrestrial cellular BS, UAV-based BS usually operates at heights of hundreds of meters, where only a few multipath components corresponding to the main scatterers can be used to establish reliable communication links with ground UEs. Therefore, the space-to-ground mmWave channel exhibits inherent sparsity in both the angular domain and the delay domain, i.e., the double sparsity in the angular-delay domain. The proposed scheme in this chapter can obtain super-resolution estimation of angles/delays based on array signal processing [99, 100] by carefully designing the transmit precoder and the receive combiner, which yields better channel parameter estimation performance than traditional CS-based schemes.

2.2 Overview of the Close-Loop Sparse Channel Estimation

Figure 2.1 shows the flow chart of the proposed closed-loop sparse CE scheme, which includes the stages of downlink CE and subsequent uplink CE. The channel reciprocity in time division duplexing (TDD) systems is utilized and the frame structure in this scheme is further illustrated in Fig. 2.2 [88, 102, 103]. Specifically, at the downlink CE stage, each UE estimates the azimuth and elevation AoAs of MPCs and then feeds back the estimated angles to the BS with limited quantization resolution through the feedback link. At this stage, we design a random transmit precoding matrix at the BS to broadcast the omnidirectional channel sounding signal to all UEs. Meanwhile, each UE designs its receive combining matrix to equivalently visualize the high-dimensional hybrid array into a low-dimensional digital array, which facilitates the use of estimating signal parameters via rotational invariance techniques (ESPRIT) algorithm to estimate the channel parameters.

Similarly, at the uplink CE stage, the BS can use the multi-dimensional unitary ESPRIT (MDU-ESPRIT) algorithm to estimate the azimuth angles, elevation angles,

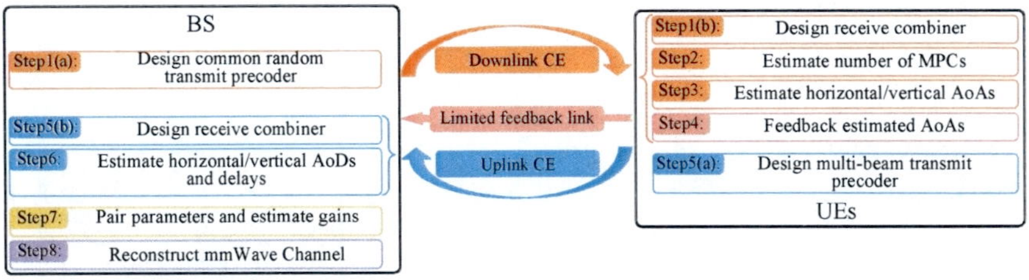

Fig. 2.1 Procedure of the proposed closed-loop sparse CE solution, where the limited feedback is realized via the low-frequency control link [101]

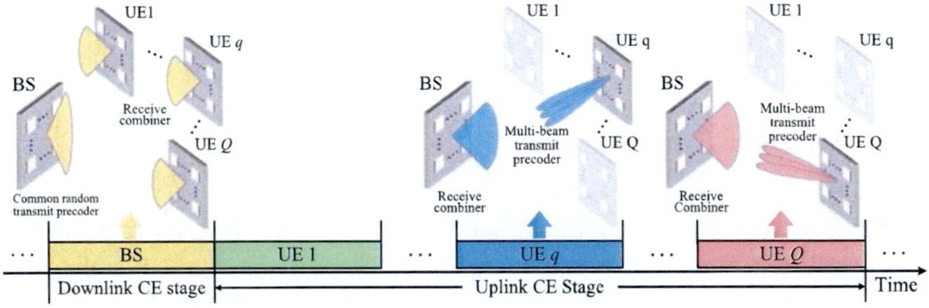

Fig. 2.2 Frame structure of the proposed closed-loop sparse CE solution [101]

and path delays corresponding to different UEs. In addition, by exploiting the channel reciprocity in TDD systems, each UE can utilize the estimated AoAs to design the multi-beam transmit precoding matrix to improve the received SNR at the BS. Next, a maximum likelihood (ML) method is adopted at the BS to pair the acquired channel parameters from two stages, and then the channel gain of each MPC can be obtained through the least squares (LS) estimator. Finally, the high-dimensional mmWave MIMO channels associated with each UE can be reconstructed based on the estimated channel parameters above.

2.3 Downlink CE Stage

As shown in Fig. 2.3, we consider a mmWave FD-MIMO system with hybrid analog-digital beamforming, where the BS and each UE are all equipped with UPA. Orthogonal frequency division multiplexing (OFDM) system with K subcarriers is used, K subcarriers, and each subcarrier transmits N_s^d independent signal streams. The BS(UE) employs $N_{\mathrm{BS}} = N_{\mathrm{BS}}^{\mathrm{h}} N_{\mathrm{BS}}^{\mathrm{v}}$ ($N_{\mathrm{UE}} = N_{\mathrm{UE}}^{\mathrm{h}} N_{\mathrm{UE}}^{\mathrm{v}}$) antennas and $N_{\mathrm{BS}}^{\mathrm{RF}} \ll N_{\mathrm{BS}}$($N_{\mathrm{UE}}^{\mathrm{RF}} \ll N_{\mathrm{UE}}$) RF chains, where $N_{\mathrm{BS}}^{\mathrm{h}}$($N_{\mathrm{UE}}^{\mathrm{h}}$) and $N_{\mathrm{BS}}^{\mathrm{v}}$($N_{\mathrm{UE}}^{\mathrm{v}}$) are the numbers of antennas in horizontal and vertical directions at the BS(UE), respectively. The mmWave FD-MIMO-OFDM system utilizes a fully-connected phase shift network (PSN) to form the analog RF network, as shown in the figure.

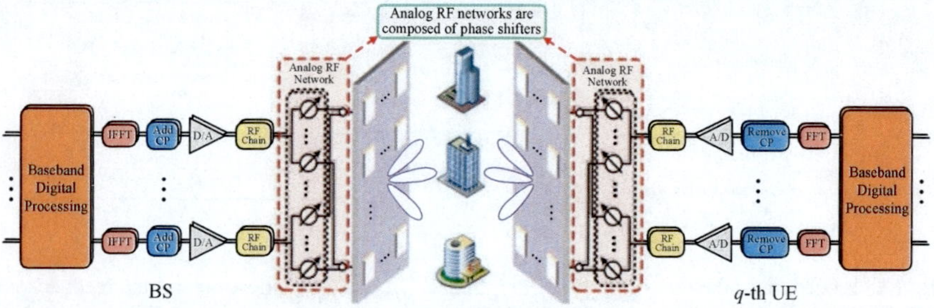

Fig. 2.3 Transceiver diagram of mmWave FD-MIMO-OFDM system with hybrid beamforming architecture, where FFT, IFFT, CP, D/A, and A/D stand for fast Fourier transform, inverse FFT, cyclic prefix, digital-to-analogue, and analogue-to-digital, respectively

2.3.1 Problem Formulation

The downlink CE stage constitutes of N_d time slots, and each time slot contains N_{o}^d OFDM symbols. The signal $\boldsymbol{y}_q[k,i,m] \in \mathbb{C}^{N_{\mathrm{s}}^d}$ received by the q-th UE at the k-th subcarrier of the i-th OFDM symbol in the m-th time slot can be expressed as

$$\boldsymbol{y}_q[k,i,m] = \boldsymbol{W}_{d,q}^{\mathrm{H}}[k,m]\boldsymbol{H}_q[k]\boldsymbol{F}_d[k,m]\boldsymbol{s}[k,i,m] + \boldsymbol{W}_{d,q}^{\mathrm{H}}[k,m]\boldsymbol{n}_q[k,i,m], \tag{2.1}$$

where $1 \le q \le Q$, $0 \le k \le K-1$, $1 \le i \le N_{\mathrm{o}}^d$, and $1 \le m \le N_d$. In Eq. (2.1), the UE's combing matrix can be divided into $\boldsymbol{W}_{d,q}[k,m] = \boldsymbol{W}_{\mathrm{RF},q}[m]\boldsymbol{W}_{\mathrm{BB},q}[k,m] \in \mathbb{C}^{N_{\mathrm{UE}} \times N_{\mathrm{s}}^d}$, where $\boldsymbol{W}_{\mathrm{RF},q}[m] \in \mathbb{C}^{N_{\mathrm{UE}} \times N_{\mathrm{UE}}^{\mathrm{RF}}}$ and $\boldsymbol{W}_{\mathrm{BB},q}[k,m] \in \mathbb{C}^{N_{\mathrm{UE}}^{\mathrm{RF}} \times N_{\mathrm{s}}^d}$ are the analog and digital combining matrices, respectively. Similarly, the BS's transmit precoding matrix can be divided into $\boldsymbol{F}_d[k,m] = \boldsymbol{F}_{\mathrm{RF},d}[m]\boldsymbol{F}_{\mathrm{BB},d}[k,m] \in \mathbb{C}^{N_{\mathrm{BS}} \times N_{\mathrm{s}}^d}$, where $\boldsymbol{F}_{\mathrm{RF},d}[m] \in \mathbb{C}^{N_{\mathrm{BS}} \times N_{\mathrm{BS}}^{\mathrm{RF}}}$ and $\boldsymbol{F}_{\mathrm{BB},d}[k,m] \in \mathbb{C}^{N_{\mathrm{BS}}^{\mathrm{RF}} \times N_{\mathrm{s}}^d}$ are the analog and digital transmit precoding matrices, respectively. Besides, $\boldsymbol{H}_q[k] \in \mathbb{C}^{N_{\mathrm{UE}} \times N_{\mathrm{BS}}}$ is the downlink channel matrix, $\boldsymbol{s}[k,i,m] \in \mathbb{C}^{N_{\mathrm{s}}^d}$ is the transmitted signal with $\mathbb{E}\left(\boldsymbol{s}[k,i,m]\boldsymbol{s}^{\mathrm{H}}[k,i,m]\right) = \frac{1}{N_{\mathrm{s}}^d}\boldsymbol{I}_{N_{\mathrm{s}}^d}$, and $\boldsymbol{n}_q[k,i,m] \in \mathbb{C}^{N_{\mathrm{UE}}}$ is the complex additive white Gaussian noise (AWGN) vector with its covariance matrix given by $\sigma_n^2 \boldsymbol{I}_{N_{\mathrm{UE}}}$, i.e., $\boldsymbol{n}_q[k,i,m] \sim \mathcal{CN}\left(\boldsymbol{0}_{N_{\mathrm{UE}}}, \sigma_n^2 \boldsymbol{I}_{N_{\mathrm{UE}}}\right)$. Constrained by the constant modulus of PSN, the analog precoder and combiner matrices satisfy $\{\boldsymbol{F}_{\mathrm{RF},d}[m]\}_{j_1,j_2} = \frac{1}{\sqrt{N_{\mathrm{BS}}}} e^{\mathrm{j}\vartheta_{1,j_1,j_2}}$ and $\{\boldsymbol{W}_{\mathrm{RF},q}[m]\}_{j_1,j_2} = \frac{1}{\sqrt{N_{\mathrm{UE}}}} e^{\mathrm{j}\vartheta_{2,j_1,j_2}}$, respectively, where $\vartheta_{1,j_1,j_2}, \vartheta_{2,j_1,j_2} \in \mathcal{A}$ and $\mathcal{A}$ represents the quantized phase set of the PSN with a resolution of N_q^{ps}, given by

$$\mathcal{A} = \left\{-\pi, -\pi + \frac{2\pi}{2^{N_q^{\mathrm{ps}}}}, -\pi + 2 \cdot \frac{2\pi}{2^{N_q^{\mathrm{ps}}}}, \ldots, \pi - \frac{2\pi}{2^{N_q^{\mathrm{ps}}}}\right\}. \tag{2.2}$$

Besides, the transmit precoder matrix satisfies the power constraint $\|\boldsymbol{F}_d[k,m]\|_F^2 \le N_{\mathrm{BS}}^{\mathrm{RF}}$ [28].

Due to the large bandwidth of mmWave communications, each MPC has a clearly distinguishable delay spread. Then according to the typical mmWave MIMO channel model, [40–45, 104], the downlink delay-domain channel matrix $\boldsymbol{H}_q(\tau) \in \mathbb{C}^{N_{\mathrm{UE}} \times N_{\mathrm{BS}}}$ with L_q MPCs can be given by

$$\boldsymbol{H}_q(\tau) = \beta_q \sum_{l=1}^{L_q} \boldsymbol{H}_{q,l}\, p\left(\tau - \tau_{q,l}\right), \tag{2.3}$$

where $\beta_q = \sqrt{N_{\mathrm{UE}} N_{\mathrm{BS}} / L_q}$ is the normalization factor, $\tau_{q,l}$ is the path delay of l-th MPC, and $p(\tau)$ represents the equivalent pulse shaping filter (PSF). The complex gain matrix $\boldsymbol{H}_{q,l} \in \mathbb{C}^{N_{\mathrm{UE}} \times N_{\mathrm{BS}}}$ is given by

$$\boldsymbol{H}_{q,l} = \alpha_{q,l} \boldsymbol{a}_{\mathrm{UE}}\left(\mu_{q,l}^{\mathrm{UE}}, \nu_{q,l}^{\mathrm{UE}}\right) \boldsymbol{a}_{\mathrm{BS}}^{\mathrm{H}}\left(\mu_{q,l}^{\mathrm{BS}}, \nu_{q,l}^{\mathrm{BS}}\right), \tag{2.4}$$

where $\alpha_{q,l} \sim CN(0, \sigma_\alpha^2)$ is the complex path gain, $\mu_{q,l}^{\mathrm{UE}} = \pi \sin\left(\theta_{q,l}^{\mathrm{UE}}\right) \cos\left(\varphi_{q,l}^{\mathrm{UE}}\right)$ ($\mu_{q,l}^{\mathrm{BS}} = \pi \sin\left(\theta_{q,l}^{\mathrm{BS}}\right) \cos\left(\varphi_{q,l}^{\mathrm{BS}}\right)$) and $\nu_{q,l}^{\mathrm{UE}} = \pi \sin\left(\varphi_{q,l}^{\mathrm{UE}}\right)$ ($\nu_{q,l}^{\mathrm{BS}} = \pi \sin\left(\varphi_{q,l}^{\mathrm{BS}}\right)$) denote the horizontal and vertical spatial frequencies with the half-wavelength antenna space at the q-th UE (the BS), respectively. Here, the array response vector at the UE side is given by $\boldsymbol{a}_{\mathrm{UE}}\left(\mu_{q,l}^{\mathrm{UE}}, \nu_{q,l}^{\mathrm{UE}}\right) = \boldsymbol{a}_{\mathrm{v}}\left(\nu_{q,l}^{\mathrm{UE}}\right) \otimes \boldsymbol{a}_{\mathrm{h}}\left(\mu_{q,l}^{\mathrm{UE}}\right) \in \mathbb{C}^{N_{\mathrm{UE}}}$ [97, 99, 105], where

$$\boldsymbol{a}_{\mathrm{h}}\left(\mu_{q,l}^{\mathrm{UE}}\right) = \frac{1}{\sqrt{N_{\mathrm{UE}}^{\mathrm{h}}}} \left[1\ e^{\mathrm{j}\mu_{q,l}^{\mathrm{UE}}} \ldots e^{\mathrm{j}(N_{\mathrm{UE}}^{\mathrm{h}}-1)\mu_{q,l}^{\mathrm{UE}}}\right]^{\mathrm{T}} \in \mathbb{C}^{N_{\mathrm{UE}}^{\mathrm{h}}}, \tag{2.5}$$

$$\boldsymbol{a}_{\mathrm{v}}\left(\nu_{q,l}^{\mathrm{UE}}\right) = \frac{1}{\sqrt{N_{\mathrm{UE}}^{\mathrm{v}}}} \left[1\ e^{\mathrm{j}\nu_{q,l}^{\mathrm{UE}}} \ldots e^{\mathrm{j}(N_{\mathrm{UE}}^{\mathrm{v}}-1)\nu_{q,l}^{\mathrm{UE}}}\right]^{\mathrm{T}} \in \mathbb{C}^{N_{\mathrm{UE}}^{\mathrm{v}}}, \tag{2.6}$$

are the steering vectors associated with the horizontal and vertical directions, respectively. Similarly, the array response vector at the BS side is $\boldsymbol{a}_{\mathrm{BS}}\left(\mu_{q,l}^{\mathrm{BS}}, \nu_{q,l}^{\mathrm{BS}}\right) = \boldsymbol{a}_{\mathrm{v}}\left(\nu_{q,l}^{\mathrm{BS}}\right) \otimes \boldsymbol{a}_{\mathrm{h}}\left(\mu_{q,l}^{\mathrm{BS}}\right) \in \mathbb{C}^{N_{\mathrm{BS}}}$, where the steering vectors $\boldsymbol{a}_{\mathrm{h}}\left(\mu_{q,l}^{\mathrm{BS}}\right) \in \mathbb{C}^{N_{\mathrm{BS}}^{\mathrm{h}}}$ and $\boldsymbol{a}_{\mathrm{h}}\left(\nu_{q,l}^{\mathrm{BS}}\right) \in \mathbb{C}^{N_{\mathrm{BS}}^{\mathrm{v}}}$ can be derived by substituting $\mu_{q,l}^{\mathrm{UE}}$ and $N_{\mathrm{UE}}^{\mathrm{h}}$ with $\mu_{q,l}^{\mathrm{BS}}$ and $N_{\mathrm{BS}}^{\mathrm{h}}$ in Eq. (2.5) as well as by substituting $\nu_{q,l}^{\mathrm{UE}}$ and $N_{\mathrm{UE}}^{\mathrm{v}}$ with $\nu_{q,l}^{\mathrm{BS}}$ and $N_{\mathrm{BS}}^{\mathrm{v}}$ in Eq. (2.6).

The frequency domain channel at the k-th ($0 \le k \le K-1$) subcarrier can then be expressed by

$$\boldsymbol{H}_q[k] = \beta_q \sum_{l=1}^{L_q} \alpha_{q,l} \boldsymbol{a}_{\mathrm{UE}}\left(\mu_{q,l}^{\mathrm{UE}}, \nu_{q,l}^{\mathrm{UE}}\right) \boldsymbol{a}_{\mathrm{BS}}^{\mathrm{H}}\left(\mu_{q,l}^{\mathrm{BS}}, \nu_{q,l}^{\mathrm{BS}}\right) e^{-\mathrm{j}\frac{2\pi k f_s \tau_{q,l}}{K}}, \tag{2.7}$$

where $f_s = 1/T_s$ denotes the system bandwidth and T_s is the sampling period. The derivation of the Eq. (2.7) is shown in the Appendix A. It can be observed from Eq. (2.7) that $\boldsymbol{H}_q[k]$ does not depends on the PSF, and it exhibits delay-domain sparsity due to the small L_q but large normalized delay spread. Existing CS-based solutions have to estimate the effective delay-domain channel impulse response (CIR) that includes the PSF. However, the delay-domain sparsity would be corrupted when the order of PSF is too large. For simplicity of expression, $\boldsymbol{H}_q[k]$ is written as

$$\boldsymbol{H}_q[k] = \boldsymbol{A}_{\mathrm{UE},q}\boldsymbol{D}_q[k]\boldsymbol{A}_{\mathrm{BS},q}^{\mathrm{H}}, \tag{2.8}$$

where $\boldsymbol{D}_q[k]=\mathrm{diag}\left(\boldsymbol{d}_q[k]\right)\in\mathbb{C}^{L_q\times L_q}$ is a diagonal matrix with $\boldsymbol{d}_q[k]=\mathrm{diag}\left(\boldsymbol{\alpha}_q\right)$ $\boldsymbol{\tau}_q[k]$, where $\boldsymbol{\alpha}_q=\beta_q\left[\alpha_{q,1}\ldots\alpha_{q,L_q}\right]^{\mathrm{T}}$ and $\boldsymbol{\tau}_q[k]=\left[e^{-\mathrm{j}2\pi kf_s\tau_{q,1}/K}\ldots e^{-\mathrm{j}2\pi kf_s\tau_{q,L_q}/K}\right]^{\mathrm{T}}$. In Eq. (2.8), $\boldsymbol{A}_{\mathrm{UE},q}\in\mathbb{C}^{N_{\mathrm{UE}}\times L_q}$ is the array response matrix associated to the q-th UE's AoAs, which can be expressed by $\boldsymbol{A}_{\mathrm{UE},q}=\boldsymbol{A}_{\mathrm{UE},q}^{\nu}\odot\boldsymbol{A}_{\mathrm{UE},q}^{\mu}$. Here, $\boldsymbol{A}_{\mathrm{UE},q}^{\mu}=\left[\boldsymbol{a}_{\mathrm{h}}(\mu_{q,1}^{\mathrm{UE}})\ldots\boldsymbol{a}_{\mathrm{h}}(\mu_{q,L_q}^{\mathrm{UE}})\right]\in\mathbb{C}^{N_{\mathrm{UE}}^{\mathrm{h}}\times L_q}$ and $\boldsymbol{A}_{\mathrm{UE},q}^{\nu}=\left[\boldsymbol{a}_{\mathrm{v}}(\nu_{q,1}^{\mathrm{UE}})\ldots\boldsymbol{a}_{\mathrm{v}}(\nu_{q,L_q}^{\mathrm{UE}})\right]\in\mathbb{C}^{N_{\mathrm{UE}}^{\mathrm{v}}\times L_q}$ are the steering matrices corresponding to the horizontal and vertical spatial frequencies, respectively. Similarly, the array response matrix associated to the BS's AoDs is denoted by $\boldsymbol{A}_{\mathrm{BS},q}=\boldsymbol{A}_{\mathrm{BS},q}^{\nu}\odot\boldsymbol{A}_{\mathrm{BS},q}^{\mu}\in\mathbb{C}^{N_{\mathrm{BS}}\times L_q}$, where the steering matrices $\boldsymbol{A}_{\mathrm{BS},q}^{\mu}\in\mathbb{C}^{N_{\mathrm{BS}}^{\mathrm{h}}\times L_q}$ and $\boldsymbol{A}_{\mathrm{BS},q}^{\nu}\in\mathbb{C}^{N_{\mathrm{BS}}^{\mathrm{v}}\times L_q}$ have the similar forms with $\boldsymbol{A}_{\mathrm{UE},q}^{\mu}$ and $\boldsymbol{A}_{\mathrm{UE},q}^{\nu}$, respectively.

2.3.2 *Horizontal/Vertical AoAs Acquisition at UE*

The downlink CE stage corresponds to the Step 1 to Step 4 of Fig. 2.1. In this subsection, the horizontal and vertical AoAs at the UE side are estimated. We first assume the designed training signal $\boldsymbol{s}[i,m]$ is independent of subcarriers, and its j_1-th element is $\{\boldsymbol{s}[i,m]\}_{j_1}=\frac{1}{N_{\mathrm{s}}^d}e^{\mathrm{j}2\pi\phi_{j_1}}$, where $\phi_{j_1}\sim\mathcal{U}(0,1)$. Then, a predefined frequency-domain scrambling sequence $\boldsymbol{x}_d\in\mathbb{C}^K$ is utilized to avoid high peak-to-average power ratio (PAPR). Here, the k-th $(0\leq k\leq K-1)$ element of $\boldsymbol{x}_d$ is denoted by $\boldsymbol{s}[k,i,m]=x_d[k]\boldsymbol{s}[i,m]$. Actually, the received signal can be first descrambled by multiplying the conjugate code $\boldsymbol{x}_d^*$, which indicates $\boldsymbol{x}_d$ will not influence the subsequent signal processing. Moreover, each subcarrier adopts the same digital transmit precoding and receive combining matrices, i.e., $\boldsymbol{F}_{\mathrm{BB},d}[k,m]=\boldsymbol{F}_{\mathrm{BB},d}[m]$ and $\boldsymbol{W}_{\mathrm{BB},q}[k,m]=\boldsymbol{W}_{\mathrm{BB},q}[m]$, $\forall 0\leq k\leq K-1$. Next, we can utilize multiple OFDM symbols to convert a high-dimensional hybrid beamforming MIMO array of size $N_{\mathrm{UE}}^{\mathrm{h}}\times N_{\mathrm{UE}}^{\mathrm{v}}$ into a equivalent low-dimensional fully-digital array of size $M_{\mathrm{UE}}^{\mathrm{h}}\times M_{\mathrm{UE}}^{\mathrm{v}}$, where $M_{\mathrm{UE}}^{\mathrm{h}}$ and $M_{\mathrm{UE}}^{\mathrm{v}}$ are the numbers of antenna in the horizontal and vertical directions, respectively. Since the number of independent signal streams associated with each subcarrier in an OFDM symbol satisfies $N_{\mathrm{s}}^d\leq N_{\mathrm{UE}}^{\mathrm{RF}}$, given $N_{\mathrm{UE}}^{\mathrm{sub}}=M_{\mathrm{UE}}^{\mathrm{h}}M_{\mathrm{UE}}^{\mathrm{v}}$, the BS only needs $N_d=\lceil N_{\mathrm{UE}}^{\mathrm{sub}}/N_{\mathrm{s}}^d\rceil$ slots to broadcast training signal, with each slot containing N_{o}^d OFDM symbols. Considering larger $M_{\mathrm{UE}}^{\mathrm{h}}$, $M_{\mathrm{UE}}^{\mathrm{v}}$, and N_{o}^d can lead to better estimation performance but higher training overhead, we need to make a trade off on the choices of these parameters.[1] Since the received signal and subsequent signal processing have similar forms for all UEs, in the next, we only focus on the q-th UE. For notation simplicity, the index q can be omitted from $\boldsymbol{y}_q[k,i,m]$, $\boldsymbol{W}_{d,q}[m]$, $\boldsymbol{H}_q[k]$, $\boldsymbol{n}_q[k,i,m]$, $\boldsymbol{A}_{\mathrm{UE},q}$, $\boldsymbol{D}_q[k]$, $\boldsymbol{A}_{\mathrm{BS},q}$ and other variables in the following parts.

[1] The training overhead is defined as the required number of OFDM symbols at the CE stage. At the downlink CE stage, the training duration is $N_d N_{\mathrm{o}}^d$ OFDM symbols.

By collecting all N_{o}^d OFDM symbols of the received signals associated with k-th subcarrier and m-th slot, as shown in Eq. (2.1), into a matrix $\boldsymbol{Y}_m[k] \in \mathbb{C}^{N_{\mathrm{s}}^d \times N_{\mathrm{o}}^d}$, we have

$$\begin{aligned} \boldsymbol{Y}_m[k] &= \left[\boldsymbol{y}[k,1,m] \ldots \boldsymbol{y}[k,N_{\mathrm{o}}^d,m]\right] \\ &= x_d^*[k]\boldsymbol{W}_d^{\mathrm{H}}[m]\boldsymbol{H}[k]\boldsymbol{F}_d[m]\boldsymbol{S}_d[k,m] + \boldsymbol{W}_d^{\mathrm{H}}[m]\boldsymbol{N}_m[k], \end{aligned} \tag{2.9}$$

where $\boldsymbol{S}_d[k,m]=\left[\boldsymbol{s}[k,1,m]\ldots\boldsymbol{s}[k,N_{\mathrm{o}}^d,m]\right]=x_d[k]\boldsymbol{S}_d[m]$, $\boldsymbol{S}_d[m]=[\boldsymbol{s}[1,m]\ldots \boldsymbol{s}[N_{\mathrm{o}}^d,m]]\in\mathbb{C}^{N_{\mathrm{s}}^d\times N_{\mathrm{o}}^d}$, and $\boldsymbol{N}_m[k]=\left[\boldsymbol{n}[k,1,m]\ldots\boldsymbol{n}[k,N_{\mathrm{o}}^d,m]\right]\in\mathbb{C}^{N_{\mathrm{UE}}\times N_{\mathrm{o}}^d}$. Since the BS transmits a public random signal $\boldsymbol{F}_d[m]\boldsymbol{S}_d[m]$, the transmit precoding matrix $\boldsymbol{F}_d[m]=\boldsymbol{F}_{\mathrm{RF},d}[m]\boldsymbol{F}_{\mathrm{BB},d}[m]$ should be a random matrix. Specifically, we can design $\boldsymbol{F}_{\mathrm{RF},d}[m]$ and $\boldsymbol{F}_{\mathrm{BB},d}[m]$ as $\{\boldsymbol{F}_{\mathrm{RF},d}[m]\}_{j_1,j_2}=\frac{1}{\sqrt{N_{\mathrm{BS}}}}e^{\mathrm{j}\vartheta_{3,j_1,j_2}}$ and $\{\boldsymbol{F}_{\mathrm{BB},d}[m]\}_{j_1,j_2} = e^{\mathrm{j}2\pi a_{j_1,j_2}}$, respectively, where ϑ_{3,j_1,j_2} is randomly sampled from the set $\mathcal{A}$ and $a_{j_1,j_2}\sim\mathcal{U}[0,1]$. Hence, the BS can use the same precoding matrix $\boldsymbol{F}_d=\boldsymbol{F}_d[m]$ to transmit the same sounding signal $\boldsymbol{S}_d=\boldsymbol{S}_d[m]$ in continuous N_d slots. Then, each UE can stack N_d received signal matrices $\{\boldsymbol{Y}_m[k]\}_{m=1}^{N_d}$ into $\widetilde{\boldsymbol{Y}}_d[k]\in\mathbb{C}^{N_dN_{\mathrm{s}}^d\times N_{\mathrm{o}}^d}$, which is given by

$$\begin{aligned} \widetilde{\boldsymbol{Y}}_d[k] &= \left[\boldsymbol{Y}_1^{\mathrm{T}}[k]\ldots\boldsymbol{Y}_{N_d}^{\mathrm{T}}[k]\right]^{\mathrm{T}} \\ &= \widetilde{\boldsymbol{W}}_d^{\mathrm{H}}\boldsymbol{A}_{\mathrm{UE}}\boldsymbol{D}[k]\boldsymbol{A}_{\mathrm{BS}}^{\mathrm{H}}\boldsymbol{F}_d\boldsymbol{S}_d + \mathrm{Bdiag}\left(\check{\boldsymbol{W}}_d\right)\widetilde{\boldsymbol{N}}_d[k], \end{aligned} \tag{2.10}$$

where $\widetilde{\boldsymbol{W}}_d=[\boldsymbol{W}_d[1]\ldots\boldsymbol{W}_d[N_d]]\in\mathbb{C}^{N_{\mathrm{UE}}\times N_dN_{\mathrm{s}}^d}$ collects N_d downlink receive combining matrices, $\mathrm{Bdiag}\left(\check{\boldsymbol{W}}_d\right)=\mathrm{Bdiag}\left(\left[\boldsymbol{W}_d^{\mathrm{H}}[1]\ldots\boldsymbol{W}_d^{\mathrm{H}}[N_d]\right]\right)\in\mathbb{C}^{N_dN_{\mathrm{s}}^d\times N_dN_{\mathrm{UE}}}$, and $\widetilde{\boldsymbol{N}}_d[k]=\left[\boldsymbol{N}_1^{\mathrm{T}}[k]\ldots\boldsymbol{N}_{N_d}^{\mathrm{T}}[k]\right]^{\mathrm{T}}\in\mathbb{C}^{N_dN_{\mathrm{UE}}\times N_{\mathrm{o}}^d}$ is the corresponding noise matrix.

After designing a selection matrix $\boldsymbol{J}_d=\left[\boldsymbol{I}_{N_{\mathrm{UE}}^{\mathrm{sub}}}\ \boldsymbol{O}_{N_{\mathrm{UE}}^{\mathrm{sub}}\times(N_dN_{\mathrm{s}}^d-N_{\mathrm{UE}}^{\mathrm{sub}})}\right]\in\mathbb{R}^{N_{\mathrm{UE}}^{\mathrm{sub}}\times N_dN_{\mathrm{s}}^d}$ and multiplying it with $\widetilde{\boldsymbol{Y}}_d[k]$, we can acquire the final received signal matrix $\bar{\boldsymbol{Y}}_d\in\mathbb{C}^{N_{\mathrm{UE}}^{\mathrm{sub}}\times KN_{\mathrm{o}}^d}$ by collecting all K subcarriers, as is given by

$$\bar{\boldsymbol{Y}}_d = \left[\boldsymbol{J}_d\widetilde{\boldsymbol{Y}}_d[0]\ \boldsymbol{J}_d\widetilde{\boldsymbol{Y}}_d[1]\ldots\boldsymbol{J}_d\widetilde{\boldsymbol{Y}}_d[K-1]\right] = \bar{\boldsymbol{A}}_{\mathrm{UE}}\bar{\boldsymbol{S}}_d + \bar{\boldsymbol{N}}_d, \tag{2.11}$$

where $\bar{\boldsymbol{N}}_d=\boldsymbol{J}_d\mathrm{Bdiag}\left(\check{\boldsymbol{W}}_d\right)\left[\widetilde{\boldsymbol{N}}_d[0]\ \widetilde{\boldsymbol{N}}_d[1]\ldots\widetilde{\boldsymbol{N}}_d[K-1]\right]$, $\bar{\boldsymbol{A}}_{\mathrm{UE}}=\boldsymbol{J}_d\widetilde{\boldsymbol{W}}_d^{\mathrm{H}}\boldsymbol{A}_{\mathrm{UE}}$, $\bar{\boldsymbol{S}}_d=\left[\bar{\boldsymbol{S}}_d[0]\ \bar{\boldsymbol{S}}_d[1]\ldots\bar{\boldsymbol{S}}_d[K-1]\right]$, and $\bar{\boldsymbol{S}}_d[k]=\boldsymbol{D}[k]\boldsymbol{A}_{\mathrm{BS}}^{\mathrm{H}}\boldsymbol{F}_d\boldsymbol{S}_d$. Observe from Eqs. (2.10) and (2.11), we cannot directly apply robust array signal processing techniques [99, 106] on $\bar{\boldsymbol{Y}}_d$ to estimate the horizontal and vertical angles at the UE. This is because the shift-invariance structure of the array response matrix $\boldsymbol{A}_{\mathrm{UE}}$ no longer holds in hybrid array [100, 107]. To this end, we design an appropriate aggregated receive combining matrix $\widetilde{\boldsymbol{W}}_d$ in the next subsection, by which the high-dimensional hybrid array can be visualized as a low-dimensional digital array. Then the shift-variance structure of the array response can be reconstructed and super-resolution spectrum estimation techniques can be applied for CE.

Algorithm 2.1: Proposed Receive Combining Matrix Design

Input: N_d, N_{s}^d, $N_{\mathrm{UE}}^{\mathrm{RF}}$, $N_{\mathrm{UE}}^{\mathrm{h}}$, $N_{\mathrm{UE}}^{\mathrm{v}}$, $M_{\mathrm{UE}}^{\mathrm{h}}$, $M_{\mathrm{UE}}^{\mathrm{v}}$
Output: $\widetilde{\boldsymbol{W}}_d$

Generate unitary matrix $\boldsymbol{U}_{N_{\mathrm{UE}}^{\mathrm{RF}}} = \left[\boldsymbol{u}_1 \ldots \boldsymbol{u}_{N_{\mathrm{UE}}^{\mathrm{RF}}}\right]$;
Construct index matrix $\boldsymbol{\Xi}_d$ in Eq. (2.12) ;
for $m = 1, 2, \ldots, N_d$ **do**
    $\boldsymbol{W}_{\mathrm{BB}}[m] = \boldsymbol{U}_{N_{\mathrm{UE}}^{\mathrm{RF}}[:,1:N_{\mathrm{s}}^d]}$, and initialize $\boldsymbol{W}_{\mathrm{RF}}[m] = \boldsymbol{1}_{N_{\mathrm{UE}}} \otimes \boldsymbol{u}_{N_{\mathrm{UE}}^{\mathrm{RF}}}^{\mathrm{H}}$;
    Extract $\boldsymbol{\Xi}_{d,m}^{\mathrm{sub}} = \boldsymbol{\Xi}_{d[:,(m-1)N_{\mathrm{s}}^d+1:mN_{\mathrm{s}}^d]}$, and obtain $\bar{\boldsymbol{\xi}}_{d,m} = \mathrm{vec}(\boldsymbol{\Xi}_{d,m}^{\mathrm{sub}})$;
    Obtain ordered index set $\mathcal{I}_m = \mathrm{mod}\left(\mathrm{find}\left(\bar{\boldsymbol{\xi}}_{d,m} \neq 0\right), N_{\mathrm{UE}}\right)$;
    Replace $\boldsymbol{W}_{\mathrm{RF}}[m]_{[\mathcal{I}_m,:]} \leftarrow \boldsymbol{W}_{\mathrm{BB}}^{\mathrm{H}}[m]$;
    Quantize phase values of $\boldsymbol{W}_{\mathrm{RF}}[m]$ based on $\mathcal{A}$;
    $\boldsymbol{W}_d[m] = \boldsymbol{W}_{\mathrm{RF}}[m]\boldsymbol{W}_{\mathrm{BB}}[m]$;
end
Return: $\widetilde{\boldsymbol{W}}_d = [\boldsymbol{W}_d[1] \ldots \boldsymbol{W}_d[N_d]]$

2.3.3 *Design Combining Matrix at UE*

Without loss of generality, we consider $N_{\mathrm{s}}^d = N_{\mathrm{UE}}^{\mathrm{RF}} - 1$ independent data streams. First, we utilize the unitary matrix $\boldsymbol{U}_{N_{\mathrm{UE}}^{\mathrm{RF}}} = \left[\boldsymbol{u}_1 \ldots \boldsymbol{u}_{N_{\mathrm{UE}}^{\mathrm{RF}}}\right] \in \mathbb{C}^{N_{\mathrm{UE}}^{\mathrm{RF}} \times N_{\mathrm{UE}}^{\mathrm{RF}}}$ to design the digital receive combining matrix $\boldsymbol{W}_{\mathrm{BB}}[m] \in \mathbb{C}^{N_{\mathrm{UE}}^{\mathrm{RF}} \times N_{\mathrm{s}}^d}$ in $\boldsymbol{W}_d[m] = \boldsymbol{W}_{\mathrm{RF}}[m]\boldsymbol{W}_{\mathrm{BB}}[m]$ at the m-th ($1 \leq m \leq N_d$) slot, i.e., $\boldsymbol{W}_{\mathrm{BB}}[m] = \boldsymbol{U}_{N_{\mathrm{UE}}^{\mathrm{RF}}\{:,1:N_{\mathrm{s}}^d\}}$. For analog receive combining matrix, we first construct the matrix $\boldsymbol{\Xi}_d \in \mathbb{R}^{N_{\mathrm{UE}} \times N_d N_{\mathrm{s}}^d}$ as

$$\boldsymbol{\Xi}_d = \begin{bmatrix} \boldsymbol{I}_{M_{\mathrm{UE}}^{\mathrm{v}}+1} \otimes \boldsymbol{B} \\ \boldsymbol{O}_{(N_{\mathrm{UE}} - N_{\mathrm{UE}}^{\mathrm{h}}(M_{\mathrm{UE}}^{\mathrm{v}}+1)) \times M_{\mathrm{UE}}^{\mathrm{h}}(M_{\mathrm{UE}}^{\mathrm{v}}+1)} \end{bmatrix} \begin{bmatrix} \boldsymbol{I}_{N_{\mathrm{s}}^d N_d} \\ \boldsymbol{O}_{(M_{\mathrm{UE}}^{\mathrm{h}}(M_{\mathrm{UE}}^{\mathrm{v}}+1) - N_{\mathrm{s}}^d N_d) \times N_{\mathrm{s}}^d N_d} \end{bmatrix}, \tag{2.12}$$

where $\boldsymbol{B} = \left[\boldsymbol{I}_{M_{\mathrm{UE}}^{\mathrm{h}}} \; \boldsymbol{O}_{M_{\mathrm{UE}}^{\mathrm{h}} \times (N_{\mathrm{UE}}^{\mathrm{h}} - M_{\mathrm{UE}}^{\mathrm{h}})}\right]^{\mathrm{T}} \in \mathbb{R}^{N_{\mathrm{UE}}^{\mathrm{h}} \times M_{\mathrm{UE}}^{\mathrm{h}}}$. Then we select a submatrix $\boldsymbol{\Xi}_{d,m}^{\mathrm{sub}} = \boldsymbol{\Xi}_{d\{:,(m-1)N_{\mathrm{s}}^d+1:mN_{\mathrm{s}}^d\}} \in \mathbb{R}^{N_{\mathrm{UE}} \times N_{\mathrm{s}}^d}$ and define $\bar{\boldsymbol{\xi}}_{d,m} = \mathrm{vec}(\boldsymbol{\Xi}_{d,m}^{\mathrm{sub}})$ to construct an ordered index set $\mathcal{D}_m = \mathrm{find}\left(\bar{\boldsymbol{\xi}}_{d,m} \neq 0\right)$, with its cardinality given by $|\mathcal{D}_m|_c = N_{\mathrm{s}}^d$. Next, we acquire an ordered index set $\mathcal{I}_m = \mathrm{mod}(\mathcal{D}_m, N_{\mathrm{UE}})$ by performing the modulo-N_{UE} operation on $\mathcal{D}_m$. Then, the rows of $\boldsymbol{W}_{\mathrm{RF}}[m]$ whose indices corresponds to $\mathcal{I}_m$ can be determined as $\boldsymbol{W}_{\mathrm{RF}}[m]_{\{\mathcal{I}_m,:\}} = \boldsymbol{W}_{\mathrm{BB}}^{\mathrm{H}}[m]$, while remaining $(N_{\mathrm{UE}} - N_{\mathrm{s}}^d)$ rows are constituted by the identical vector $\boldsymbol{u}_{N_{\mathrm{UE}}^{\mathrm{RF}}}^{\mathrm{H}}$. We denote the phase of arbitrary entry of $\boldsymbol{W}_{\mathrm{RF}}[m]$ by ϑ_d, whose value can be acquired by quantizing it to the closest ϑ using a minimum Euclidean distance rule, i.e., $\vartheta_d = \arg\min_{\vartheta \in \mathcal{A}} \|\vartheta_d - \vartheta\|_2$. Then, the m-th receive combining matrix can be obtained as $\boldsymbol{W}_d[m] = \boldsymbol{W}_{\mathrm{RF}}[m]\boldsymbol{W}_{\mathrm{BB}}[m]$. The overall design process of the receive combining matrix $\widetilde{\boldsymbol{W}}_d$ is summarized in Algorithm 2.1.

Since the number of RF chains is usually the power of 2, we can utilize Hadamard matrix (for $N_q^{\mathrm{ps}} \geq 1$) or discrete Fourier transform (DFT) matrix (for $N_q^{\mathrm{ps}} \geq 2$) when constructing $\boldsymbol{U}_{N_{\mathrm{UE}}^{\mathrm{RF}}}$.[2] The above analysis clearly state that the design of analog combining matrix can be applied to the PSN with arbitrary resolution N_q^{ps} (even for $N_q^{\mathrm{ps}} = 1$). With the designed $\widetilde{\boldsymbol{W}}_d$, the $\bar{\boldsymbol{A}}_{\mathrm{UE}} = \boldsymbol{J}_d \widetilde{\boldsymbol{W}}_d^{\mathrm{H}} \boldsymbol{A}_{\mathrm{UE}}$ in Eq. (2.11) can be further expressed by

$$\bar{\boldsymbol{A}}_{\mathrm{UE}} = \boldsymbol{J}_d \widetilde{\boldsymbol{W}}_d^{\mathrm{H}} \left(\boldsymbol{A}_{\mathrm{UE}}^{\nu} \odot \boldsymbol{A}_{\mathrm{UE}}^{\mu}\right) = \bar{\boldsymbol{A}}_{\mathrm{UE}}^{\nu} \odot \bar{\boldsymbol{A}}_{\mathrm{UE}}^{\mu}. \tag{2.13}$$

It can be seen from Eq. (2.13) that $\bar{\boldsymbol{A}}_{\mathrm{UE}}^{\mu} \in \mathbb{C}^{M_{\mathrm{UE}}^{\mathrm{h}} \times L}$ ($\bar{\boldsymbol{A}}_{\mathrm{UE}}^{\nu} \in \mathbb{C}^{M_{\mathrm{UE}}^{\mathrm{v}} \times L}$) is constituted by the first $M_{\mathrm{UE}}^{\mathrm{h}}$ ($M_{\mathrm{UE}}^{\mathrm{v}}$) rows of $\boldsymbol{A}_{\mathrm{UE}}^{\mu}$ ($\boldsymbol{A}_{\mathrm{UE}}^{\nu}$). Therefore, $\bar{\boldsymbol{A}}_{\mathrm{UE}}$ preserves the double shift-invariance structure of the original array response matrix $\boldsymbol{A}_{\mathrm{UE}}$ for horizontal and vertical AoAs. Accordingly, the $\widetilde{\boldsymbol{W}}_d$ designed in Algorithm 2.1 transforms the high-dimensional hybrid array into a equivalent low-dimensional fully-digital array. Furthermore, by utilizing the MDU-ESPRIT algorithm detailed in Sect. 2.5, the super-resolution estimates of the horizontal/vertical AoAs at UE can be acquired. Since the ESPRIT algorthms [99, 100, 106, 107] need the known number of MPCs, the next task is to estimate the number of MPCs at the receiver, i.e., Step 2 of Fig. 2.1.

2.3.4 EVD-Based Estimates for the Number of MPCs

In OFDM systems, the channels of adjacent subcarriers are highly correlated within the coherent bandwidth. If the maximum delay spread is $\tau_{\max} = N_{\mathrm{c}} T_{\mathrm{s}}$, i.e., the channel has N_{c} delay taps, then the coherent bandwidth is $B_{\mathrm{c}} \approx \frac{1}{\tau_{\max}} = \frac{f_{\mathrm{s}}}{N_{\mathrm{c}}}$. Therefore, we can jointly utilize $P \leq B_{\mathrm{c}}/\Delta f = \frac{K}{N_{\mathrm{c}}}$ adjacent subcarriers to estimate the number of MPCs, where $\Delta f = f_{\mathrm{s}}/K$ is the bandwidth of each subcarrier. Specifically, by dividing K signal matrices $\{\boldsymbol{J}_d \widetilde{\boldsymbol{Y}}_d[k]\}_{k=0}^{K-1}$ into $N_P = \lfloor K/P \rfloor$ groups, then the average of measurements $\check{\boldsymbol{Y}}_d[n_p] \in \mathbb{C}^{N_{\mathrm{UE}}^{\mathrm{sub}} \times N_{\mathrm{o}}^d}$ in the n_p-th ($1 \leq n_p \leq N_P$) group can be obtained as

$$\check{\boldsymbol{Y}}_d[n_p] = \frac{1}{P} \sum_{k=(n_p-1)P}^{n_p P-1} \boldsymbol{J}_d \widetilde{\boldsymbol{Y}}_d[k]. \tag{2.14}$$

By collecting N_P averaged measurements as $\check{\boldsymbol{Y}}_d = \left[\check{\boldsymbol{Y}}_d[1] \ldots \check{\boldsymbol{Y}}_d[N_P]\right] \in \mathbb{C}^{N_{\mathrm{UE}}^{\mathrm{sub}} \times N_{\mathrm{o}}^d N_P}$, we can obtain the covariance matrix $\boldsymbol{R}_d = \frac{1}{N_{\mathrm{o}}^d N_P} \check{\boldsymbol{Y}}_d \check{\boldsymbol{Y}}_d^{\mathrm{H}}$. After applying EVD to $\boldsymbol{R}_d$, we obtain $\boldsymbol{R}_d = [\boldsymbol{U}_s \; \boldsymbol{U}_n] \operatorname{diag}(\boldsymbol{\lambda}_d) [\boldsymbol{U}_s \; \boldsymbol{U}_n]^{\mathrm{H}}$, where $\boldsymbol{\lambda}_d = [\lambda_1 \ldots \lambda_L \; \lambda_{L+1} \ldots \lambda_{N_{\mathrm{UE}}^{\mathrm{sub}}}]^{\mathrm{T}} = [\boldsymbol{\lambda}_s^{\mathrm{T}} \; \boldsymbol{\lambda}_n^{\mathrm{T}}]^{\mathrm{T}}$ is the eigenvalue vector with its eigenvalues arranged in a descending order, $\boldsymbol{U}_s$ and $\boldsymbol{U}_n$ are the eigen matrices corresponding to the signal and noise subspace, respectively, and $\boldsymbol{\lambda}_s = [\lambda_1 \ldots \lambda_L]^{\mathrm{T}}$ and $\boldsymbol{\lambda}_n = [\lambda_{L+1} \ldots \lambda_{N_{\mathrm{UE}}^{\mathrm{sub}}}]^{\mathrm{T}}$ are the eigenvalue vec-

[2] Note that the quantized phase value of each entry in Hadamard matrix or DFT matrix still belongs to set $\mathcal{A}$, therefore the columns of $\boldsymbol{U}_{N_{\mathrm{UE}}^{\mathrm{RF}}}$ are ensured to be mutually orthogonal.

tors corresponding to $\boldsymbol{U}_s$ and $\boldsymbol{U}_n$, respectively. Besides, the number of MPCs L is the dimension of $\boldsymbol{\lambda}_s$.

To acquire an accurate estimate of L, we first construct $\widetilde{\boldsymbol{\lambda}} = \left[\boldsymbol{\lambda}_s^{\mathrm{T}}\ \mathbf{0}^{\mathrm{T}}_{N_{\mathrm{UE}}^{\mathrm{sub}}-L}\right]^{\mathrm{T}} \in \mathbb{C}^{N_{\mathrm{UE}}^{\mathrm{sub}}}$, then the optimal solution of $\widetilde{\boldsymbol{\lambda}}$ can be solved from the following optimization problem,

$$\widetilde{\boldsymbol{\lambda}}^{\star} = \arg \min_{\widetilde{\boldsymbol{\lambda}} \geq \mathbf{0}_{N_{\mathrm{UE}}^{\mathrm{sub}}}} \frac{1}{2} \left\| \widetilde{\boldsymbol{\lambda}} - \boldsymbol{\lambda}_d \right\|_2^2 + \varepsilon \left\| \widetilde{\boldsymbol{\lambda}} \right\|_1, \tag{2.15}$$

where ε is a threshold parameter related to the power of AWGN, and it can be chosen empirically. It is obvious that the optimal solution to Eq. (2.15) is [108]

$$\widetilde{\lambda}_i^{\star} = \begin{cases} \lambda_i - \varepsilon, & \lambda_i \geq \varepsilon, \\ 0, & \lambda_i < \varepsilon, \end{cases} \tag{2.16}$$

where $\widetilde{\lambda}_i^{\star}$ is the i-th element of $\widetilde{\boldsymbol{\lambda}}^{\star}$. From the estimated $\widetilde{\boldsymbol{\lambda}}^{\star}$, we obtain the estimate of the number of MPCs $\widehat{L}$, which can be further used as the input of the MDU-ESPRIT algorithm to estimate $\widehat{L}$ pairs of horizontal and vertical AoAs at the UE side.

Finally, the estimated $\{\widehat{\theta}_l^{\mathrm{UE}}, \widehat{\varphi}_l^{\mathrm{UE}}\}_{l=1}^{\widehat{L}}$ can be quantized to $\{\bar{\theta}_l^{\mathrm{UE}}, \bar{\varphi}_l^{\mathrm{UE}}\}_{l=1}^{\widehat{L}}$ within the range of $[-\pi/2,\ \pi/2]$ according to N_q^{ang} quantization angles. Since only very little data of quantized angle estimates need to be fed back to BS, the feedback overhead of this stage can be neglected in our proposed closed-loop sparse CE scheme.[3]

2.4 Uplink CE Stage

2.4.1 Estimate Horizontal/Vertical AoDs and Delays at BS

At the uplink CE stage, the BS needs to jointly estimate the horizontal/vertical AoAs and delays of each UE. According to the channel reciprocity of TDD systems [88, 102, 103], the uplink channel matrix of the q-th UE is $\boldsymbol{H}^{\mathrm{T}}[k] = \boldsymbol{A}_{\mathrm{BS}}^{*} \boldsymbol{D}[k] \boldsymbol{A}_{\mathrm{UE}}^{\mathrm{T}} \in \mathbb{C}^{N_{\mathrm{BS}} \times N_{\mathrm{UE}}}$, here we omit the UE index q again. For uplink CE, we employ $N_s^u = N_{\mathrm{BS}}^{\mathrm{RF}} - 1$ independent data streams and a equivalent low-dimensional $M_{\mathrm{BS}}^{\mathrm{h}} \times M_{\mathrm{BS}}^{\mathrm{v}}$ fully-digital array visualized from a $N_{\mathrm{BS}}^{\mathrm{h}} \times N_{\mathrm{BS}}^{\mathrm{v}}$ high-dimensional hybrid array, where $M_{\mathrm{BS}}^{\mathrm{h}}$ and $M_{\mathrm{BS}}^{\mathrm{v}}$ are the number of antennas in the horizontal and vertical directions, respectively. Then the overall antennas number of this array is $N_{\mathrm{BS}}^{\mathrm{sub}} = M_{\mathrm{BS}}^{\mathrm{h}} M_{\mathrm{BS}}^{\mathrm{v}}$. Each UE employs $N_u = \lceil N_{\mathrm{BS}}^{\mathrm{sub}} / N_s^u \rceil$ slots for transmitting training signal, with each slot consisting of N_{o}^u symbols. Hence, the training overhead of Q UEs at the uplink

[3] In the open-loop CE schemes [42, 43], the estimated support sets and channel gains on each subcarrier at the receiver also need be fed back to the transmitter for the following beamforming or channel equalization [3, 16]. Compared with these schemes, the proposed close-loop scheme only needs to feed back/forward the estimated channel parameters to each other, therefore the relevant costs are negligible.

CE stage is $QN_uN_o^u$, and the overall overhead training overhead of the proposed close-loop sparse CE scheme is $T_{CE}=N_dN_o^d+QN_uN_o^u$. Similar to Eq. (2.10), after frequency domain scrambling/descrambling, the signal $\widetilde{\boldsymbol{Y}}_u[k]\in\mathbb{C}^{N_uN_s^u\times N_o^u}$ received at the k-th subcarrier over the N_u slots can be expressed as

$$\widetilde{\boldsymbol{Y}}_u[k] = \widetilde{\boldsymbol{W}}_u^{\mathrm{H}}\boldsymbol{A}_{\mathrm{BS}}^*\boldsymbol{D}[k]\boldsymbol{A}_{\mathrm{UE}}^{\mathrm{T}}\boldsymbol{F}_u\boldsymbol{S}_u + \mathrm{Bdiag}\left(\bar{\boldsymbol{W}}_u\right)\widetilde{\boldsymbol{N}}_u[k], \tag{2.17}$$

where $\widetilde{\boldsymbol{W}}_u=[\boldsymbol{W}_u[1]\ldots\boldsymbol{W}_u[N_u]]\in\mathbb{C}^{N_{\mathrm{BS}}\times N_uN_s^u}$ with $\boldsymbol{W}_u[n]\in\mathbb{C}^{N_{\mathrm{BS}}\times N_s^u}$ being the uplink receive combining matrix at the BS in the n-th ($1\le n\le N_u$) slot. In addition, $\bar{\boldsymbol{W}}_u=\left[\boldsymbol{W}_u^{\mathrm{H}}[1]\ldots\boldsymbol{W}_u^{\mathrm{H}}[N_u]\right]$, $\boldsymbol{F}_u\in\mathbb{C}^{N_{\mathrm{UE}}\times N_s^u}$ is the multi-beam transmit precoding matrix at the UE side, $\boldsymbol{S}_u\in\mathbb{C}^{N_s^u\times N_o^u}$ and $\widetilde{\boldsymbol{N}}_u[k]$ are the uplink training signal matrix and noise matrix, respectively. Afterwards, $\widetilde{\boldsymbol{Y}}_u[k]$ is multiplied by a designed selection matrix $\boldsymbol{J}_u=\left[\boldsymbol{I}_{N_{\mathrm{BS}}^{\mathrm{sub}}}\ \boldsymbol{O}_{N_{\mathrm{BS}}^{\mathrm{sub}}\times(N_s^uN_u-N_{\mathrm{BS}}^{\mathrm{sub}})}\right]\in\mathbb{R}^{N_{\mathrm{BS}}^{\mathrm{sub}}\times N_s^uN_u}$, after performing vectorization operation $\widetilde{\boldsymbol{y}}_u[k]=\mathrm{vec}\left((\boldsymbol{J}_u\widetilde{\boldsymbol{Y}}_u[k])^{\mathrm{T}}\right)\in\mathbb{C}^{N_{\mathrm{BS}}^{\mathrm{sub}}N_o^u}$, we have

$$\widetilde{\boldsymbol{y}}_u[k] = \left(\bar{\boldsymbol{A}}_{\mathrm{BS}}\odot\left(\boldsymbol{A}_{\mathrm{UE}}^{\mathrm{T}}\boldsymbol{F}_u\boldsymbol{S}_u\right)^{\mathrm{T}}\right)\mathrm{diag}(\boldsymbol{\alpha})\boldsymbol{\tau}[k] + \widetilde{\boldsymbol{n}}_u[k], \tag{2.18}$$

where $\bar{\boldsymbol{A}}_{\mathrm{BS}}=\boldsymbol{J}_u\widetilde{\boldsymbol{W}}_u^{\mathrm{H}}\boldsymbol{A}_{\mathrm{BS}}^*$, and $\widetilde{\boldsymbol{n}}_u[k]$ is the corresponding noise vector. Equation (2.18) has utilized the property $\mathrm{vec}(\boldsymbol{ABC})=\left(\boldsymbol{C}^{\mathrm{T}}\odot\boldsymbol{A}\right)\boldsymbol{b}$, with $\boldsymbol{B}=\mathrm{diag}(\boldsymbol{b})$ [109]. Furthermore, by aggregating $\{\widetilde{\boldsymbol{y}}_u[k]\}_{k=1}^K$ at K subcarriers, we acquire the collected matrix $\widetilde{\boldsymbol{Y}}_u\in\mathbb{C}^{N_{\mathrm{BS}}^{\mathrm{sub}}N_o^u\times K}$, as given by

$$\begin{aligned}\widetilde{\boldsymbol{Y}}_u &= \left[\widetilde{\boldsymbol{y}}_u[0]\ \widetilde{\boldsymbol{y}}_u[1]\ldots\widetilde{\boldsymbol{y}}_u[K-1]\right]\\ &= \left(\bar{\boldsymbol{A}}_{\mathrm{BS}}\odot\left(\boldsymbol{A}_{\mathrm{UE}}^{\mathrm{T}}\boldsymbol{F}_u\boldsymbol{S}_u\right)^{\mathrm{T}}\right)\mathrm{diag}\left(\boldsymbol{\alpha}\right)\boldsymbol{A}_{\boldsymbol{\tau}}^{\mathrm{T}} + \widetilde{\boldsymbol{N}}_u,\end{aligned} \tag{2.19}$$

where $\boldsymbol{A}_{\boldsymbol{\tau}}=[\boldsymbol{\tau}[0]\ \boldsymbol{\tau}[1]\ldots\boldsymbol{\tau}[K-1]]^{\mathrm{T}}\in\mathbb{C}^{K\times L}$, and $\widetilde{\boldsymbol{N}}_u$ is the corresponding noise matrix. According to the channel model, $\boldsymbol{\tau}[k]=\left[e^{-\mathrm{j}2\pi kf_s\tau_1/K}\ \ldots\ e^{-\mathrm{j}2\pi kf_s\tau_L/K}\right]^{\mathrm{T}}$, then the steering matrix can be further expressed by $\boldsymbol{A}_{\boldsymbol{\tau}}=\left[\boldsymbol{a}_\tau\left(\mu_1^\tau\right)\ldots\boldsymbol{a}_\tau\left(\mu_L^\tau\right)\right]$, where $\boldsymbol{a}_\tau\left(\mu_l^\tau\right)=\left[1\ e^{\mathrm{j}\mu_l^\tau}\ldots e^{\mathrm{j}(K-1)\mu_l^\tau}\right]^{\mathrm{T}}\in\mathbb{C}^K$, with $\mu_l^\tau=-2\pi f_s\tau_l/K$. Taking the vectorization of $\widetilde{\boldsymbol{Y}}_u$, i.e., $\check{\boldsymbol{y}}_u=\mathrm{vec}(\widetilde{\boldsymbol{Y}}_u)\in\mathbb{C}^{KN_{\mathrm{BS}}^{\mathrm{sub}}N_o^u}$, we obtain

$$\check{\boldsymbol{y}}_u = \left(\left(\boldsymbol{A}_{\boldsymbol{\tau}}\odot\bar{\boldsymbol{A}}_{\mathrm{BS}}\right)\odot\left(\boldsymbol{A}_{\mathrm{UE}}^{\mathrm{T}}\boldsymbol{F}_u\boldsymbol{S}_u\right)^{\mathrm{T}}\right)\boldsymbol{\alpha} + \check{\boldsymbol{n}}_u, \tag{2.20}$$

where Eq. (2.20) has utilized the identity transformation $\boldsymbol{A}\odot(\boldsymbol{B}\odot\boldsymbol{C})=(\boldsymbol{A}\odot\boldsymbol{B})\odot\boldsymbol{C}$ [109], and $\check{\boldsymbol{n}}_u=\mathrm{vec}(\widetilde{\boldsymbol{N}}_u)$. By further reshaping $\check{\boldsymbol{y}}_u$ into a matrix $\check{\boldsymbol{Y}}_u=\mathrm{mat}\left(\check{\boldsymbol{y}}_u;\ N_o^u,\ KN_{\mathrm{BS}}^{\mathrm{sub}}\right)\in\mathbb{C}^{N_o^u\times KN_{\mathrm{BS}}^{\mathrm{sub}}}$, we arrive

$$\check{\boldsymbol{Y}}_u = \left(\boldsymbol{A}_{\mathrm{UE}}^{\mathrm{T}}\boldsymbol{F}_u\boldsymbol{S}_u\right)^{\mathrm{T}}\mathrm{diag}(\boldsymbol{\alpha})\left(\boldsymbol{A}_{\boldsymbol{\tau}}\odot\bar{\boldsymbol{A}}_{\mathrm{BS}}\right)^{\mathrm{T}} + \check{\boldsymbol{N}}_u, \tag{2.21}$$

where $\check{\boldsymbol{N}}_u=\mathrm{mat}\left(\check{\boldsymbol{n}}_u;\ N_o^u,\ KN_{\mathrm{BS}}^{\mathrm{sub}}\right)$. Hence, $\bar{\boldsymbol{Y}}_u=\check{\boldsymbol{Y}}_u^{\mathrm{T}}\in\mathbb{C}^{KN_{\mathrm{BS}}^{\mathrm{sub}}\times N_o^u}$ can be written as

$$\bar{\boldsymbol{Y}}_u = \left(\boldsymbol{A}_\tau \odot \bar{\boldsymbol{A}}_{\text{BS}}\right) \text{diag}(\boldsymbol{\alpha}) \left(\boldsymbol{A}_{\text{UE}}^{\text{T}} \boldsymbol{F}_u \boldsymbol{S}_u\right) + \check{\boldsymbol{N}}_u^{\text{T}}. \tag{2.22}$$

From $\bar{\boldsymbol{A}}_{\text{BS}} = \boldsymbol{J}_u \widetilde{\boldsymbol{W}}_u^{\text{H}} \boldsymbol{A}_{\text{BS}}^*$ in Eq. (2.18), we observe that $\widetilde{\boldsymbol{W}}_u$ may destroy the shift-invariance structure of $\boldsymbol{A}_{\text{BS}}$. Similar to the downlink CE stage, we can also use Algorithm 2.1 to design $\widetilde{\boldsymbol{W}}_u$, where the input parameters N_d, N_{s}^d, $N_{\text{UE}}^{\text{RF}}$, N_{UE}^{h}, N_{UE}^{v}, M_{UE}^{h}, and M_{UE}^{v} need to be replaced by N_u, N_{s}^u, $N_{\text{BS}}^{\text{RF}}$, N_{BS}^{h}, N_{BS}^{v}, M_{BS}^{h}, and M_{BS}^{v} correspondingly. Next, by substituting the designed $\boldsymbol{W}_u$ into Eq. (2.22), we have

$$\bar{\boldsymbol{Y}}_u = \boldsymbol{A}_{\tau\text{BS}} \bar{\boldsymbol{S}}_u + \check{\boldsymbol{N}}_u^{\text{T}}, \tag{2.23}$$

where $\bar{\boldsymbol{S}}_u = \text{diag}(\boldsymbol{\alpha}) \left(\boldsymbol{A}_{\text{UE}}^{\text{T}} \boldsymbol{F}_u \boldsymbol{S}_u\right)$, and $\boldsymbol{A}_{\tau\text{BS}} \in \mathbb{C}^{K M_{\text{BS}}^{\text{v}} M_{\text{BS}}^{\text{h}} \times L}$ is given by

$$\begin{aligned} \boldsymbol{A}_{\tau\text{BS}} &= \boldsymbol{A}_\tau \odot \left(\boldsymbol{J}_u \widetilde{\boldsymbol{W}}_u^{\text{H}} \left((\boldsymbol{A}_{\text{BS}}^{\nu})^* \odot (\boldsymbol{A}_{\text{BS}}^{\mu})^*\right)\right) \\ &= \boldsymbol{A}_\tau \odot \bar{\boldsymbol{A}}_{\text{BS}}^{\nu} \odot \bar{\boldsymbol{A}}_{\text{BS}}^{\mu}. \end{aligned} \tag{2.24}$$

In Eq. (2.24), $\bar{\boldsymbol{A}}_{\text{BS}}^{\mu} \in \mathbb{C}^{M_{\text{BS}}^{\text{h}} \times L}$ $\left(\bar{\boldsymbol{A}}_{\text{BS}}^{\nu} \in \mathbb{C}^{M_{\text{BS}}^{\text{v}} \times L}\right)$ is composed by the first M_{BS}^{h} (M_{BS}^{v}) rows of $\left(\boldsymbol{A}_{\text{BS}}^{\mu}\right)^*$ $\left(\left(\boldsymbol{A}_{\text{BS}}^{\nu}\right)^*\right)$. In this way, the designed $\widetilde{\boldsymbol{W}}_u$ transforms the high-dimensional hybrid beamforming matrix into a low-dimensional fully-digital array, which makes $\boldsymbol{A}_{\tau\text{BS}}$ preserve the triple shift-invariance structure [107] of horizontal/vertical AoDs and delays in $\boldsymbol{A}_{\text{BS}}$. Finally, by utilizing the MDU-ESPRIT algorithm in Sect. 2.5, we can obtain the super-resolution estimates of horizontal/vertical AoDs and delays, $\{\widehat{\theta}_l^{\text{BS}}, \widehat{\varphi}_l^{\text{BS}}, \widehat{\tau}_l\}_{l=1}^{\widehat{L}}$.

2.4.2 Design Multi-beam Transmit Precoding Matrix at UE

At the uplink CE stage, the transmit power-limited UE can utilize the estimated horizontal/vertical AoAs to design the uplink transmit precoding matrix $\boldsymbol{F}_u = \boldsymbol{F}_{\text{RF},u} \boldsymbol{F}_{\text{BB},u}$, which helps to improve the received SNR at BS by transmitting directional multi-beam signals.

First, we consider the design of analog transmit precoding matrix $\boldsymbol{F}_{\text{RF},u} \in \mathbb{C}^{N_{\text{UE}} \times N_{\text{UE}}^{\text{RF}}}$. Given estimated horizontal/vertical angles $\{\widehat{\theta}_l^{\text{UE}}, \widehat{\varphi}_l^{\text{UE}}\}_{l=1}^{\widehat{L}}$, we can construct the array response matrix as $\widehat{\boldsymbol{A}}_{\text{UE}}$. To fully utilize the acquired $\{\widehat{\theta}_l^{\text{UE}}, \widehat{\varphi}_l^{\text{UE}}\}_{l=1}^{\widehat{L}}$, the designed multi-beam transmit precoding matrix should align $\widehat{L}$ beams to $\widehat{L}$ estimated AoAs. Specifically, we first divide the phase shifters of the PSN into $\widehat{L}$ groups as equally as possible, then, the designs of $\boldsymbol{F}_{\text{RF},u}$ can be classified as the following three scenarios according to $\widehat{L}$ and $N_{\text{UE}}^{\text{RF}}$.

- *Case 1:* $\widehat{L} > N_{\text{UE}}^{\text{RF}}$. In this case, the number of beams transmitted by UE is larger than the number of RF chains. For a hybrid MIMO architecture adopting fully-connected PSN, the total number of phase shifters is $N_{\text{PS}} = N_{\text{UE}}^{\text{RF}} N_{\text{UE}}$. First, by assigning $n_{\text{ps},l}$ phase shifters to the l-th group. Then, we introduce a $\widehat{L}$-length vector $\boldsymbol{v}_{\text{ps}}$ as

$$\boldsymbol{v}_{\text{ps}} = \left[n_{\text{ps},1} \ldots n_{\text{ps},\widehat{L}}\right]^{\text{T}} = n_{\text{ps}}\mathbf{1}_{\widehat{L}} + \left[\mathbf{1}_{n_{\text{re}}}^{\text{T}}\ \mathbf{0}_{\widehat{L}-n_{\text{re}}}^{\text{T}}\right]^{\text{T}}, \tag{2.25}$$

where $n_{\text{ps}} = \lfloor N_{\text{PS}}/\widehat{L}\rfloor$ and $n_{\text{re}} = \text{mod}(N_{\text{PS}}, \widehat{L})$. Defining a index vector $\boldsymbol{p} = [1\ 2 \ldots N_{\text{PS}}]^{\text{T}}$, then the ordered index set corresponding to l-th group can be represented by $\mathcal{P}_l = \boldsymbol{p}_{\left\{\sum_{i=1}^{l-1} n_{\text{ps},i}+1:\sum_{i=1}^{l} n_{\text{ps},i}\right\}}$, and the cardinality of it is $|\mathcal{P}_l|_c = n_{\text{ps},l}$. We define the vector $\boldsymbol{f} = [\boldsymbol{f}_1^{\text{T}} \ldots \boldsymbol{f}_{\widehat{L}}^{\text{T}}]^{\text{T}} \in \mathbb{C}^{N_{\text{PS}}}$, where $\boldsymbol{f}_l = \widehat{\boldsymbol{A}}^*_{\text{UE}\{\text{mod}(\mathcal{P}_l, N_{\text{UE}}), l\}} \in \mathbb{C}^{n_{\text{ps},l}}$, $\forall 1 \le l \le \widehat{L}$. Finally, we obtain $\boldsymbol{F}_{\text{RF},u} = \text{mat}\left(\boldsymbol{f};\ N_{\text{UE}}, N_{\text{UE}}^{\text{RF}}\right)$.

- *Case 2*: $\widehat{L} \le N_{\text{UE}}^{\text{RF}}$, *and* $N_{\text{UE}}^{\text{RF}}$ *can be exactly divided by* $\widehat{L}$. The $N_{\text{UE}}^{\text{RF}}$ RF chains can be equally allocated to $\widehat{L}$ groups. By defining $N_{\text{rep}} = N_{\text{UE}}^{\text{RF}}/\widehat{L}$, then we have $\boldsymbol{F}_{\text{RF},u} = \mathbf{1}_{N_{\text{rep}}}^{\text{T}} \otimes \widehat{\boldsymbol{A}}^*_{\text{UE}}$.
- *Case 3*: $\widehat{L} \le N_{\text{UE}}^{\text{RF}}$, *and* $N_{\text{UE}}^{\text{RF}}$ *can not be exactly divided by* $\widehat{L}$. In this case, we first define $\boldsymbol{F}_{\text{RF},u} = \left[\boldsymbol{F}_{\text{RF},u}^1\ \boldsymbol{F}_{\text{RF},u}^2\right]$, where $\boldsymbol{F}_{\text{RF},u}^1 \in \mathbb{C}^{N_{\text{UE}} \times \widehat{L} N_{\text{rep}}}$ with $N_{\text{rep}} = \lfloor N_{\text{UE}}^{\text{RF}}/\widehat{L}\rfloor$, and $\boldsymbol{F}_{\text{RF},u}^2 \in \mathbb{C}^{N_{\text{UE}} \times N_{\text{UE,re}}^{\text{RF}}}$, with $N_{\text{UE,re}}^{\text{RF}} = \text{mod}(N_{\text{UE}}^{\text{RF}}, \widehat{L})$. The design of $\boldsymbol{F}_{\text{RF},u}^1$ is similar to that in *Case 2*, i.e., $\boldsymbol{F}_{\text{RF},u}^1 = \mathbf{1}_{N_{\text{rep}}}^{\text{T}} \otimes \widehat{\boldsymbol{A}}^*_{\text{UE}}$. Besides, the design of $\boldsymbol{F}_{\text{RF},u}^2$ is similar to that in *Case 1*, i.e., $\boldsymbol{F}_{\text{RF},u}^2 = \text{mat}\left([\widetilde{\boldsymbol{f}}_1^{\text{T}} \ldots \widetilde{\boldsymbol{f}}_{\widehat{L}}^{\text{T}}]^{\text{T}};\ N_{\text{UE}}, N_{\text{UE,re}}^{\text{RF}}\right)$, where the $\widetilde{\boldsymbol{f}}_l$ can be acquired through replacing $N_{\text{UE}}^{\text{RF}}$ by $N_{\text{UE,re}}^{\text{RF}}$.

Due to the limited resolution of the phase shifters in the PSN, each element of $\boldsymbol{F}_{\text{RF},u}$ needs to be quantized to the nearest value in phase set $\mathcal{A}$. For digital transmit precoding matrix $\boldsymbol{F}_{\text{BB},u} \in \mathbb{C}^{N_{\text{UE}}^{\text{RF}} \times N_s^u}$, each element of it can be assigned as $\{\boldsymbol{F}_{\text{BB},u}\}_{j_1,j_2} = e^{\text{j}2\pi b_{j_1,j_2}}$, where $b_{j_1,j_2} \sim \mathcal{U}(0, 1)$. Finally, the multi-beam transmit precoding matrix at the UE side is acquired as $\boldsymbol{F}_u = \boldsymbol{F}_{\text{RF},u}\boldsymbol{F}_{\text{BB},u}$.

To give an intuitive comparison of the designed $\boldsymbol{F}_d$ at the downlink CE stage side and the $\boldsymbol{F}_u$ at the uplink CE stage, Fig. 2.4 plots three kinds of beam patterns, where we consider both the BS and UE are equipped with $N_{\text{BS}}^{\text{RF}} = N_{\text{UE}}^{\text{RF}} = 4$ RF chains, and the 5 MPCs of the channel are assumed to be known at the UE side. Specifically, Fig. 2.4a plots the beam pattern transmitted by the random transmit precoding matrix $\boldsymbol{F}_d$ in an 8×8 UPA. On the contrary, Fig. 2.4b, c depict the beam patterns transmitted by a well-designed multi-beam transmit precoding matrix $\boldsymbol{F}_u$ in an 8×8 and 16×16 UPA, respectively. Compared with the beam pattern in Fig. 2.4a, the beam pattern in Fig. 2.4b clearly exhibits 5 main lobes corresponding to 5 MPCs in the channel, thus improving the received SNR obviously. In addition, by comparing Fig. 2.4b, c, it can be found that the sidelobes of the multi-beam signals can be further suppressed by enlarging the size of the array. In short, when the transmit power is limited at the UE, by leveraging the designed multi-beam precoding matrix to align the multi-beam directional signals to the estimated AoA directions, the uplink CE performance can be improved.

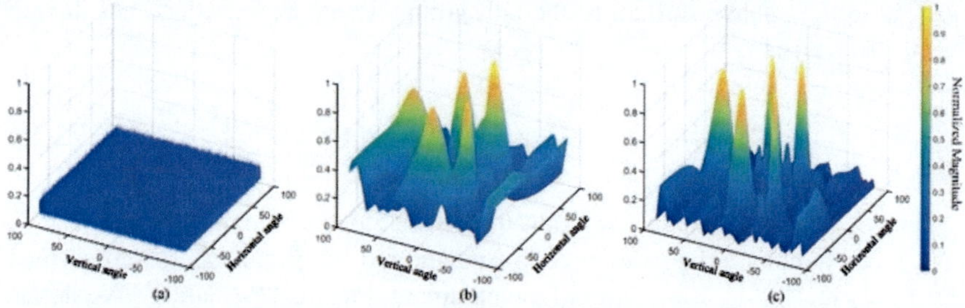

Fig. 2.4 Comparison of beam patterns, where $N_{\text{BS}}^{\text{RF}} = N_{\text{UE}}^{\text{RF}} = 4$ and the AoAs of the 5 MPCs are known to the UE: **a** random transmit precoding matrix with 8×8 UPA, **b** multi-beam transmit precoding matrix with 8×8 UPA, and **c** multi-beam transmit precoding matrix with 16×16 UPA [101]

2.5 MDU-ESPRIT Algorithm

Since the designed receive combining matrix $\widetilde{\boldsymbol{W}}_d$ ($\widetilde{\boldsymbol{W}}_u$) reconstructs the double (triple) shift-invariance structure of the array response at the UE (BS), we can use the R-dimensional ($R \geq 2$) unitary ESPRIT algorithm detailed in this section to estimate the channel parameters. Without loss of generality, we define the following general signal transmission model that consists of L MPCs and R spatial frequencies.

$$\boldsymbol{Y} = \boldsymbol{A}\boldsymbol{S} + \boldsymbol{N}, \tag{2.26}$$

where $\boldsymbol{Y} \in \mathbb{C}^{M \times N}$ is aggregated data matrix received at N snapshots. $M = \prod_{r=1}^{R} M_r$, and M_r is the dimension of the parameter vector that corresponds to r-th ($1 \leq r \leq R$) spatial frequency. Besides, $\boldsymbol{S} \in \mathbb{C}^{L \times N}$ and $\boldsymbol{N} \in \mathbb{C}^{M \times N}$ are the transmit signal and noise matrix, respectively. The array response array $\boldsymbol{A} \in \mathbb{C}^{M \times L}$ is given by

$$\begin{aligned} \boldsymbol{A} &= \boldsymbol{A}_{\mu_R} \odot \cdots \odot \boldsymbol{A}_{\mu_2} \odot \boldsymbol{A}_{\mu_1} \\ &= \left[\boldsymbol{a}\left(\mu_1^1, \mu_1^2, \ldots, \mu_1^R\right) \ldots \boldsymbol{a}\left(\mu_L^1, \mu_L^2, \ldots, \mu_L^R\right)\right]. \end{aligned} \tag{2.27}$$

In Eq. (2.27), $\boldsymbol{A}_{\mu_r} = \left[\boldsymbol{a}(\mu_1^r) \ldots \boldsymbol{a}(\mu_L^r)\right] \in \mathbb{C}^{M_r \times L}$ is the steering matrix associated with r-th set of spatial frequencies $\{\mu_l^r\}_{l=1}^{L}$, and $\boldsymbol{a}\left(\mu_l^r\right) = \left[1\ e^{\mathrm{j}\mu_l^r} \ldots e^{\mathrm{j}(M_r-1)\mu_l^r}\right]^{\mathrm{T}} \in \mathbb{C}^{M_r}$ is the l-th steering vector. Furthermore, the array response vector associated to l-th MPC can be expressed by

$$\boldsymbol{a}\left(\mu_l^1, \mu_l^2, \ldots, \mu_l^R\right) = \boldsymbol{a}\left(\mu_l^R\right) \otimes \cdots \otimes \boldsymbol{a}\left(\mu_l^2\right) \otimes \boldsymbol{a}\left(\mu_l^1\right). \tag{2.28}$$

The proposed MDU-ESPRIT algorithm, which can acquire R sets of super-resolution estimates of spatial frequencies $\left\{\widehat{\mu}_l^r\right\}_{l=1}^{L}$ ($1 \leq r \leq R$) from Eq. (2.26), mainly contains the following five steps.

Step 1: R-D Spatial Smoothing Preprocessing (SSP)
Since the limited training overhead will lead to insufficient observation dimension N, spatial smoothing preprocessing technique [100] is exploited to preprocess the original data matrix $\boldsymbol{Y}$ in Eq. (2.26). This preprocessing method can mitigate the influence of other coherent signals and avoid the rank-deficiency of covariance matrix $\boldsymbol{S}$, thus increasing the robustness of the estimation. Specifically, we first define R spatial smoothing parameters $\{G_r\}_{r=1}^{R}$ $(1 \le G_r \le M_r)$, then the sub-dimension $\{M_r^{\text{sub}}\}_{r=1}^{R}$ corresponding to $\{M_r\}_{r=1}^{R}$ is $M_r^{\text{sub}} = M_r - G_r + 1$ $(1 \le r \le R)$. Therefore, the total dimension is $M_{\text{sub}} = \prod_{r=1}^{R} M_r^{\text{sub}}$. To acquire R-D selection matrix, we first define the g_r-th $(1 \le g_r \le G_r)$ 1-D selection matrix as $\boldsymbol{J}^{(g_r)} = \left[\boldsymbol{O}_{M_r^{\text{sub}} \times (g_r - 1)}\ \boldsymbol{I}_{M_r^{\text{sub}}}\ \boldsymbol{O}_{M_r^{\text{sub}} \times (G_r - g_r)}\right] \in \mathbb{R}^{M_r^{\text{sub}} \times M_r}$, then we obtain $G = \prod_{r=1}^{R} G_r$ R-D selection matrices, where the $(g_1, g_2, \ldots, g_R)$-th matrix is denoted by $\boldsymbol{J}_{g_1, g_2, \ldots, g_R} = \boldsymbol{J}^{(g_R)} \otimes \cdots \otimes \boldsymbol{J}^{(g_2)} \otimes \boldsymbol{J}^{(g_1)} \in \mathbb{R}^{M_{\text{sub}} \times M}$. By applying the R-D selection matrices to $\boldsymbol{Y}$, we acquire the smoothed complex-valued data matrix $\bar{\boldsymbol{Y}} \in \mathbb{C}^{M_{\text{sub}} \times NG}$ as

$$\bar{\boldsymbol{Y}} = \left[\left(\boldsymbol{J}_{1,1,\ldots,1,1}\boldsymbol{Y}\right) \ldots \left(\boldsymbol{J}_{1,1,\ldots,1,G_R}\boldsymbol{Y}\right) \left(\boldsymbol{J}_{1,1,\ldots,2,1}\boldsymbol{Y}\right) \ldots \left(\boldsymbol{J}_{G_1,G_2,\ldots,G_{R-1},G_R}\boldsymbol{Y}\right)\right]. \tag{2.29}$$

Step 2: Real-Valued Processing (RVP)
To reduce the computational complexity, we employ the forward-backward averaging technique [99] to transform the $\bar{\boldsymbol{Y}}$ into a real-value matrix $\bar{\boldsymbol{Y}}_{\text{re}} \in \mathbb{R}^{M_{\text{sub}} \times 2NG}$, as given by

$$\bar{\boldsymbol{Y}}_{\text{re}} = \boldsymbol{Q}_{M_{\text{sub}}}^{\text{H}} \left[\bar{\boldsymbol{Y}}\ \left(\boldsymbol{\Pi}_{M_{\text{sub}}} \bar{\boldsymbol{Y}}^* \boldsymbol{\Pi}_{NG}\right)\right] \boldsymbol{Q}_{2NG}, \tag{2.30}$$

where $\boldsymbol{\Pi}_n$ is a $n \times n$ exchanging matrix (obtained by permuting the order of $\boldsymbol{I}_n$'s rows), and $\boldsymbol{Q}_n \in \mathbb{C}^{n \times n}$ is a sparse unitary matrix [110] which satisfies $\boldsymbol{\Pi}_n \boldsymbol{Q}_n^{\text{H}} = \boldsymbol{Q}_n$, where

$$\boldsymbol{Q}_2 n = \frac{1}{\sqrt{2}} \begin{bmatrix} \boldsymbol{I}_n & \text{j}\boldsymbol{I}_n \\ \boldsymbol{\Pi}_n & -\text{j}\boldsymbol{\Pi}_n \end{bmatrix}, \boldsymbol{Q}_{2n+1} = \frac{1}{\sqrt{2}} \begin{bmatrix} \boldsymbol{I}_n & \boldsymbol{0}_n & \text{j}\boldsymbol{I}_n \\ \boldsymbol{0}_n^{\text{T}} & \sqrt{2} & \boldsymbol{0}_n^{\text{T}} \\ \boldsymbol{\Pi}_n & \boldsymbol{0}_n & -\text{j}\boldsymbol{\Pi}_n \end{bmatrix}. \tag{2.31}$$

Step 3: Signal Subspace Approximation (SSA)
To extra spatial-frequency information from the real-valued matrix $\bar{\boldsymbol{Y}}_{\text{re}}$, we introduce a transforming matrix $\boldsymbol{K}$, which satisfies [99]

$$\Re\left\{\boldsymbol{Q}_{m_r}^{\text{H}} \boldsymbol{J}_r \boldsymbol{Q}_{M_{\text{sub}}}\right\} \boldsymbol{K} \boldsymbol{\Lambda}_r = \Im\left\{\boldsymbol{Q}_{m_r}^{\text{H}} \boldsymbol{J}_r \boldsymbol{Q}_{M_{\text{sub}}}\right\} \boldsymbol{K}, \tag{2.32}$$

where $1 \le r \le R$, $m_r = M_{\text{sub}}\left(M_r^{\text{sub}} - 1\right)/M_r^{\text{sub}}$, $\boldsymbol{\Lambda}_r = \text{diag}\left(\left[\tan\left(\mu_1^r/2\right) \ldots \tan\left(\mu_L^r/2\right)\right]^{\text{T}}\right)$ is a real-valued diagonal matrix which contains the target spatial frequencies $\{\mu_l^r\}_{l=1}^{L}$, and $\boldsymbol{J}_r = \boldsymbol{I}_{\prod_{i=r+1}^{R} M_i^{\text{sub}}} \otimes \widetilde{\boldsymbol{J}}^{(r)} \otimes \boldsymbol{I}_{\prod_{i=1}^{r-1} M_i^{\text{sub}}} \in \mathbb{R}^{m_r \times M_{\text{sub}}}$ with $\widetilde{\boldsymbol{J}}^{(r)} = \left[\boldsymbol{0}_{M_r^{\text{sub}}-1}\ \boldsymbol{I}_{M_r^{\text{sub}}-1}\right]$. Note that $\boldsymbol{K}$ is related to the approximated signal subspace matrix $\boldsymbol{E}_{\text{s}} \in \mathbb{R}^{M_{\text{sub}} \times L}$. Specifically, since the columns of $\boldsymbol{K}$ and $\boldsymbol{E}_{\text{s}}$ span the same L-

dimensional signal subspace [100, 107], $\boldsymbol{K} = \boldsymbol{E}_s \boldsymbol{T}$, where $\boldsymbol{T} \in \mathbb{R}^{L \times L}$ is a non-singular matrix. According to the definition of the signal subspace, $\boldsymbol{E}_s$ can be determined by the left singular vectors corresponding to the L largest singular values of $\bar{\boldsymbol{Y}}_{\text{re}}$. Therefore, by applying singular value decomposition (SVD) to the real part of $\bar{\boldsymbol{Y}}_{\text{re}}$, $\bar{\boldsymbol{Y}}_{\text{re}} = \boldsymbol{U}_{\text{re}} \boldsymbol{\Sigma}_{\text{re}} \boldsymbol{V}_{\text{re}}^{\text{H}}$, we obtain $\boldsymbol{E}_s = \boldsymbol{U}_{\text{re}\{:,1:L\}}$.

Step 4: Shift-Invariance Equation Solving (SIES)
After obtaining the approximated signal subspace $\boldsymbol{E}_s$, we can obtain the following R shift-invariance equations by substituting $\boldsymbol{K} = \boldsymbol{E}_s \boldsymbol{T}$ into Eq. (2.32)

$$\Re\left\{\boldsymbol{Q}_{m_r}^{\text{H}} \boldsymbol{J}_r \boldsymbol{Q}_{M_{\text{sub}}}\right\} \boldsymbol{E}_s \boldsymbol{\Phi}_r = \Im\left\{\boldsymbol{Q}_{m_r}^{\text{H}} \boldsymbol{J}_r \boldsymbol{Q}_{M_{\text{sub}}}\right\} \boldsymbol{E}_s, \ 1 \le r \le R, \tag{2.33}$$

where $\boldsymbol{\Phi}_r = \boldsymbol{T} \boldsymbol{\Lambda}_r \boldsymbol{T}^{-1} \in \mathbb{R}^{L \times L}$. To estimate R diagonal matrices $\{\boldsymbol{\Lambda}_r\}_{r=1}^R$, we exploit LS or total least squares (TLS) estimators to solve R shift-invariance equations in (2.33), from which we acquire R real-valued estimates of $\{\boldsymbol{\Phi}_r\}_{r=1}^R$, denoted by $\{\widehat{\boldsymbol{\Phi}}_r\}_{r=1}^R$.

Step 5: R-D Joint Diagonalization (JD)
For $1 \le r \le R$, we first define the estimate of $\boldsymbol{\Lambda}_r$ as $\widehat{\boldsymbol{\Lambda}}_r$, then the estimated $\widehat{\boldsymbol{\Phi}}_r$ can be further expressed by $\widehat{\boldsymbol{\Phi}}_r = \boldsymbol{T} \widehat{\boldsymbol{\Lambda}}_r \boldsymbol{T}^{-1}$. Next, we perform the R-D joint diagonalization of $\{\widehat{\boldsymbol{\Lambda}}_r\}_{r=1}^R$ to acquire paired estimates of spatial frequencies $\{\widehat{\mu}_l^r\}_{l=1}^L$. Specifically, we consider the cases of $R=2$ and $R \ge 3$. For the 2-D case $R=2$, since both $\widehat{\boldsymbol{\Phi}}_1$ and $\widehat{\boldsymbol{\Phi}}_2$ have the same eigen-matrix $\boldsymbol{T}$, we can obtain $\widehat{\boldsymbol{\Lambda}}_1$ and $\widehat{\boldsymbol{\Lambda}}_2$ by calculating the EVD of the complex-valued matrix $\boldsymbol{\Psi} = \widehat{\boldsymbol{\Phi}}_1 + \text{j}\widehat{\boldsymbol{\Phi}}_2$. Specifically, denote the EVD of $\boldsymbol{\Psi}$ as $\boldsymbol{\Psi} = \boldsymbol{T} \boldsymbol{\Delta} \boldsymbol{T}^{-1}$, where $\boldsymbol{\Delta} = \widehat{\boldsymbol{\Lambda}}_1 + \text{j}\widehat{\boldsymbol{\Lambda}}_2$, then we have $\widehat{\boldsymbol{\Lambda}}_1 = \Re\{\boldsymbol{\Delta}\}$ and $\widehat{\boldsymbol{\Lambda}}_2 = \Im\{\boldsymbol{\Delta}\}$. For $R \ge 3$, the R matrices $\{\widehat{\boldsymbol{\Phi}}_r\}_{r=1}^R$ polluted by noise do not always share the same $\boldsymbol{T}$. Therefore, we resort to the simultaneous Schur decomposition (SSD) algorithm [106] derived from real-valued Schur decomposition [109] to estimate and pair the multiple parameters. By exploiting SSD algorithms, we can obtain R approximated upper-triangular matrices $\{\boldsymbol{\Gamma}_r\}_{r=1}^R$, from which we can further determine $\{\widehat{\boldsymbol{\Lambda}}_r\}_{r=1}^R$ according to the diagonal elements of $\{\boldsymbol{\Gamma}_r\}_{r=1}^R$, i.e., $\widehat{\boldsymbol{\Lambda}}_r = \text{diag}\left(\text{vdiag}(\boldsymbol{\Gamma}_r)\right), \forall 1 \le r \le R$. Finally, we obtain R sets of paired super-resolution of spatial frequencies $\{\widehat{\mu}_l^r\}_{l=1}^L$ based on $\{\widehat{\boldsymbol{\Lambda}}_r\}_{r=1}^R$, i.e., $\widehat{\mu}_l^r = 2\arctan([\widehat{\boldsymbol{\Lambda}}_r]_{l,l}), \forall 1 \le l \le L, 1 \le r \le R$.

The above MDU-ESPRIT algorithm is summarized in Algorithm 2.2. At the downlink CE stage, we can obtain the $\{\widehat{\mu}_l^{\text{UE}}, \widehat{\nu}_l^{\text{UE}}\}_{l=1}^{\widehat{L}}$ by applying the 2DU-ESPRIT ($R = 2$) algorithm to $\bar{\boldsymbol{Y}}_d$ in Eq. (2.11), while at the uplink CE stage, we can obtain $\{\widehat{\mu}_l^{\text{BS}}, \widehat{\nu}_l^{\text{BS}}, \widehat{\mu}_l^{\tau}\}_{l=1}^{\widehat{L}}$ by applying the 3DU-ESPRIT ($R = 3$) algorithm to $\bar{\boldsymbol{Y}}_u$ in Eq. (2.23). Hence, the spatial smoothing parameters of Algorithm 2.2 are $\{G_1^d, G_2^d\}$ and $\{G_1^u, G_2^u, G_3^u\}$ in 2D and 3D cases, respectively. The total sub-dimensions at the UE and BS are $M_{\text{sub}}^{\text{UE}} = (M_{\text{UE}}^{\text{h}} - G_1^d + 1)(M_{\text{UE}}^{\text{v}} - G_2^d + 1)$ and $M_{\text{sub}}^{\text{BS}} = (M_{\text{BS}}^{\text{h}} - G_1^u + 1)(M_{\text{BS}}^{\text{v}} - G_2^u + 1)(K - G_3^u + 1)$, respectively.

Algorithm 2.2: MDU-ESPRIT Algorithm

Input: Data matrix $\boldsymbol{Y}$, number of MPCs L, sub-dimensions $\{M_r\}_{r=1}^{R}$, and spatial smoothing parameters $\{G_r\}_{r=1}^{R}$

Output: Estimates of spatial frequencies $\left\{\widehat{\mu}_l^r\right\}_{l=1}^{L}$, $1 \leq r \leq R$

Obtain $\bar{\boldsymbol{Y}}$ in Eq. (2.29) using R-D spatial smoothing preprocessing ;
Obtain $\bar{\boldsymbol{Y}}_{\mathrm{re}}$ in Eq. (2.30) using forward backward averaging ;
Determine approximate signal subspace matrix $\boldsymbol{E}_{\mathrm{s}}$ through SVD ;
Solve Eq. (2.33) to obtain R real-valued matrices $\left\{\widehat{\boldsymbol{\Phi}}_r\right\}_{r=1}^{R}$;
if $R = 2$ **then**
  Use EVD of $\boldsymbol{\Psi} = \boldsymbol{T}\boldsymbol{\Delta}\boldsymbol{T}^{-1}$ to obtain $\widehat{\boldsymbol{\Lambda}}_1 = \Re\{\boldsymbol{\Delta}\}$ and $\widehat{\boldsymbol{\Lambda}}_2 = \Im\{\boldsymbol{\Delta}\}$
else if $R \geq 3$ **then**
  Obtain R diagonal matrices $\left\{\widehat{\boldsymbol{\Lambda}}_r\right\}_{r=1}^{R}$ via SSD algorithm
end
Extract R paired $\left\{\widehat{\mu}_l^r\right\}_{l=1}^{L}$ from $\left\{\widehat{\boldsymbol{\Lambda}}_r\right\}_{r=1}^{R}$

2.6 ML-Based Parameters Pairing and Channel Gains Estimation

At the downlink CE stage, the BS has acquired the estimated horizontal/vertical AoAs $\left\{\bar{\theta}_l^{\mathrm{UE}}, \bar{\varphi}_l^{\mathrm{UE}}\right\}_{l=1}^{\widehat{L}}$ fed back by UE. Then, the BS acquires the estimates of horizontal/vertical AoDs and delays $\{\widehat{\theta}_l^{\mathrm{BS}}, \widehat{\varphi}_l^{\mathrm{BS}}, \widehat{\tau}_l\}_{l=1}^{\widehat{L}}$ at the uplink CE stage. Since the $\left\{\bar{\theta}_l^{\mathrm{UE}}, \bar{\varphi}_l^{\mathrm{UE}}\right\}_{l=1}^{\widehat{L}}$ and $\{\widehat{\theta}_l^{\mathrm{BS}}, \widehat{\varphi}_l^{\mathrm{BS}}, \widehat{\tau}_l\}_{l=1}^{\widehat{L}}$ are acquired by two different devices, we need to pair these parameters. Besides, we also need to estimate the path gain vector $\boldsymbol{\alpha}$. We propose to use the ML method to pair the channel parameters and estimate the path gains, which corresponds to the Step 7 in Fig. 2.1.

Specifically, according to Eq. (2.24), we first construct the steering matrix $\widehat{\boldsymbol{A}}_{\tau \mathrm{BS}}$ associated to $\{\widehat{\theta}_l^{\mathrm{BS}}, \widehat{\varphi}_l^{\mathrm{BS}}, \widehat{\tau}_l\}_{l=1}^{\widehat{L}}$, where $\{\widehat{\tau}_l\}_{l=1}^{\widehat{L}}$ is arranged in a ascending order. Then, similar to the construction method of $\boldsymbol{F}_u$ in Sect. 2.4.2, we utilize the received $\left\{\bar{\theta}_l^{\mathrm{UE}}, \bar{\varphi}_l^{\mathrm{UE}}\right\}_{l=1}^{\widehat{L}}$ fed back by the UE to construct the multi-beam transmit precoding matrix $\widehat{\boldsymbol{F}}_u$. It is obvious that there are a total of $J_{\mathrm{c}} = \widehat{L}!$ possible ordered paired methods. Equivalently, there are J_{c} possible pairing methods between $\widehat{\boldsymbol{A}}_{\tau \mathrm{BS}}(\{\widehat{\theta}_l^{\mathrm{BS}}, \widehat{\varphi}_l^{\mathrm{BS}}, \widehat{\tau}_l\}_{l=1}^{\widehat{L}})$ and $\left\{\bar{\theta}_l^{\mathrm{UE}}, \bar{\varphi}_l^{\mathrm{UE}}\right\}_{l=1}^{\widehat{L}}$, where $j \in \mathcal{J}$ and $|\mathcal{J}|_{\mathrm{c}} = J_{\mathrm{c}}$. For each $\{\bar{\theta}_{l_j}^{\mathrm{UE}}, \bar{\varphi}_{l_j}^{\mathrm{UE}}\}_{l_j=1}^{\widehat{L}}$, we can construct the corresponding array response matrix $\boldsymbol{A}_{\mathrm{UE}}$, which is denoted as $\widehat{\boldsymbol{A}}_{\mathrm{UE},j}$. Then for each pair of the parameters at the UE/BS, we can obtain a corresponding $\widehat{\boldsymbol{A}}_{\tau \mathrm{BS}}$, $\widehat{\boldsymbol{A}}_{\mathrm{UE},j}$ and $\widehat{\boldsymbol{F}}_u$. By substituting them into Eq. (2.20), we have $\check{\boldsymbol{y}}_u = \widehat{\boldsymbol{A}}_j \boldsymbol{\alpha}_j + \check{\boldsymbol{n}}_u$, where $\widehat{\boldsymbol{A}}_j = \widehat{\boldsymbol{A}}_{\tau \mathrm{BS}} \odot (\widehat{\boldsymbol{A}}_{\mathrm{UE},j}^{\mathrm{T}} \widehat{\boldsymbol{F}}_u \boldsymbol{S}_u)^{\mathrm{T}}$ and $\boldsymbol{\alpha}_j$ is the channel gain corresponding to the j-th pair of the AoDs and AoAs with $j \in \mathcal{J}$. Hence, the LS estimation of $\boldsymbol{\alpha}_j$ can be easily obtained by

$$\widehat{\boldsymbol{\alpha}}_j = \left(\widehat{\boldsymbol{A}}_j^{\mathrm{H}} \widehat{\boldsymbol{A}}_j\right)^{-1} \widehat{\boldsymbol{A}}_j^{\mathrm{H}} \check{\boldsymbol{y}}_u. \tag{2.34}$$

From the estimated $\widehat{\boldsymbol{\alpha}}_j$ in Eq. (2.34) and the approximated relationships $\check{\boldsymbol{y}}_u \approx \widehat{\boldsymbol{A}}_j \widehat{\boldsymbol{\alpha}}_j$, we have $\widehat{\check{\boldsymbol{y}}}_{u,j} = \widehat{\boldsymbol{A}}_j \widehat{\boldsymbol{\alpha}}_j$, where the residual between $\check{\boldsymbol{y}}_u$ and $\widehat{\check{\boldsymbol{y}}}_{u,j}$ is $\|\check{\boldsymbol{y}}_u - \widehat{\check{\boldsymbol{y}}}_{u,j}\|_2^2$. Then optimal pairing index can be acquired from the following optimization problem

$$j^{\star} = \arg\min_{j \in \mathcal{J}} \left\| \check{\boldsymbol{y}}_u - \widehat{\check{\boldsymbol{y}}}_{u,j} \right\|_2^2. \tag{2.35}$$

Hence, the optimal estimation of path gains can be represented by $\widehat{\boldsymbol{\alpha}} = \widehat{\boldsymbol{\alpha}}_{j^\star} = \beta \left[\widehat{\alpha}_1 \ldots \widehat{\alpha}_{\widehat{L}}\right]^{\mathrm{T}}$, and the optimal ordered parameters of mmWave channel are given as $\{\bar{\theta}_l^{\mathrm{UE}}, \bar{\varphi}_l^{\mathrm{UE}}, \widehat{\theta}_l^{\mathrm{BS}}, \widehat{\varphi}_l^{\mathrm{BS}}, \widehat{\tau}_l, \widehat{\alpha}_l\}_{l=1}^{\widehat{L}}$.

Finally, by substituting the acquired channel parameters $\{\bar{\theta}_l^{\mathrm{UE}}, \bar{\varphi}_l^{\mathrm{UE}}, \widehat{\theta}_l^{\mathrm{BS}}, \widehat{\varphi}_l^{\mathrm{BS}}, \widehat{\tau}_l, \widehat{\alpha}_l\}_{l=1}^{\widehat{L}}$ into Eq. (2.7), we obtain the optimally-estimated frequency domain channel $\widehat{\boldsymbol{H}}[k]$ of the k-th subcarrier, as given by

$$\widehat{\boldsymbol{H}}[k] = \beta \sum_{l=1}^{\widehat{L}} \widehat{\alpha}_l \boldsymbol{a}_{\mathrm{UE}}\left(\bar{\mu}_l^{\mathrm{UE}}, \bar{\nu}_l^{\mathrm{UE}}\right) \boldsymbol{a}_{\mathrm{BS}}^{\mathrm{H}}\left(\widehat{\mu}_l^{\mathrm{BS}}, \widehat{\nu}_l^{\mathrm{BS}}\right) e^{-\mathrm{j}2\pi \frac{k f_s}{K} \widehat{\tau}_l}, \tag{2.36}$$

where $\bar{\mu}_l^{\mathrm{UE}} = \pi \sin(\bar{\theta}_l^{\mathrm{UE}}) \cos(\bar{\varphi}_l^{\mathrm{UE}})$, $\bar{\nu}_l^{\mathrm{UE}} = \pi \sin(\bar{\varphi}_l^{\mathrm{UE}})$, $\widehat{\mu}_l^{\mathrm{BS}} = \pi \sin(\widehat{\theta}_l^{\mathrm{BS}}) \cos(\widehat{\varphi}_l^{\mathrm{BS}})$, and $\widehat{\nu}_l^{\mathrm{BS}} = \pi \sin(\widehat{\varphi}_l^{\mathrm{BS}})$.

2.7 Performance Evaluation

This section conducts an extensive simulation investigation to evaluate the performance of the proposed closed-loop space CE scheme and the corresponding computational complexity. During simulations, the carrier frequency is $f_c = 30$ GHz, the system bandwidth is $f_s = 200$ MHz, the number of RF chains at BS and UE are both $N_{\mathrm{BS}}^{\mathrm{RF}} = N_{\mathrm{UE}}^{\mathrm{RF}} = 4$, the antenna number at both horizontal and vertical directions at the BS and UE are both $N_{\mathrm{BS}}^{\mathrm{h}} = N_{\mathrm{BS}}^{\mathrm{v}} = N_{\mathrm{UE}}^{\mathrm{h}} = N_{\mathrm{UE}}^{\mathrm{v}} = 12$, the quantization resolution of PSN is $N_q^{\mathrm{ps}} = 3$ bits, and the quantization resolution of the feedback angle at UE side is $N_q^{\mathrm{ang}} = 10$ bits. Without loss generality, we assume only single UE is considered (i.e., $Q = 1$) during simulations. It can be seen from Fig. 2.2 that the uplink training overhead is $Q N_u N_{\mathrm{o}}^u$ when $Q > 1$ rather than $N_u N_{\mathrm{o}}^u$ in $Q = 1$ case. The channel parameters are as follows. The channel gains follow $CN(0, 1)$, and other channel parameters $\{\tau_l, \theta_l^{\mathrm{UE}}, \theta_l^{\mathrm{BS}}, \varphi_l^{\mathrm{UE}}, \varphi_l^{\mathrm{BS}}\}_{l=1}^{L}$ follow uniform distribution. Specifically, for the l-th MPC, $\tau_l \sim \mathcal{U}(0, \tau_{\max})$, and $\theta_l^{\mathrm{UE}}, \theta_l^{\mathrm{BS}}, \varphi_l^{\mathrm{UE}}, \varphi_l^{\mathrm{BS}} \sim \mathcal{U}(-\pi/3, \pi/3)$. The maximum multipath delay is set to $\tau_{\max} = 16 T_s$, i.e., $N_{\mathrm{c}} = 16$. The subcarrier number is set to $K = 128$ with the length of cyclic prefix (CP) being 32. In the proposed scheme, the high-dimensional hybrid beamforming array is equivalently visualized as a fully-digital array with sizes of $M_{\mathrm{BS}}^{\mathrm{h}} = M_{\mathrm{BS}}^{\mathrm{v}} = M_{\mathrm{UE}}^{\mathrm{h}} = M_{\mathrm{UE}}^{\mathrm{v}} = 8$. Therefore, given the number of independent signal streams $N_s^d = N_{\mathrm{UE}}^{\mathrm{RF}} - 1 = 3$ and $N_s^u = N_{\mathrm{BS}}^{\mathrm{RF}} - 1 = 3$ at the downlink and uplink CE stages, respectively, the number of time slots needed at the downlink and uplink CE stage are $N_d = 22$ and $N_u = 22$, respectively. In addition, $P = 8$ adjacent subcarriers are jointly employed to estimate the number of MPCs, and the threshold parameters ε are set to 1.54, 0.50, 0.16, 0.05, 0.016, and 0.005, respectively, at the SNR of -15 dB -10 dB, -5 dB, 0 dB, 5 dB, and 10 dB. The spatial smoothing parameters are set to $G_1^d = G_2^d = G_1^u = G_2^u = 2$ and $G_3^u = K/2$ in Algorithm 2.2. The downlink and uplink SNRs are both defined as $\rho\sigma_\alpha^2/\sigma_n^2$, where ρ and σ_n^2 are transmit power and the variance of received noise, respectively.

We consider orthogonal matching pursuit (OMP)-based [42] frequency domain CE scheme[4] and the simultaneous weighted-OMP (SW-OMP)-based [43] CE scheme as two benchmarks. To keep consistence with [42, 43], the digital transmit precoding/receive combing matrix in their corresponding schemes are replaced by identity matrices, while the analog designs are similar to that of $\boldsymbol{F}_{\mathrm{RF},d}[m]$ in Sect. 2.3.2. The size of quantized angular grids corresponding to the horizontal and vertical directions $G_{\mathrm{BS}}^{\mathrm{h}}$, $G_{\mathrm{BS}}^{\mathrm{v}}$, $G_{\mathrm{UE}}^{\mathrm{h}}$, and $G_{\mathrm{UE}}^{\mathrm{v}}$ are twice the number of antennas in the horizontal and vertical directions of UPA, respectively, i.e., $G_{\mathrm{BS}} = G_{\mathrm{BS}}^{\mathrm{h}} \times G_{\mathrm{BS}}^{\mathrm{v}} = 2N_{\mathrm{BS}}^{\mathrm{h}} \times 2N_{\mathrm{BS}}^{\mathrm{v}} = 24 \times 24$ and $G_{\mathrm{UE}} = G_{\mathrm{UE}}^{\mathrm{h}} \times G_{\mathrm{UE}}^{\mathrm{v}} = 2N_{\mathrm{UE}}^{\mathrm{h}} \times 2N_{\mathrm{UE}}^{\mathrm{v}} = 24 \times 24$. Additionally, to assuring the fairness, all the CE schemes employ the same training overhead, which is equal to the number of training frames in [42, 43].

2.7.1 Evaluation of CE Performance

First, we utilize the normalized mean square error (NMSE) as a metric to evaluate the CE performance, with its expression given by

$$\mathrm{NMSE} = 10\log_{10}\left(\mathbb{E}\Big(\sum_{k=0}^{K-1}\|\boldsymbol{H}[k] - \widehat{\boldsymbol{H}}[k]\|_F^2 \Big/ \sum_{k=0}^{K-1}\|\boldsymbol{H}[k]\|_F^2\Big)\right) \tag{2.37}$$

Figure 2.5 compares the NMSE performance of the proposed close-loop sparse CE scheme with the OMP-based and SW-OMP-based schemes for different SNRs, given the numbers of MPCs $L = 3$ and $L = 5$. For the proposed scheme, the number of OFDM symbols in each slot of the downlink and uplink CE stages are $N_o^d = N_o^u = 3$. Hence, the total training overhead of the proposed scheme is $T_{\mathrm{CE}} = N_d N_o^d + N_u N_o^u = 132$. In Fig. 2.5, the NMSE curve labeled 'Proposed Close-Loop' is our proposed closed-loop CE scheme, which also estimates the number of MPCs. The 'Benchmark of Closed-Loop' is the proposed closed-loop CE scheme with a known MPCs number. By comparing these two curves, it can be found that our scheme can approach the

[4] The redundant dictionary of the OMP-based time-domain method in [42] is designed by using redundant grids in delay-angular domains, and directly employing massive MIMO in mmWave systems will incur unaffordable computational complexity and storage demand. Therefore, we only consider the frequency-domain scheme in the simulations.

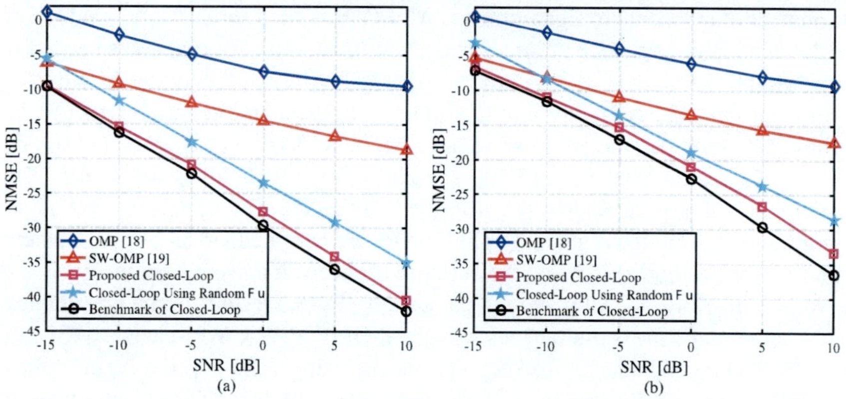

Fig. 2.5 NMSE performance comparisons of different CE schemes versus SNRs with the same training overhead $T_{\text{CE}} = 132$: **a** the number of MPCs $L = 3$; and **b** $L = 5$ [101]

lower bound when the number of MPCs L is unknown, which verifies that our scheme can acquire the super-resolution estimates of the channel parameters. In addition, Fig. 2.5 also compares the scheme that random transmit precoding matrix $\boldsymbol{F}_u$ is employed at the downlink and uplink CE stages. The results exhibit that our well-designed transmit precoding matrix can acquire a performance gain of 5 dB and 3 dB in $L = 3$ and $L = 5$ cases, respectively, compared with the random design. This clearly shows that by fully exploiting the estimated horizontal/vertical AoAs at the UE to design the precoding matrix, a better SNR can be obtained at the BS for an improved CE performance. Besides, the proposed parameterized CE scheme performs better than the other two CS-based schemes. Restricted by limited quantization resolution of grids, OMP-based scheme and SW-OMP-based scheme [42, 43] will lead to the NMSE floor at the high SNR regime. By employing denser discrete angular domain grids to achieve a higher dimensional CS dictionary, the performance of CS-based schemes can improve. However, this also leads to a significant increase in the computational complexity and the storage demand, which is unaffordable for an FD-MIMO system with massive antenna elements.

Figure 2.6 compares the NMSE performance of different schemes against different SNRs, given the same number of MPCs $L = 4$. Figure 2.6a, b employ a training overhead of $T_{\text{CE}} = 88$ and $T_{\text{CE}} = 176$ for $N_{\text{o}}^d = N_{\text{o}}^u = 2$ and $N_{\text{o}}^d = N_{\text{o}}^u = 4$, respectively. Similar conclusions can be drawn from Fig. 2.6 as those in Fig. 2.5. In particular, our closed-cloop CE scheme outperforms the other CE-based schemes significantly.

Figure 2.7 exhibits the NMSE performance of different schemes against the the number of MPCs L, given two SNRs of 0 dB and 10 dB as well as two training overheads $T_{\text{CE}} = 88$ and $T_{\text{CE}} = 176$. Except for the number of MPCs L, all the simulation parameters are the same as those in Fig. 2.6. It can be observed from Fig. 2.7 that the proposed multi-beam transmit precoding matrix design and EVD-based MPCs number estimates contribute to good NMSE performance. Meanwhile, the proposed scheme acquires better performance in the mmWave channel when the

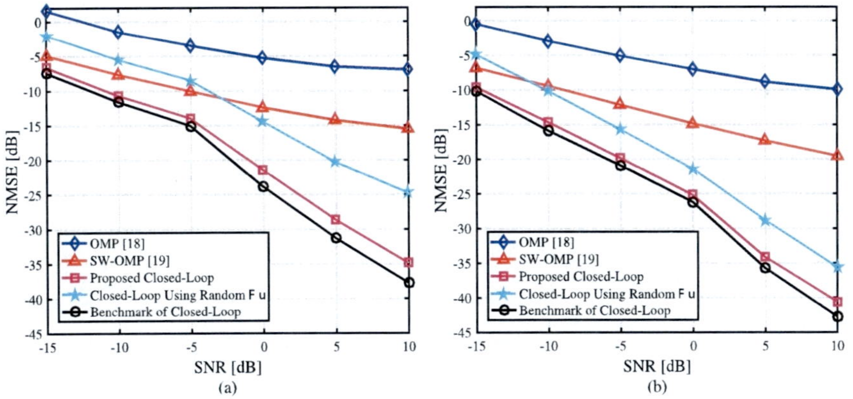

Fig. 2.6 NMSE performance comparisons of different CE schemes versus SNRs with the same number of MPCs $L = 4$: **a** training overhead $T_{\text{CE}} = 88$; and **b** $T_{\text{CE}} = 176$ [101]

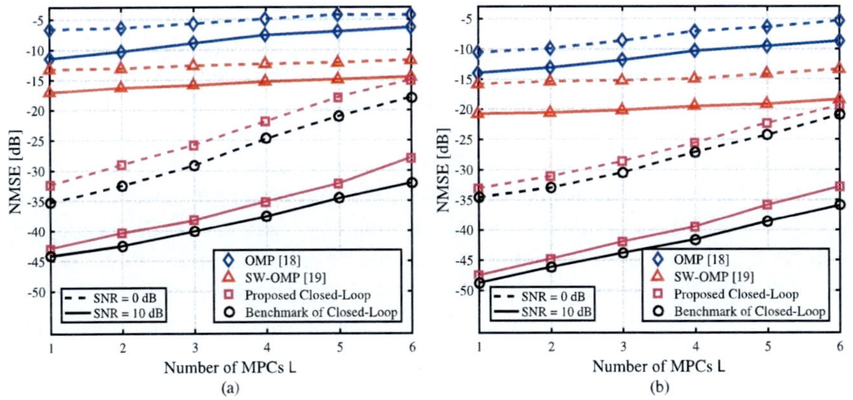

Fig. 2.7 NMSE performance comparisons of different CE schemes versus the number of MPCs L with SNR = 0 dB and 10 dB: **a** training overhead $T_{\text{CE}} = 88$; and **b** $T_{\text{CE}} = 176$ [101]

number of MPCs is fewer. In addition, although the performance gap between the proposed scheme and other schemes decreases with the increase of L in the lower SNR region, our proposed scheme can still harvest considerable performance gain in the high SNR region. Hence, the proposed scheme is suitable for sparse mmWave channels, and more performance gain can be obtained for sparser channels.

In the next, we consider the average spectrum efficiency (ASE) metric [111], which is defined as

$$\text{ASE} = \frac{1}{K}\sum_{k=0}^{K-1}\log_2\det\left(\boldsymbol{I}_{N_s} + \tfrac{1}{N_s}\boldsymbol{R}_n^{-1}[k]\boldsymbol{W}_c^{\text{H}}[k]\boldsymbol{H}[k]\boldsymbol{F}_p[k]\boldsymbol{F}_p^{\text{H}}[k]\boldsymbol{H}^{\text{H}}[k]\boldsymbol{W}_c[k]\right), \tag{2.38}$$

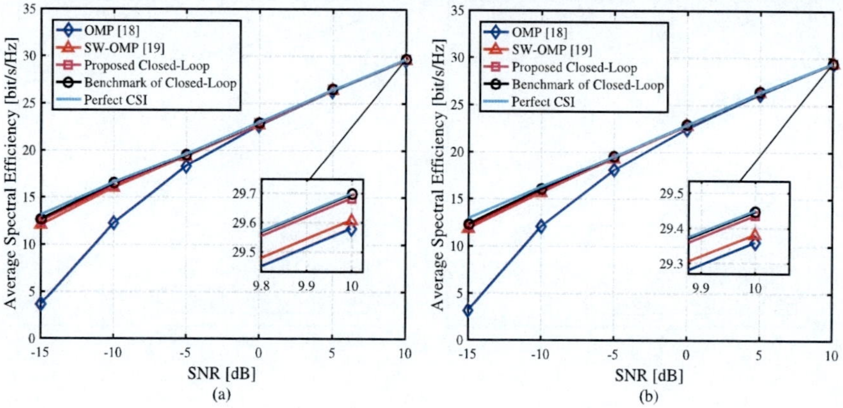

Fig. 2.8 ASE performance comparison of different CE schemes versus SNRs: **a** the number of MPCs $L = 3$; and **b** $L = 5$ [101]

where $\boldsymbol{R}_n[k] = \sigma_n^2 \boldsymbol{W}_c^{\mathrm{H}}[k]\boldsymbol{W}_c[k]$, $\boldsymbol{F}_p[k] = \boldsymbol{F}_{\mathrm{RF},p}\boldsymbol{F}_{\mathrm{BB},p}[k]$ and $\boldsymbol{W}_c[k] = \boldsymbol{W}_{\mathrm{RF},c}$ $\boldsymbol{W}_{\mathrm{BB},c}[k]$ are the transmit precoding matrix and receive combining matrix the data transmission phase, respectively, and N_{s} is the number of data streams. The principle component analysis (PCA)-based hybrid beamforming method in [111] can be used to evaluate the ASE performance, where the CSI is the estimated channel. Besides, the PCA-based hybrid beamforming scheme with a perfectly known CSI is adopted as the performance upper bound. Figure 2.8 compares the ASE of different schemes against different SNRs. The simulation parameters are the same as those in Fig. 2.5, and $N_{\mathrm{s}} = 2$. It can be seen from the figure that ASE performance based on estimated CSI can approach the performance upper bound using perfectly known CSI. In addition, in the high SNR region, the proposed scheme obtains a SE gain of 0.1 bit/s/Hz compared with the other two CS-based schemes. In the low SNR region, this performance gain becomes larger. It is worth noting that when SNR ≤ 0 dB, OMP-based schemes can only acquire poor CE results, which lead to degraded ASE compared with other schemes.

2.7.2 *Computational Complexity*

Table 2.1 exhibits the computational complexity of the proposed closed-loop sparse CE scheme in detail. For steps in Fig. 2.1, the complexities of Step 1(a) and Step 5(a) are omitted since they are far less than other steps. In addition, Step 4 does not involve computations. It can be seen from Table 2.1 that, Step 6 and Step 7 contribute most to the overall complexity, where Step 6 is corresponding to the case of $R = 3$ in Algorithm 2.2. Besides, the complexity of Step 6 and Step 7 is proportional to $\widehat{L}^4$ and $\widehat{L}!$, respectively, which makes the complexity of the proposed scheme increase rapidly with the increase of the estimated number of MPCs.

Table 2.1 Computational complexity of proposed closed-loop scheme [101]

Operation	Complexity
Step 1(b) and 5(b)	$\mathcal{O}\left(N_d N_{\mathrm{UE}} N_{\mathrm{UE}}^{\mathrm{RF}}(1+N_{\mathrm{s}}^{d}) + N_u N_{\mathrm{BS}} N_{\mathrm{BS}}^{\mathrm{RF}}(1+N_{\mathrm{s}}^{u})\right)$
Step 2	$\mathcal{O}\left((N_{\mathrm{UE}}^{\mathrm{sub}})^3 + (N_{\mathrm{UE}}^{\mathrm{sub}})^2 N_{\mathrm{o}}^{d} N_P\right)$
Step 3 ($R=2$)	$\mathcal{O}\Big(\underbrace{M_{\mathrm{sub}}^{\mathrm{UE}} K N_{\mathrm{o}}^{d} G_1^d G_2^d}_{\text{2-D SSP}} + \underbrace{8 M_{\mathrm{sub}}^{\mathrm{UE}} K N_{\mathrm{o}}^{d} G_1^d G_2^d}_{\text{RVP}} + \underbrace{\tfrac{1}{4} M_{\mathrm{sub}}^{\mathrm{UE}} \widehat{L}^2}_{\text{SSA}} + \underbrace{\tfrac{1}{2}\left(\widehat{L}^3 + 2\widehat{L}^2 M_{\mathrm{sub}}^{\mathrm{UE}} + 2(\widehat{L}+1)(M_{\mathrm{sub}}^{\mathrm{UE}})^2\right)}_{\text{SIES}} + \underbrace{\tfrac{1}{4}\widehat{L}^3}_{\text{2-D JD}}\Big)$
Step 6 ($R=3$)	$\mathcal{O}\Big(\underbrace{M_{\mathrm{sub}}^{\mathrm{BS}} N_{\mathrm{o}}^{u} G_1^u G_2^u G_3^u}_{\text{3-D SSP}} + \underbrace{8 M_{\mathrm{sub}}^{\mathrm{BS}} N_{\mathrm{o}}^{u} G_1^u G_2^u G_3^u}_{\text{RVP}} + \underbrace{\tfrac{1}{4} M_{\mathrm{sub}}^{\mathrm{BS}} \widehat{L}^2}_{\text{SSA}} + \underbrace{\tfrac{3}{4}\left(\widehat{L}^3 + 2\widehat{L}^2 M_{\mathrm{sub}}^{\mathrm{BS}} + 2(\widehat{L}+1)(M_{\mathrm{sub}}^{\mathrm{BS}})^2\right)}_{\text{SIES}} + \underbrace{\tfrac{3}{4}\widehat{L}^4}_{\text{3-D JD}}\Big)$
Step 7	$\mathcal{O}\left(\widehat{L}!(\widehat{L}^3 + 2\widehat{L}^2 K N_{\mathrm{BS}}^{\mathrm{sub}} N_{\mathrm{o}}^{u})\right)$
Step 8	$\mathcal{O}\left(K \widehat{L} N_{\mathrm{BS}} N_{\mathrm{UE}}\right)$

Table 2.2 exhibits the complexity of each step in two CS-based CE schemes, where the number of iterations of the k-th subcarrier is denoted by I_k and I for OMP algorithm and SW-OMP algorithm, respectively, and for I_k can be different for different subcarriers. It can be seen from Table 2.2 that the computational complexities of these two CS-based schemes increase rapidly with the increasing number of the quantized grids G_{BS} and G_{UE}. The complexity of the OMP-based scheme is about K times of that in the SW-OMP-based scheme. This is because all the channels of K subcarriers are estimated independently in the OMP-based scheme, while the SW-OMP-based scheme estimates K subchannels jointly. Moreover, due to the mismatch between the discrete angular domain dictionary adopted in CS-based schemes and the continuously distributed AoAs/AoDs of channels, severe power leakage will happen, leading to a greater number of effective MPCs represented in the dictionary than L. Hence, the I_k in OMP algorithm and I in SW-OMP algorithm are usually not fixed and greater than L. Here, we use $I = I_k = L$ to calculate the lower bound of the complexities of these two algorithms.

Figure 2.9 compares the complexity of the proposed closed-loop sparse CE scheme and those of two CS-based schemes, given a fixed training overhead of $T_{\mathrm{CE}} = 88$ (corresponding to $N_{\mathrm{o}}^{d} = N_{\mathrm{o}}^{u} = 2$ in our scheme). It can be seen from Fig. 2.9a that the complexity of the proposed scheme increases slightly with the increase of L. In the mmWave MIMO channels, the number of MPCs L is usually small due to the sparsity. Besides, when the sizes of UPAs at BS and UE are both 12×12, the complexity of our scheme is 3 orders of magnitude lower than SW-OMP scheme and 5 orders of magnitude lower than OMP scheme. The results in Fig. 2.9b exhibit that the complexity of the proposed scheme is almost irrelevant to the size of the UPAs at the BS and UE. By contrast, the complexity of CS-based schemes increases

Table 2.2 Computational complexity of two CS-based CE schemes [101]

Operation	OMP-based scheme [42]	SW-OMP-based scheme [43]
Measurement matrix	$\mathsf{O}\left(K T_{\mathrm{CE}} N_{\mathrm{BS}}^{\mathrm{RF}} N_{\mathrm{BS}} N_{\mathrm{UE}} G_{\mathrm{BS}} G_{\mathrm{UE}}\right)$	$\mathsf{O}\left(T_{\mathrm{CE}} N_{\mathrm{BS}}^{\mathrm{RF}} N_{\mathrm{BS}} N_{\mathrm{UE}} G_{\mathrm{BS}} G_{\mathrm{UE}}\right)$
Whitening	NA	$\mathsf{O}\left(T_{\mathrm{CE}} N_{\mathrm{BS}}^{\mathrm{RF}}\left((T_{\mathrm{CE}} N_{\mathrm{BS}}^{\mathrm{RF}})^2 + K + G_{\mathrm{BS}} G_{\mathrm{UE}}\right)\right)$
Correlation	$\mathsf{O}(T_{\mathrm{CE}} N_{\mathrm{BS}}^{\mathrm{RF}} G_{\mathrm{BS}} G_{\mathrm{UE}} (\sum_{k=1}^{K} I_k))$	$\mathsf{O}\left(T_{\mathrm{CE}} N_{\mathrm{BS}}^{\mathrm{RF}} G_{\mathrm{BS}} G_{\mathrm{UE}} K I\right)$
Project subspace	$\mathsf{O}(\sum_{k=1}^{K} (\frac{1}{4} I_k^2 (I_k+1)^2 +$ $\frac{1}{3} I_k (I_k+1)(2I_k+1) T_{\mathrm{CE}} N_{\mathrm{BS}}^{\mathrm{RF}}))$	$\mathsf{O}(\frac{1}{4} I^2 (I+1)^2 + \frac{1}{3} K I (I+1)(2I+1) T_{\mathrm{CE}} N_{\mathrm{BS}}^{\mathrm{RF}})$
Update residual	$\mathsf{O}(T_{\mathrm{CE}} N_{\mathrm{BS}}^{\mathrm{RF}} (\sum_{k=1}^{K} \frac{I_k}{2} (I_k+1)))$	$\mathsf{O}(T_{\mathrm{CE}} N_{\mathrm{BS}}^{\mathrm{RF}} K \frac{I}{2} (I+1))$
Compute MSE	$\mathsf{O}(T_{\mathrm{CE}} N_{\mathrm{BS}}^{\mathrm{RF}} (\sum_{k=}^{K} I_k))$	$\mathsf{O}\left(T_{\mathrm{CE}} N_{\mathrm{BS}}^{\mathrm{RF}} K^2 I\right)$
Reestablishment	$\mathsf{O}(N_{\mathrm{BS}} N_{\mathrm{UE}} (\sum_{k=1}^{K} I_k))$	$\mathsf{O}(N_{\mathrm{BS}} N_{\mathrm{UE}} K I)$

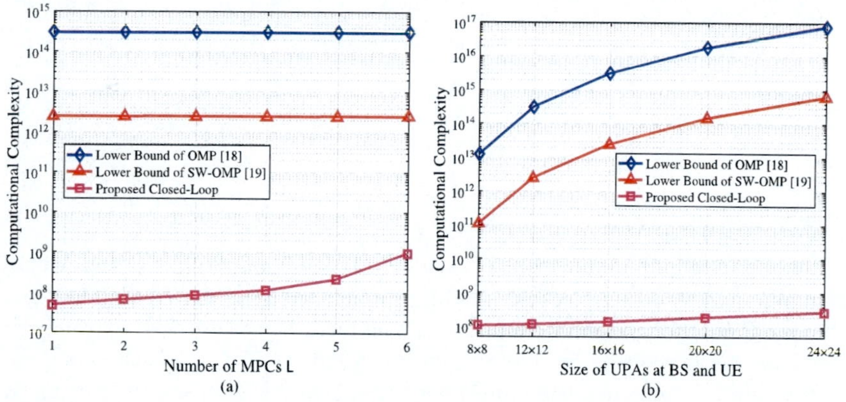

Fig. 2.9 Computational complexity comparisons of different CE schemes given the training overhead $T_{\mathrm{CE}} = 88$: **a** sizes of UPAs at BS and UE are both 12×12; and **b** number of MPCs $L = 4$ [101]

dramatically with the size of UPAs. Recalling that CS-based schemes need a higher dimensional redundant dictionary to suppress the power leakage, this will leads to unaffordable storage demands with the increase of the number of antennas. Obviously, for FD-MIMO systems equipped with a large number of antennas, the proposed closed-loop solution is significantly superior to the CS-based schemes in terms of both computational complexity and storage requirements.

The advantages and disadvantages of the proposed closed-loop sparse CE scheme and the other two CS-based CE schemes are analyzed in detail in Table 2.3, where the training/feedback overhead, storage requirements, computational complexity and received SNR at the CE stage are compared.

Table 2.3 Comparisons of advantages and disadvantages of different CE schemes [101]

		Proposed closed-loop scheme	OMP [18] and SW-OMP [19] schemes	
			CE in Downlink	CE in Uplink
Overhead	Training overhead	**Medium**; Q users share downlink CE but perform respective uplink CE	**Small**; Q users share downlink CE	**Large**; Q users have to perform respective uplink CE
	Feedback overhead	**Small**; Quantized dominant channel parameters are fed back/forward	**Large**; Support sets and channel gains for every subcarrier are fed back	**Large**; Support sets and channel gains for every subcarrier are fed forward
Storage requirements		**Small** storage at UDs in downlink; **Medium** storage required at BS in uplink	**Excessive high** storage of large redundant dictionary required at UDs	**High** storage of large redundant dictionary required at BS
Computational complexity (CC)		**Lower** CC imposed on UDs in downlink; **Medium** CC imposed on BS in uplink	**Excessive high** CC caused by large matrix operations at UDs	**High** CC due to large matrix operations at BS
Received SNR at CE		**High**; Large transmit power in downlink; Directional multi-beam design in uplink	**High**; Large transmit power at BS	**Low**; Limited transmit power of UDs

2.8 Summary

This chapter proposed a closed-loop sparse CE scheme based on multi-dimensional array signal processing techniques, which can be applied to multi-user wideband mmWave FD-MIMO systems with hybrid beamforming architectures. By exploiting the double sparsity of the mmWave MIMO channel in the angular domain and delay domain, the proposed scheme can visualize the high-dimensional hybrid beamforming MIMO as an equivalent low-dimensional fully-digital array, which facilitates the utilization of MDU-ESPRIT to obtain the super-resolution estimation of the horizontal/vertical AoAs and delays at the transmitter and receiver. Specifically, at the downlink CE stage, the common transmit precoding matrix at the BS and the receive combining matrix of each UE are designed to estimate the AoAs corresponding to each sparse MPC. At the uplink CE stage, the horizontal/vertical AoAs and delays at the BS can be estimated simultaneously by using the designed receive combining matrix at the BS. In addition, each UE utilizes the estimated AoAs to design respective uplink multi-beam transmit precoding matrix, which can enhance the CE performance. Finally, the BS utilizes an ML method to match the estimated parameters at the above two stages, and then estimate the corresponding path gains. Simulation results exhibit that, compared with the CS-based CE schemes, the proposed closed-loop CE scheme can significantly improve the accuracy of the estimated CSI, and exhibits considerable advantages in reducing the computational complexity and storage requirements.

Chapter 3
Compressive Sensing Based Channel Estimation for MmWave Full-Dimensional Lens Antenna Array

Abstract This chapter studies the channel estimation (CE) for the mmWave full-dimensional (FD) massive MIMO with lens antenna array. The lens-array using antenna switching network (ASN) simplifies the hardware, but the associated CE is a challenging task due to the constraint imposed by ASN. To tackle the problem, we first propose a dedicated approach of pilot training under FD lens-array with ASN, and formulate the associated CE problem as a sparse signal recovery problem. Then, a redundant dictionary is tailored for FD lens antenna array to combat the power leakage caused by the continuous angles of multipath components. Further, we design the baseband pilot signals to minimize the total mutual coherence of the measurement matrix based on compressive sensing (CS) theory for more reliable CE performance. Our solution provides a framework for applying CS techniques to lens antenna array using simple and practical ASN.

3.1 Introduction

In order to strike a balance between the system performance, hardware complexity and power consumption in massive MIMO, various advanced forms of antenna arrays have appeared, among which the lens antenna array is regarded as one of the most promising antenna forms. By utilizing the energy focusing property of the lens antenna array and the angular sparsity of the mmWave channels, the mmWave MIMO with lens antenna array can achieve near-optimal capacity and greatly reduce the processing complexity and power consumption [27]. In this chapter, we focus our attention on the pilot design and channel estimation (CE) in mmWave full-dimensional (FD) MIMO with lens antenna arrays. Note that, as mentioned in Sect. 1.3.1, most of the current channel estimation works on lens antenna arrays [47–50] employ the large-scale phase shifter network (PSN) to assist channel estimation, which contradicts the key idea of using an electromagnetic (EM) lens to replace the bulky PSN, and thus cannot make full use of the high energy efficiency brought by the lens antenna array. As an alternative solution, we consider replacing the complex and power-hungry PSN with a simple antenna switching network (ASN) to achieve energy-efficient CE. We propose a pilot training scheme for FD lens antenna array

Z. Gao et al., *Millimeter-Wave/Sub-Terahertz Ultra-Massive MIMO Transmission Technology*, https://doi.org/10.1007/978-981-99-2388-5_3

under the use of ASN, which is able to sound the FD spatial range with reduced pilot overhead. Also, we propose to use the redundant dictionary to formulate the CE problem as the sparse signal recovery problem, which can be solved by advanced compressive sensing (CS) techniques. Compared with the traditional estimation algorithm, the proposed CE scheme can greatly reduce the pilot overhead, and can mitigate the power leakage caused by the angle mismatch.

3.2 The Principle of Lens Antenna Array

We first introduce the general model of a lens antenna array, and accordingly derive its array response, which lays the foundation for the further channel modeling. Consider a three-dimensional (3D) coordinate system and a FD lens antenna array which consists of an EM lens with negligible thickness on the yz plane, and the antenna elements located on the focal sphere of the EM lens, as shown in Fig. 3.1. Note that the center of the focal sphere is also the center of the EM lens. In practice, EM lenses can be commonly implemented with three different techniques [27]:

- Dielectric lenses (e.g., convex lenses) made of dielectric materials with carefully designed front and/or rear surfaces.
- Traditional planar lenses consisting of arrays of transmitting and receiving antennas connected via transmission lines with different lengths.
- Modern compact planar lenses composed of sub-wavelength spatial phase shifters with periodic inductive and capacitive structures.

No matter which implementation method is used, the principle of the EM lens is to add a designed phase shift to the EM signals passing through it, so that the EM signals can be constructively superposed at a certain point on the focal sphere, realizing the focusing of EM energy. Let the physical size of the EM lens in Fig. 3.1 be D^{h} along the horizontal direction and be D^{v} along the vertical direction, and its corresponding electrical sizes (physical sizes normalized by the wavelength of EM signal) are denoted as $\tilde{D}^{\mathrm{h}}$ and $\tilde{D}^{\mathrm{v}}$. The focal length of the EM lens is F. Then, based on the 3D coordinate system in Fig. 3.1, the coordinate of the m-th antenna element on the focal sphere can be written as $p_m = (F \cos\theta_m \cos\phi_m, F \cos\theta_m \sin\phi_m, F \sin\theta_m)$, where $\theta_m, \phi_m \in [-\pi/2, \pi/2]$ are the azimuth angle and the elevation angle, respectively. The following Lemma 3.2.1 gives the array response of such lens antenna array.

Lemma 3.2.1 *When $F \gg D^{\mathrm{h}}$ and $F \gg D^{\mathrm{v}}$, the array response at the m-th antenna element aroused by the incoming EM signal with the azimuth angle θ and elevation angle ϕ can be represented as*

$$a_m(\theta, \phi) \approx \mathrm{sinc}\left[\tilde{D}^{\mathrm{v}} (\sin\theta_m - \sin\theta)\right] \mathrm{sinc}\left[\tilde{D}^{\mathrm{h}} (\cos\theta_m \sin\phi_m - \cos\theta \sin\phi)\right] \tag{3.1}$$

where $\mathrm{sinc}(x) = \sin(\pi x)/(\pi x)$ *is the normalized sinc function.*

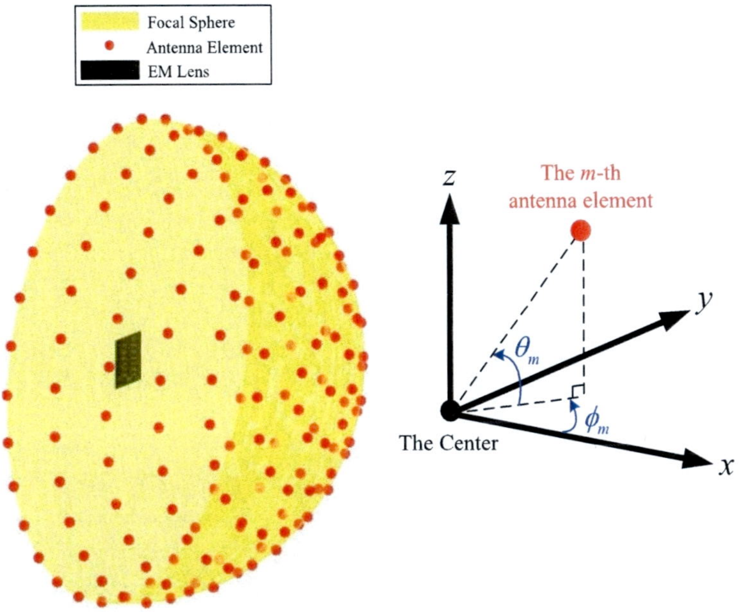

Fig. 3.1 The model of a FD lens antenna array

Proof See Appendix B. □

We demonstrate some examples of the array responses of lens antenna array in Fig. 3.2. For clarity, we assume $\phi = 0$ and only demonstrate the array responses in those antenna elements with $\phi_m = 0$. It is readily seen from Fig. 3.2 that for the incident signal from the direction (θ, ϕ), only those antenna elements that satisfy $\theta_m \approx \theta$ ($\phi_m \approx \phi$) have the array responses with significant amplitude, while other elements have weak and even negligible array responses, which embodies the energy focusing property of the lens antenna array. On the other hand, Fig. 3.2 provides insight in the angular resolution of lens antenna array. Specifically, for two incident signals from two different directions (θ, ϕ) and $\left(\theta', \phi'\right)$, if $\left|\sin\theta - \sin\theta'\right| \geq \frac{1}{\tilde{D}^{\mathrm{v}}}$ or $\left|\cos\theta \sin\phi - \cos\theta' \sin\phi'\right| \geq \frac{1}{\tilde{D}^{\mathrm{h}}}$, their corresponding mainlobes of array responses do not affect each other. As long as two different antenna elements on the focal sphere are properly selected for reception, the two incident signals can be received separately without inter-path interference. Note that the elimination of inter-path interference in lens antenna arrays does not require complex PSN as in traditional massive MIMO systems. Furthermore, considering the reversibility of the receiving and transmitting path, the lens antenna array can also be used as the transmitter, where only a few antenna elements need to be selected and fed into the signal to be transmitted. After passing through the EM lens, the transmitted signals form the directional beams

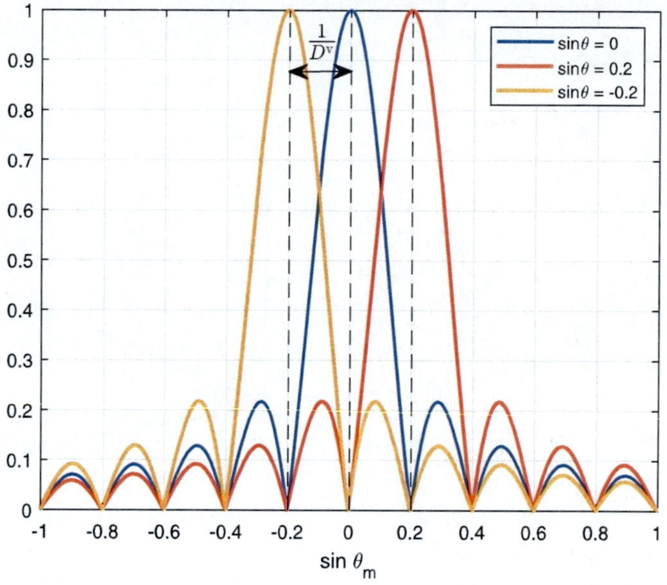

Fig. 3.2 The examples of the array responses of the lens antenna array $\tilde{D}^{\mathrm{v}} = 5$

towards the spatial directions corresponding to the selected antenna elements. The transmit beam pattern can also be described by (3.1). In a word, the massive MIMO based on lens antenna array can realize beamforming without PSN, improving the energy efficiency of communication systems.

3.3 An Overview of Compressive Sensing Theory

Almost all previous wireless cellular networks have relied upon the classic Nyquist sampling theorem, stating that any bandwidth-limited signal can be perfectly reconstructed when the sampling rate is higher than twice the signal's highest frequency [103]. However, with the development of mobile communication, the bandwidth, number of antennas, density of base stations and users, and other key performance indicators of the communication system increase significantly. In the upcoming 6G communications, these indicators will also reach unprecedented peaks (see Fig. 1.1). Obviously, the signal processing in 6G systems based on Nyquist sampling theorem will face severe challenges such as the excessive cost for signal storage and feedback, and the unaffordable complexity of processing algorithm. Fortunately, the *compressive sensing* (CS) theory, which was founded by Donoho et al. in the 2000s [112–116], has pointed out an effective way to reduce the cost of signal transmission and the complexity of signal processing. Basically, by using some sparse transformation to re-represent the original signal as the sparse signal (i.e., only a few samplings contain most of the useful information), the original signal can be recovered from the carefully-designed compressed data set via various advanced CS-based algorithms. In the context of CS, the required sampling rate for signal recovery no longer depends on the signal bandwidth, but depends on the intrinsic structure of the signal. At present, the CS theory has been widely applied to CE, beamforming design, random access and other promising scenarios in modern wireless communications. This chapter introduces the basic principle of CS. In particular, we focus our attention on how the sensing matrix in CS influences the performance of sparse signal recovery, providing a theoretical basis for the pilot design in the FD lens antenna array that we will discuss later.

3.3.1 Compressive Sensing Modeling

Although the CS theory was originally proposed to break through the limitation of Nyquist sampling theorem, its mathematical model is different from the traditional signal sampling and reconstruction. In general, the objective of a CS problem is to recover the original high-dimensional signal $\boldsymbol{s} \in \mathbb{C}^N$ from its under-determined linear observation $\boldsymbol{y} \in \mathbb{C}^M$, $M \ll N$. Specifically,

$$\boldsymbol{y} = \boldsymbol{\Phi s} \tag{3.2}$$

where $\boldsymbol{\Phi} \in \mathbb{C}^{M \times N}$ is the measurement matrix with rank $(\boldsymbol{\Phi}) = M$. Since $M \ll N$, recovering $\boldsymbol{y}$ from $\boldsymbol{s}$ is equivalent to solving the under-determined linear equations. If no further prior information is given, there are infinite solutions of $\boldsymbol{s}$ that satisfy (3.2), and it is impossible to determine which the original signal is. In order to find the unique solution of $\boldsymbol{s}$ via (3.2), we re-represent $\boldsymbol{s}$ and exploit its structural feature. Consider a basis (also known as *dictionary*) $\left\{\boldsymbol{\psi}_i \in \mathbb{C}^N\right\}_{i=1}^G$, where $G \geq N$, and accordingly we re-represent $\boldsymbol{s}$ as

$$\boldsymbol{s} = \sum_{i=1}^{G} x_i \boldsymbol{\psi}_i = \boldsymbol{\Psi x} \tag{3.3}$$

where $\boldsymbol{\Psi} = \left[\boldsymbol{\psi}_1, \boldsymbol{\psi}_2, \ldots, \boldsymbol{\psi}_G\right] \in \mathbb{C}^{N \times G}$ is the dictionary matrix, and $\boldsymbol{x} = \left[x_1, x_2, \ldots, x_G\right]^{\mathrm{T}} \in \mathbb{C}^G$ is the representation of $\boldsymbol{s}$ under the basis $\left\{\boldsymbol{\psi}_i\right\}_{i=1}^G$. Note that when $G > N$ or $\boldsymbol{\Psi}$ is rank-insufficient, there will also be infinite possible representations $\boldsymbol{x}$ corresponding to $\boldsymbol{s}$. We assume that for a certain $\boldsymbol{s}$, we can always select a basis $\left\{\boldsymbol{\psi}_i\right\}_{i=1}^G$ so that the representation $\boldsymbol{x}$ is *sparse* (or nearly sparse), namely the number of non-zero elements (or the elements with significant amplitudes) in $\boldsymbol{x}$ is very small (typically it should be much smaller than M). This is also called the *compressibility* of $\boldsymbol{s}$. In fact, the signal with compressibility is common in nature, since most signals from the real world exhibit some inherent redundancy or correlation [103]. As an example, Fig. 3.3

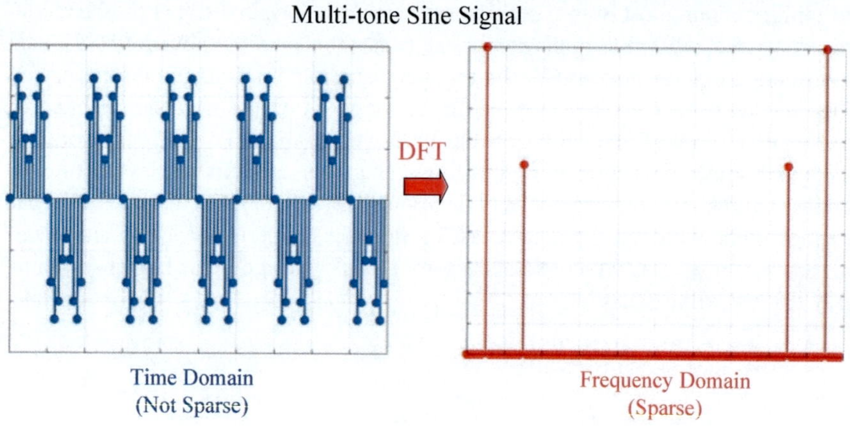

Fig. 3.3 The sparse representation of a multi-tone sine signal via DFT

demonstrates that a multi-tone sine signal in the time-domain can be represented as the frequency-domain sparse signal via the discrete Fourier transform (DFT).

Combining (3.2) and (3.3), we have

$$\boldsymbol{y} = \boldsymbol{\Phi}\boldsymbol{\Psi}\boldsymbol{x} = \boldsymbol{A}\boldsymbol{x} \tag{3.4}$$

where $\boldsymbol{A} = \boldsymbol{\Phi}\boldsymbol{\Psi} \in \mathbb{C}^{M \times G}$ is the *sensing matrix*. Although recovering $\boldsymbol{x}$ from $\boldsymbol{y}$ in (3.4) is also an under-determined problem, the number of unknown elements in $\boldsymbol{x}$ has been significantly reduced since $\boldsymbol{x}$ is sparse, which makes it possible to recover $\boldsymbol{x}$ uniquely and then reconstruct $\boldsymbol{s}$ based on (3.3). Note that in practice, we are only interested on the "sparsest" solution of $\boldsymbol{x}$ that satisfies (3.4). Therefore, we formulate the CS problem as the following optimization problem

$$\begin{cases} \min\limits_{\boldsymbol{x}} \|\boldsymbol{x}\|_0 \\ \text{s.t.} \;\; \boldsymbol{y} = \boldsymbol{A}\boldsymbol{x} \end{cases} \tag{3.5}$$

where the l_0-norm stands for the number of non-zero elements in a vector.

For the practical application of (3.5), there are two key questions need to answer: (1) how the sensing matrix $\boldsymbol{A}$ guarantees a unique sparse solution for $\boldsymbol{x}$; and (2) how to obtain the unique sparse $\boldsymbol{x}$ according to $\boldsymbol{y}$ and $\boldsymbol{A}$ with acceptable complexity. In the rest of this section, we will focus on the first problem, that is, the uniqueness of sparse recovery, which is one of the most important issues in CS. Note that the theories involved in the uniqueness of sparse recovery also provide guidance for the second problem, that is, the design of practical reconstruction algorithms.

3.3.2 Sensing Matrix

For simplicity, we assume that each column of the sensing matrix $\boldsymbol{A}$ has the normalized l_2-norm, namely $\left\|\boldsymbol{A}_{\{:,i\}}\right\|_2 = 1, \forall i$. This causes no loss of generality, since we can re-write (3.4) as $\boldsymbol{y} = \boldsymbol{A}\boldsymbol{D}\boldsymbol{D}^{-1}\boldsymbol{x} = \tilde{\boldsymbol{A}}\tilde{\boldsymbol{x}}$, where $\tilde{\boldsymbol{A}} = \boldsymbol{A}\boldsymbol{D}, \tilde{\boldsymbol{x}} = \boldsymbol{D}^{-1}\boldsymbol{x}$, and $\boldsymbol{D} \in \mathbb{C}^{G\times G}$ is a positive definite diagonal matrix normalizing the columns of $\tilde{\boldsymbol{A}}$ while keeping the sparsity of $\tilde{\boldsymbol{x}}$ unchanged.

At present, three main indicators that can evaluate the sensing matrix are widely considered in the CS theory, namely, the Spark, mutual coherence, and restricted isometric constant (RIC). Next we only focus on the Spark and mutual coherence. Interested readers are referred to [113, 116] to learn more about RIC.

Definition 3.3.1 (Spark) For the sensing matrix $\boldsymbol{A}$, Spark $(\boldsymbol{A})$ is defined as the smallest integer n such that there exists a set of n columns from $\boldsymbol{A}$ that are linearly dependent [112].

Note that although Spark $(\boldsymbol{A})$ is similar to the rank of $\boldsymbol{A}$ in some ways, they can be totally different. For example, the ranks of matrices

$$\boldsymbol{A}_1 = \begin{bmatrix} 1\,0\,0\,1 \\ 0\,1\,0\,0 \\ 0\,0\,1\,0 \end{bmatrix}, \boldsymbol{A}_2 = \begin{bmatrix} 1\,0\,0\,1 \\ 0\,1\,0\,1 \\ 0\,0\,1\,0 \end{bmatrix}, \boldsymbol{A}_3 = \begin{bmatrix} 1\,0\,0\,1 \\ 0\,1\,0\,1 \\ 0\,0\,1\,1 \end{bmatrix}$$

are all 3, while their Sparks are 2,3, and 4, respectively.

Theorem 3.3.1 shows the importance of the Spark for the uniqueness of sparse recovery.

Theorem 3.3.1 *If there exists a signal $\bar{\boldsymbol{x}}$ that satisfies $\boldsymbol{y} = \boldsymbol{A}\bar{\boldsymbol{x}}$ and $\|\bar{\boldsymbol{x}}\|_0 <$ Spark $(\boldsymbol{A})/2$, $\bar{\boldsymbol{x}}$ is necessarily the unique sparsest solution for $\boldsymbol{y} = \boldsymbol{A}\boldsymbol{x}$.*

Proof Suppose another $\boldsymbol{x}' \neq \bar{\boldsymbol{x}}$ that satisfies $\boldsymbol{y} = \boldsymbol{A}\boldsymbol{x}'$, and thus we have

$$\boldsymbol{A}\left(\bar{\boldsymbol{x}} - \boldsymbol{x}'\right) = \boldsymbol{0} \tag{3.6}$$

This indicates that there exists a set of $\left\|\bar{\boldsymbol{x}} - \boldsymbol{x}'\right\|_0$ columns from $\boldsymbol{A}$ that are linearly dependent, which renders Spark $(\boldsymbol{A}) \leq \left\|\bar{\boldsymbol{x}} - \boldsymbol{x}'\right\|_0$. On the other hand, according to the famous triangle inequality, we have

$$\text{Spark}\,(\boldsymbol{A}) \leq \left\|\bar{\boldsymbol{x}} - \boldsymbol{x}'\right\|_0 \leq \|\bar{\boldsymbol{x}}\|_0 + \left\|\boldsymbol{x}'\right\|_0 \tag{3.7}$$

that is $\left\|\boldsymbol{x}'\right\|_0 \geq$ Spark $(\boldsymbol{A}) - \|\bar{\boldsymbol{x}}\|_0 >$ Spark $(\boldsymbol{A})/2$. This proves that $\bar{\boldsymbol{x}}$ is the unique sparsest solution for $\boldsymbol{y} = \boldsymbol{A}\boldsymbol{x}$, since any $\boldsymbol{x}' \neq \bar{\boldsymbol{x}}$ that satisfies $\boldsymbol{y} = \boldsymbol{A}\boldsymbol{x}'$ has larger l_0-norm than $\bar{\boldsymbol{x}}$. □

Theorem 3.3.1 provides a method to verify or guarantee the CS (sparse) recovery. On the one hand, if we obtain a solution for $\boldsymbol{y} = \boldsymbol{A}\boldsymbol{x}$ whose sparsity is smaller than

Spark $(\boldsymbol{A})/2$, we can draw the conclusion that this is the unique sparsest solution we want. On the other hand, in most practical cases, we cannot know prior the sparsity of the signal to be recovered, therefore enlarging Spark $(\boldsymbol{A})$ will provide a broad margin for the sparsity and guarantee the high probability of signal recovery. However, calculating the Spark of a matrix requires the combinatorial complexity that is unaffordable in practice, not to mention optimizing it. Fortunately, the mutual coherence, as a more straightforward indicator defined below, provides a feasible way to optimize the sensing matrix.

Definition 3.3.2 (Mutual Coherence) Let $\boldsymbol{G} = \boldsymbol{A}^{\mathrm{H}}\boldsymbol{A}$ denotes the Gram matrix of $\boldsymbol{A}$. Note that $[\boldsymbol{G}]_{i,i} = 1$ and $\left|[\boldsymbol{G}]_{i,j}\right| \leq 1$, $\forall i \neq j$ since we assume the column normalization of $\boldsymbol{A}$. The (maximum) mutual coherence of $\boldsymbol{A}$ is defined as $\mu(\boldsymbol{A}) = \max\limits_{i \neq j} \left|[\boldsymbol{G}]_{i,j}\right|$.

Compared to the Spark, the mutual coherence of a sensing matrix is easy to calculate, which only involves the matrix multiplication and the element-wise search. More importantly, the mutual coherence gives a lower bound of the Spark, as stated in Theorem 3.3.2.

Theorem 3.3.2 (A Lower Bound of the Spark) *Given a sensing matrix* $\boldsymbol{A}$*,*

$$\text{Spark}(\boldsymbol{A}) \geq \lceil 1/\mu(\boldsymbol{A}) \rceil \tag{3.8}$$

Proof See Appendix C. □

Theorem 3.3.2 bridges the mutual coherence and the uniqueness of sparse recovery (via the Spark). Specifically, if we replace Spark $(\boldsymbol{A})$ by $\lceil 1/\mu(\boldsymbol{A}) \rceil$ in Theorem 3.3.1, the conclusion still holds. Consequently, we can optimize the mutual coherence of the sensing matrix to guarantee the performance of CS, which can be supported by existing theories and techniques. For instance, the author of [117] established an achievable lower bound of the mutual coherence, namely Welch bound.

Theorem 3.3.3 (Welch Bound) *Given a sensing matrix* $\boldsymbol{A} \in \mathbb{C}^{M \times G}$*,*

$$\mu(\boldsymbol{A}) \geq \sqrt{\frac{G - M}{M(G-1)}} \tag{3.9}$$

Proof See Appendix D. □

Many off-the-shelf techniques based on either constructive (e.g., [118]) or numerical (e.g., [119]) methods have been applied to optimize the mutual coherence of sensing matrices.

Last but not least, the indicators discussed in this section (the Spark, mutual coherence, RIC) also provide insight in the second key problem of CS, namely how to obtain the unique sparse solution for $\boldsymbol{x}$ according to $\boldsymbol{y}$ and $\boldsymbol{A}$. Instead of the intractable NP-hard combinatorial search to determine the support (the set of non-zero elements) of the sparse signal, a practical alternative is to relax the problem (3.5) by replacing the non-convex l_0-norm constraint by the convex l_1-norm constraint, that is

$$\begin{cases} \min\limits_{\boldsymbol{x}} \|\boldsymbol{x}\|_1 \\ \text{s.t. } \boldsymbol{y} = \boldsymbol{A}\boldsymbol{x} \end{cases} \tag{3.10}$$

Obviously, (3.10) is a convex optimization problem, which can be cast as a linear programming problem and solved by various reliable methods. It has been proven [112, 113] that for the sensing matrix which satisfies some strict condition (in terms of the Spark, or mutual coherence, or RIC), the solution for (3.10) may be the same as that for (3.5). Due to the page limitation, we will not go deep into the details. In the next section, we will resort to the CS theories introduced above and propose a CS-based CE scheme for FD lens antenna array.

3.4 The Proposed Channel Estimation Scheme

Based on the preparation in Sects. 3.2 and 3.3, in this section, we propose a CS-based CE scheme for the FD lens antenna array, and design the pilot signal according to the CS theory, aiming to improve the CE performance with reduced pilot overhead.

3.4.1 System Model

Consider a mmWave massive MIMO system with the FD lens antenna arrays equipped at both the transmitter and receiver, as shown in Fig. 3.4. The electrical sizes of the EM lenses at the transmitter and receiver are $\tilde{D}_{\mathrm{T}}^{\mathrm{h}} \times \tilde{D}_{\mathrm{T}}^{\mathrm{v}}$ and $\tilde{D}_{\mathrm{R}}^{\mathrm{h}} \times \tilde{D}_{\mathrm{R}}^{\mathrm{v}}$, respectively. The numbers of antenna elements at the transmitter and receiver are N_{T} and $N_{\mathrm{T}}^{\mathrm{RF}}$, respectively, while only $N_{\mathrm{T}}^{\mathrm{RF}} \ll N_{\mathrm{T}}$ ($N_{\mathrm{R}}^{\mathrm{RF}} \ll N_{\mathrm{R}}$) radio frequency (RF) chains are used at the transmitter (receiver). Unlike most of the previous works [47–50], we adopt the simple ASN, rather than complex PSN, to connect RF chains and antenna elements, where each RF chain can activate at most one antenna at a time. This architecture avoids the hardware cost, insertion loss, and power consumption brought by the PSN, taking full advantage of the high energy efficiency of lens antenna arrays.

The distribution of antenna elements on the focal sphere of the EM lens is designed based on [120]. Taking the transmitter as an example, we denote the angle coordinate (see Fig. 3.1) of the m-th antenna element as (θ_m, ϕ_m), and θ_m is drawn from the set

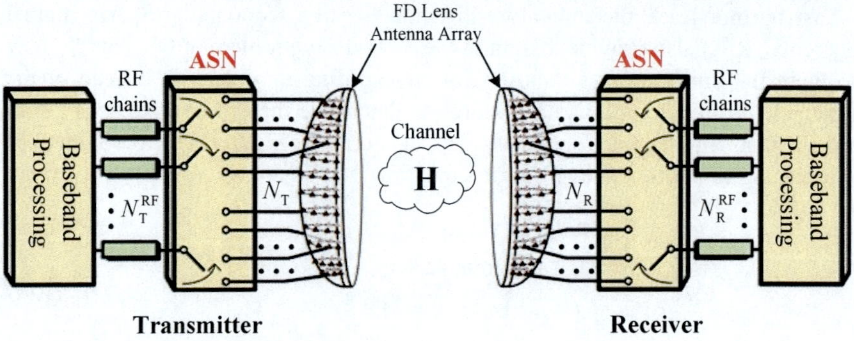

Fig. 3.4 MmWave massive MIMO with lens antenna arrays

$$\theta_m \in \left\{ \theta \left| \sin\theta = \frac{1}{2\tilde{D}_T^v} + \frac{n}{\tilde{D}_T^v}, n = -\left\lfloor \tilde{D}_T^v \right\rfloor, \ldots, -1, 0, 1, \ldots, \left\lfloor \tilde{D}_T^v \right\rfloor - 1 \right. \right\} \tag{3.11}$$

Furthermore, for a given $\theta_{m'}$, the corresponding ϕ_m is drawn from the set

$$\phi_m \in \left\{ \phi \left| \sin\phi = \frac{n}{\tilde{D}_T^h \cos\theta_{m'}}, n = 0, \pm 1, \pm 2, \ldots, \pm \left\lfloor \tilde{D}_T^h \cos\theta_{m'} \right\rfloor \right. \right\} \tag{3.12}$$

This design guarantees that for any two different antenna elements (θ_m, ϕ_m) and $(\theta_{m'}, \phi_{m'})$, $m \neq m'$, we have $|\sin\theta_m - \sin\theta_{m'}| \geq \frac{1}{\tilde{D}_T^v}$ or $\left| \cos\theta_m \sin\phi_m - \cos\theta_{m'} \sin\phi_{m'} \right| \geq \frac{1}{\tilde{D}_T^h}$, that is to say, the mainlobes of their beam patterns do not affect each other.

Based on (3.11) and (3.12), we can calculate N_T, which is the number of all possible pairs (θ_m, ϕ_m), as

$$N_T = \sum_{n=-\left\lfloor \tilde{D}_T^v \right\rfloor}^{\left\lfloor \tilde{D}_T^v \right\rfloor - 1} 2 \left\lfloor \tilde{D}_T^h \cos\left[\arcsin\left(\frac{1}{2\tilde{D}_T^v} + \frac{n}{\tilde{D}_T^v} \right) \right] \right\rfloor + 1 \tag{3.13}$$

The distribution design and the calculation of N_R at the receiver are similar to (3.11)–(3.13), so they are omitted for brevity.

The channel matrix $\boldsymbol{H} \in \mathbb{C}^{N_R \times N_T}$ between the transmitter and receiver is modeled as

$$\boldsymbol{H} = \sqrt{\frac{N_T N_R}{L}} \sum_{l=1}^{L} g_l \boldsymbol{a}_R\left(\theta_R^l, \phi_R^l\right) \boldsymbol{a}_T^H\left(\theta_T^l, \phi_T^l\right) \tag{3.14}$$

where L is number of channel paths, g_l is the complex gain of the l-th path, θ_{T}^{l} and θ_{R}^{l} are the azimuth angle of departure (AoD) and angle of arrival (AoA) of the l-th path, respectively, ϕ_{T}^{l} and ϕ_{R}^{l} are the elevation AoD and AoA of the l-th path, respectively, and $\boldsymbol{a}_{\mathrm{R}}\left(\theta_{\mathrm{R}}^{l}, \phi_{\mathrm{R}}^{l}\right) \in \mathbb{C}^{N_{\mathrm{R}}}$ and $\boldsymbol{a}_{\mathrm{T}}\left(\theta_{\mathrm{T}}^{l}, \phi_{\mathrm{T}}^{l}\right) \in \mathbb{C}^{N_{\mathrm{T}}}$ are the array response vectors of the receive and transmit lens antenna arrays, respectively. Taking $\boldsymbol{a}_{\mathrm{T}}\left(\theta_{\mathrm{T}}^{l}, \phi_{\mathrm{T}}^{l}\right)$ as an example, it can be written as

$$\left[\boldsymbol{a}_{\mathrm{T}}\left(\theta_{\mathrm{T}}^{l}, \phi_{\mathrm{T}}^{l}\right)\right]_m = \operatorname{sinc}\left[\tilde{D}_{\mathrm{T}}^{\mathrm{v}}\left(\sin\theta_m - \sin\theta_{\mathrm{T}}^{l}\right)\right] \times \operatorname{sinc}\left[\tilde{D}_{\mathrm{T}}^{\mathrm{h}}\left(\cos\theta_m \sin\phi_m - \cos\theta_{\mathrm{T}}^{l}\sin\phi_{\mathrm{T}}^{l}\right)\right] \tag{3.15}$$

where (θ_m, ϕ_m) denotes the angle coordinate of the m-th antenna element. $\boldsymbol{a}_{\mathrm{R}}\left(\theta_{\mathrm{R}}^{l}, \phi_{\mathrm{R}}^{l}\right)$ can be similarly modeled like (3.15).

3.4.2 *Pilot Training Scheme Design*

For ease of expression, we define a *time slot* as one transmission (at the transmitter) or one reception (at the receiver) of the pilot signal, and a *time block* as several successive time slot. During the CE stage, we assume that the transmitter use $N_{\mathrm{T}}^{\mathrm{P}}$ (the superscript "P" stands for "pilot") time blocks to transmit the pilot signals, and we further assume $N_{\mathrm{T}}^{\mathrm{G}} = N_{\mathrm{T}}/N_{\mathrm{T}}^{\mathrm{RF}}$ (the superscript "G" stands for "group") and $N_{\mathrm{T}}^{\mathrm{P}}/N_{\mathrm{T}}^{\mathrm{G}}$ are integers. In the m-th transmit time block, $1 \le m \le N_{\mathrm{T}}^{\mathrm{P}}$, the transmit pilot signal $\boldsymbol{s}_m \in \mathbb{C}^{N_{\mathrm{T}}}$ can be represented as the product of the RF precoder $\boldsymbol{F}_{\mathrm{RF},p} \in \mathbb{C}^{N_{\mathrm{T}} \times N_{\mathrm{T}}^{\mathrm{RF}}}$ and the baseband signal $\boldsymbol{f}_{\mathrm{RF},m} \in \mathbb{C}^{N_{\mathrm{T}}^{\mathrm{RF}}}$, which is given by

$$\boldsymbol{s}_m = \boldsymbol{F}_{\mathrm{RF},p}\boldsymbol{f}_{\mathrm{BB},m} \tag{3.16}$$

where $p = \left\lceil m N_{\mathrm{T}}^{\mathrm{G}}/N_{\mathrm{T}}^{\mathrm{P}} \right\rceil \in \{1, 2, \ldots, N_{\mathrm{T}}^{\mathrm{G}}\}$. This indicates that every $N_{\mathrm{T}}^{\mathrm{P}}/N_{\mathrm{T}}^{\mathrm{G}}$ successive baseband signals share the same RF precoder $\boldsymbol{f}_{\mathrm{BB},m}$.

At the receiver, we assume that $N_{\mathrm{R}}^{\mathrm{G}} = N_{\mathrm{R}}/N_{\mathrm{R}}^{\mathrm{RF}}$ is also an integer, and each transmit time block is divided into $N_{\mathrm{R}}^{\mathrm{G}}$ equal-length time slots. The received pilot signal of the m-th time block at the n-th time slot can be expressed as

$$\boldsymbol{y}_{n,m} = \boldsymbol{W}_{\mathrm{BB},n}^{\mathrm{H}}\boldsymbol{W}_{\mathrm{RF},n}^{\mathrm{H}}\boldsymbol{H}\boldsymbol{s}_m + \boldsymbol{n}_{n,m} \tag{3.17}$$

where $\boldsymbol{W}_{\mathrm{RF},n} \in \mathbb{C}^{N_{\mathrm{R}} \times N_{\mathrm{R}}^{\mathrm{RF}}}$ and $\boldsymbol{W}_{\mathrm{BB},n} \in \mathbb{C}^{N_{\mathrm{R}}^{\mathrm{RF}} \times N_{\mathrm{R}}^{\mathrm{RF}}}$ are the RF combiner and the baseband combiner, respectively, and $\boldsymbol{n}_{n,m} \sim \mathcal{CN}\left(\boldsymbol{0}, \sigma^2 \boldsymbol{W}_{\mathrm{BB},n}^{\mathrm{H}}\boldsymbol{W}_{\mathrm{RF},n}^{\mathrm{H}}\boldsymbol{W}_{\mathrm{RF},n}\boldsymbol{W}_{\mathrm{BB},n}\right)$ is the additive White Guassian noise (AWGN) vector after combination. Collecting $\boldsymbol{y}_{n,m}$ from all $N_{\mathrm{R}}^{\mathrm{G}}$ successive time slots, the aggregated received pilot signal of the m-th transmit time block is

$$\boldsymbol{y}_m = \boldsymbol{W}_{\mathrm{BB}}^{\mathrm{H}}\boldsymbol{W}_{\mathrm{RF}}^{\mathrm{H}}\boldsymbol{H}\boldsymbol{s}_m + \boldsymbol{n}_m \tag{3.18}$$

where

$$
\begin{aligned}
\boldsymbol{y}_m &= \left[\boldsymbol{y}_{1,m}^{\mathrm{T}}, \boldsymbol{y}_{2,m}^{\mathrm{T}}, \ldots, \boldsymbol{y}_{N_{\mathrm{R}}^{\mathrm{g}},m}^{\mathrm{T}}\right]^{\mathrm{T}} \in \mathbb{C}^{N_{\mathrm{R}}} \\
\boldsymbol{n}_m &= \left[\boldsymbol{n}_{1,m}^{\mathrm{T}}, \boldsymbol{n}_{2,m}^{\mathrm{T}}, \ldots, \boldsymbol{n}_{N_{\mathrm{R}}^{\mathrm{g}},m}^{\mathrm{T}}\right]^{\mathrm{T}} \in \mathbb{C}^{N_{\mathrm{R}}} \\
\boldsymbol{W}_{\mathrm{RF}} &= \left[\boldsymbol{W}_{\mathrm{RF},1}, \boldsymbol{W}_{\mathrm{RF},2}, \ldots, \boldsymbol{W}_{\mathrm{RF},N_{\mathrm{R}}^{\mathrm{G}}}\right] \in \mathbb{C}^{N_{\mathrm{R}} \times N_{\mathrm{R}}} \\
\boldsymbol{W}_{\mathrm{BB}} &= \mathrm{Bdiag}\left(\boldsymbol{W}_{\mathrm{BB},1}, \boldsymbol{W}_{\mathrm{BB},2}, \ldots, \boldsymbol{W}_{\mathrm{BB},N_{\mathrm{R}}^{\mathrm{G}}}\right) \in \mathbb{C}^{N_{\mathrm{R}} \times N_{\mathrm{R}}}
\end{aligned}
$$

Furthermore, collect $\boldsymbol{y}_m$ for $m = 1, 2, \ldots, N_{\mathrm{T}}^{\mathrm{P}}$ and thus the overall observation of the channel can be written as

$$
\boldsymbol{Y} = \boldsymbol{W}_{\mathrm{BB}}^{\mathrm{H}} \boldsymbol{W}_{\mathrm{RF}}^{\mathrm{H}} \boldsymbol{H} \boldsymbol{F}_{\mathrm{RF}} \boldsymbol{F}_{\mathrm{BB}} + \boldsymbol{N} \tag{3.19}
$$

where

$$
\begin{aligned}
\boldsymbol{Y} &= \left[\boldsymbol{y}_1, \boldsymbol{y}_2, \ldots, \boldsymbol{y}_{N_{\mathrm{T}}^{\mathrm{P}}}\right] \in \mathbb{C}^{N_{\mathrm{R}} \times N_{\mathrm{T}}^{\mathrm{P}}} \\
\boldsymbol{F}_{\mathrm{RF}} &= \left[\boldsymbol{F}_{\mathrm{RF},1}, \boldsymbol{F}_{\mathrm{RF},2}, \ldots, \boldsymbol{F}_{\mathrm{RF},N_{\mathrm{T}}^{\mathrm{P}}}\right] \in \mathbb{C}^{N_{\mathrm{T}} \times N_{\mathrm{T}}} \\
\boldsymbol{F}_{\mathrm{BB}} &= \mathrm{Bdiag}\left(\boldsymbol{F}_{\mathrm{BB},1}, \boldsymbol{F}_{\mathrm{BB},2}, \ldots, \boldsymbol{F}_{\mathrm{BB},N_{\mathrm{T}}^{\mathrm{G}}}\right) \in \mathbb{C}^{N_{\mathrm{T}} \times N_{\mathrm{T}}^{\mathrm{P}}} \\
\boldsymbol{F}_{\mathrm{RF},p} &= \left[\boldsymbol{f}_{\mathrm{BB},\frac{(p-1)N_{\mathrm{T}}^{\mathrm{P}}}{N_{\mathrm{T}}^{\mathrm{G}}}+1}, \boldsymbol{f}_{\mathrm{BB},\frac{(p-1)N_{\mathrm{T}}^{\mathrm{P}}}{N_{\mathrm{T}}^{\mathrm{G}}}+2}, \ldots, \boldsymbol{f}_{\mathrm{BB},\frac{pN_{\mathrm{T}}^{\mathrm{P}}}{N_{\mathrm{T}}^{\mathrm{G}}}}\right] \in \mathbb{C}^{N_{\mathrm{R}} \times \frac{N_{\mathrm{T}}^{\mathrm{P}}}{N_{\mathrm{T}}^{\mathrm{G}}}} \\
\boldsymbol{n} &= \left[\boldsymbol{n}_1, \boldsymbol{n}_2, \ldots, \boldsymbol{n}_{N_{\mathrm{T}}^{\mathrm{P}}}\right] \in \mathbb{C}^{N_{\mathrm{R}} \times N_{\mathrm{T}}^{\mathrm{P}}}
\end{aligned}
$$

The proposed pilot training scheme is illustrated in Fig. 3.5. In a nutshell, in each time block, the transmitter (receiver) generates $N_{\mathrm{T}}^{\mathrm{RF}}$ ($N_{\mathrm{R}}^{\mathrm{RF}}$) non-overlapped beams to sound the channel. By collecting the beams from multiple time blocks, the FD spatial channel can be sounded. In addition, in the simulations, we set $N_{\mathrm{T}}^{\mathrm{P}}/N_{\mathrm{T}}^{\mathrm{G}} < N_{\mathrm{T}}^{\mathrm{RF}}$ (i.e., $N_{\mathrm{T}}^{\mathrm{P}} < N_{\mathrm{T}}$), that is to say, in every transmit time block, the pilot overhead is always less than the number of non-overlapped beams. In other words, the channel observation is *compressed*. Recall that the number of effective multi-path components in mmWave channels is limited (i.e., $L \ll \min\{N_{\mathrm{T}}, N_{\mathrm{R}}\}$) due to the high path loss and severe blockage effect. This, together with the energy focusing property of lens antenna arrays, renders the sparsity of channel matrix $\boldsymbol{H}$, namely only a minority of the elements in $\boldsymbol{H}$ have significant magnitudes, while the others are close to zero. The analysis above motivates us to formulate the CE problem as a CS problem. Vectorize the expressions in (3.19) and we arrive at

$$
\boldsymbol{y} = \boldsymbol{\Phi} \boldsymbol{h} + \boldsymbol{n} \tag{3.20}
$$

where $\boldsymbol{y} = \mathrm{vec}(\boldsymbol{Y})$, $\boldsymbol{h} = \mathrm{vec}(\boldsymbol{H})$, $\boldsymbol{n} = \mathrm{vec}(\boldsymbol{N})$, $\boldsymbol{\Phi} = \left(\boldsymbol{F}_{\mathrm{BB}}^{\mathrm{T}} \boldsymbol{F}_{\mathrm{RF}}^{\mathrm{T}}\right) \otimes \left(\boldsymbol{W}_{\mathrm{BB}}^{\mathrm{H}} \boldsymbol{W}_{\mathrm{RF}}^{\mathrm{H}}\right)$.

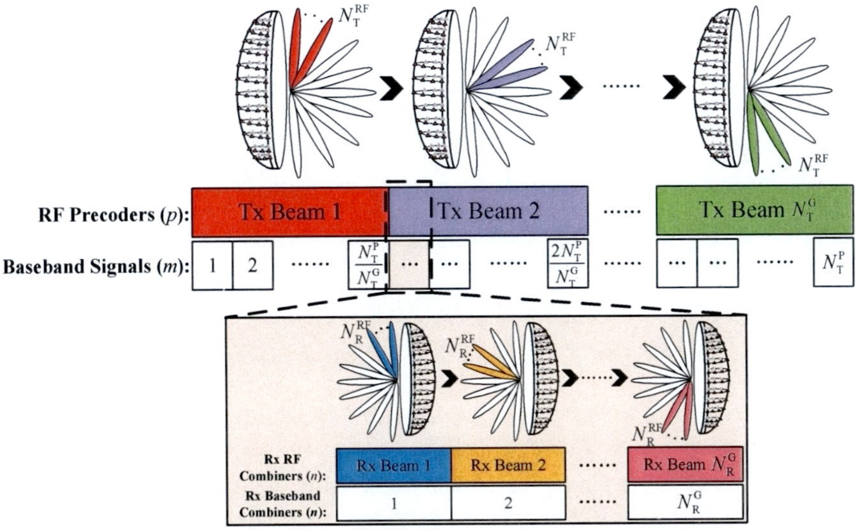

Fig. 3.5 The proposed pilot training scheme [121]

The task of CE is recovering the channel matrix $\boldsymbol{h} \in \mathbb{C}^{N_T N_R}$ via the known observation $\boldsymbol{y} \in \mathbb{C}^{N_T^P N_R}$ (corrupted by noise) and the measurement matrix $\boldsymbol{\Phi}$ (i.e., the pilot signals). Obviously, it is an under-determined problem with $N_T^P < N_T$, which is a canonical form of CS problem as shown in (3.4) and can be effectively solved by advanced CS techniques.

3.4.3 *Redundant Dictionary Design*

The accuracy of CS-based CE depends heavily on the sparsity of $\boldsymbol{h}$ in (3.20), while the power leakage caused by the mismatch between continuous AoAs/AoDs and discrete dictionary with limited resolution may weaken this sparsity [47–50]. To mitigate this behaviour, we design a redundant dictionary to re-represent the "sparse" $\boldsymbol{h}$ as a "sparser" vector. First, we quantize both the azimuth and elevation angles with a finer resolution, where the sets of the quantized virtual angles can be expressed as

$$\mathcal{A}_h = \left\{ \vartheta_g \,\middle|\, \sin\vartheta_g = -1 + \frac{2g-1}{G_h}, g = 1, 2, \ldots, G_h \right\} \tag{3.21}$$

$$\mathcal{A}_v = \left\{ \varphi_g \,\middle|\, \sin\varphi_g = -1 + \frac{2g-1}{G_v}, g = 1, 2, \ldots, G_v \right\} \tag{3.22}$$

where G_h and G_v are the quantization levels for azimuth angle and elevation angle, respectively. We assume $G_\text{h}G_\text{v} > \max\{N_\text{T}, N_\text{R}\}$, namely the quantization level is larger than the number of antenna elements. Based on (3.21) and (3.22), we re-write $\boldsymbol{H}$ as

$$\boldsymbol{H} \approx \boldsymbol{A}_\text{R}\tilde{\boldsymbol{H}}\boldsymbol{A}_\text{T}^\text{H} \tag{3.23}$$

where

$$\boldsymbol{A}_\text{R} = \left[\boldsymbol{a}_\text{R}(\vartheta_1, \varphi_1), \ldots, \boldsymbol{a}_\text{R}\left(\vartheta_1, \varphi_{G_\text{v}}\right), \boldsymbol{a}_\text{R}(\vartheta_2, \varphi_1), \ldots, \boldsymbol{a}_\text{R}\left(\vartheta_{G_\text{h}}, \varphi_{G_\text{v}}\right)\right] \in \mathbb{C}^{N_\text{R}\times G_\text{h}G_\text{v}} \tag{3.24}$$

$$\boldsymbol{A}_\text{T} = \left[\boldsymbol{a}_\text{T}(\vartheta_1, \varphi_1), \ldots, \boldsymbol{a}_\text{T}\left(\vartheta_1, \varphi_{G_\text{v}}\right), \boldsymbol{a}_\text{T}(\vartheta_2, \varphi_1), \ldots, \boldsymbol{a}_\text{T}\left(\vartheta_{G_\text{h}}, \varphi_{G_\text{v}}\right)\right] \in \mathbb{C}^{N_\text{T}\times G_\text{h}G_\text{v}} \tag{3.25}$$

and $\tilde{\boldsymbol{H}} \in \mathbb{C}^{G_\text{h}G_\text{v}\times G_\text{h}G_\text{v}}$ is the representation of $\boldsymbol{H}$ using the dictionary $\boldsymbol{A}_\text{R}$ and $\boldsymbol{A}_\text{T}$. Since $G_\text{h}G_\text{v} > \max\{N_\text{T}, N_\text{R}\}$, we call the dictionary matrices in (3.23) and (3.24) as the *redundant dictionary*. Substituting (3.23) in (3.20), we have

$$\boldsymbol{y} \approx \boldsymbol{\Phi}\text{vec}\left(\boldsymbol{A}_\text{R}\tilde{\boldsymbol{H}}\boldsymbol{A}_\text{T}^\text{H}\right) + \boldsymbol{n} = \boldsymbol{\Phi}\boldsymbol{\Psi}\tilde{\boldsymbol{h}} + \boldsymbol{n} \tag{3.26}$$

where $\boldsymbol{\Psi} = \boldsymbol{A}_\text{T}^* \otimes \boldsymbol{A}_\text{R}$ and $\tilde{\boldsymbol{h}} = \text{vec}\left(\tilde{\boldsymbol{H}}\right)$. Benefiting from the large quantization level in redundant dictionary, the angles of channels are more likely to coincide with the pre-defined angles in the dictionary, and thus the power leakage is suppressed. Compared to directly estimating $\boldsymbol{H}$, we can first estimate $\tilde{\boldsymbol{h}}$ (i.e., $\tilde{\boldsymbol{H}}$) and then obtain $\boldsymbol{H}$ based on (3.23) to improve the CE performance.

3.4.4 Pilot Design Based on CS Theory

In this section, in order to further improve the performance of CS-based CE, we resort to the CS theory introduced in previous text to optimize the sensing matrix $\boldsymbol{\Phi}\boldsymbol{\Psi}$ in (3.26). First of all, we focus on the RF precoder $\boldsymbol{F}_\text{RF}$ and RF combiner $\boldsymbol{W}_\text{RF}$. As indicated in Fig. 3.5, every element should be activated a least once during the CE stage to realize FD sounding. Given that each RF chain can activate at most one antenna element in the adopted ASN, we set

$$\boldsymbol{F}_\text{RF} = \tilde{\boldsymbol{I}}_{N_\text{T}}, \ \boldsymbol{W}_\text{RF} = \tilde{\boldsymbol{I}}_{N_\text{R}} \tag{3.27}$$

where $\tilde{\boldsymbol{I}}_{N_\text{T}}$ ($\tilde{\boldsymbol{I}}_{N_\text{R}}$) is an $N_\text{T} \times N_\text{T}$ ($N_\text{R} \times N_\text{R}$) identical matrix after column permutation. Note that the elements "1" and "0" in $\tilde{\boldsymbol{I}}_{N_\text{T}}$ or $\tilde{\boldsymbol{I}}_{N_\text{R}}$ stand for the connection and the disconnection between a RF chain and a antenna element, respectively.

Then we try to optimize the baseband pilot signals, including the transmit baseband signal $\boldsymbol{F}_{\mathrm{BB}}$ and the baseband combiner $\boldsymbol{W}_{\mathrm{BB}}$.

We have argued in Sect. 3.3 that optimizing the mutual coherence of the sensing matrix $\mu(\boldsymbol{\Phi\Psi})$ is able to improve the performance of CS recovery. However, the optimization of the mutual coherence may be still too complex, so the *total mutual coherence*, as defined below, is widely considered as an alternative optimization objective.

Definition 3.4.1 (Total Mutual Coherence) The total mutual coherence of the sensing matrix $\boldsymbol{A}$ is defined as

$$\mu_{\mathrm{total}}(\boldsymbol{A}) = \sum_{i \neq j} \left| \left[\boldsymbol{A}^{\mathrm{H}} \boldsymbol{A} \right]_{\{i,j\}} \right|^2 \tag{3.28}$$

It has been shown in literature [37, 119, 122] that the optimization of total mutual coherence can greatly simplify the problem while keeping a sufficiently good performance of the optimized sensing matrix. Therefore, we can adopt $\mu_{\mathrm{total}}(\boldsymbol{\Phi\Psi})$ as the optimization objective and formulate the optimization of baseband pilot signals as the following problem

$$\begin{cases} \left(\boldsymbol{F}_{\mathrm{BB}}^{\mathrm{opt}}, \boldsymbol{W}_{\mathrm{BB}}^{\mathrm{opt}} \right) = \underset{\boldsymbol{F}_{\mathrm{BB}}, \boldsymbol{W}_{\mathrm{BB}}}{\arg\min}\, \mu_{\mathrm{total}}(\boldsymbol{\Phi\Psi}) \\ \text{s.t.} \quad \left\| \boldsymbol{f}_{\mathrm{BB},i} \right\|_2^2 = 1, \forall i, \quad \left\| \boldsymbol{W}_{\mathrm{BB}} \right\|_F^2 = 1, \forall j \end{cases} \tag{3.29}$$

where the first constraint is due to the limited transmit power, and the second constraint is considered for ease of analysis.[1]

To further simplify the problem, we consider a lower bound of $\mu_{\mathrm{total}}(\boldsymbol{\Phi\Psi})$ in [37]

$$\mu_{\mathrm{total}}(\boldsymbol{\Phi\Psi}) \leq \mu_{\mathrm{total}}(\boldsymbol{F})\, \mu_{\mathrm{total}}(\boldsymbol{W}) \tag{3.30}$$

where $\boldsymbol{F} = \boldsymbol{F}_{\mathrm{BB}}^{\mathrm{T}} \boldsymbol{F}_{\mathrm{RF}}^{\mathrm{T}} \boldsymbol{A}_{\mathrm{T}}^{*}$, $\boldsymbol{W} = \boldsymbol{W}_{\mathrm{BB}}^{\mathrm{H}} \boldsymbol{W}_{\mathrm{RF}}^{\mathrm{H}} \boldsymbol{A}_{\mathrm{R}}$. The inequality (3.30) sheds light on how to decouple problem (3.29) to avoid difficult joint optimization. Specifically, we minimize $\mu_{\mathrm{total}}(\boldsymbol{F})$ and $\mu_{\mathrm{total}}(\boldsymbol{W})$ in terms of $\boldsymbol{F}_{\mathrm{BB}}$ and $\boldsymbol{W}_{\mathrm{BB}}$, respectively. Taking $\mu_{\mathrm{total}}(\boldsymbol{F})$ as as an example and assuming the columns of $\boldsymbol{F}$ are normalized, we have

[1] Note that the power of baseband combiner does not influence the CE performance, since the noise is combined and scaled in the same proportion as the received pilot signals. In practice, we hope the baseband combiner to be orthogonal in order not to color the noise. It will be seen that our optimized resulted can guarantee the orthogonality of the baseband combiner.

$$\begin{aligned}
\mu_{\text{total}}(\boldsymbol{F}) &= \left\| \boldsymbol{F}^{\text{H}}\boldsymbol{F} - \boldsymbol{I}_{G_{\text{h}}G_{\text{v}}} \right\|_F^2 \\
&= \text{tr}\left(\boldsymbol{F}^{\text{H}}\boldsymbol{F}\boldsymbol{F}^{\text{H}}\boldsymbol{F} - 2\boldsymbol{F}^{\text{H}}\boldsymbol{F} + \boldsymbol{I}_{G_{\text{h}}G_{\text{v}}} \right) \\
&= \text{tr}\left(\boldsymbol{F}\boldsymbol{F}^{\text{H}}\boldsymbol{F}\boldsymbol{F}^{\text{H}} - 2\boldsymbol{F}\boldsymbol{F}^{\text{H}} + \boldsymbol{I}_{N_{\text{T}}^{\text{P}}} \right) + G_{\text{h}}G_{\text{v}} - N_{\text{T}}^{\text{P}} \\
&= \left\| \boldsymbol{F}\boldsymbol{F}^{\text{H}} - \boldsymbol{I}_{N_{\text{T}}^{\text{P}}} \right\|_F^2 + G_{\text{h}}G_{\text{v}} - N_{\text{T}}^{\text{P}} \\
&= \left\| \boldsymbol{F}_{\text{BB}}^{\text{T}}\boldsymbol{F}_{\text{RF}}^{\text{T}}\boldsymbol{A}_{\text{T}}^{*}\boldsymbol{A}_{\text{T}}^{\text{T}}\boldsymbol{F}_{\text{RF}}^{*}\boldsymbol{F}_{\text{BB}}^{*} - \boldsymbol{I}_{N_{\text{T}}^{\text{P}}} \right\|_F^2 + G_{\text{h}}G_{\text{v}} - N_{\text{T}}^{\text{P}} \\
&\approx \left\| c\boldsymbol{F}_{\text{BB}}^{\text{T}}\boldsymbol{F}_{\text{BB}}^{*} - \boldsymbol{I}_{N_{\text{T}}^{\text{P}}} \right\|_F^2 + G_{\text{h}}G_{\text{v}} - N_{\text{T}}^{\text{P}} \\
&= \sum_{p=1}^{N_{\text{T}}^{\text{G}}} \left\| c\boldsymbol{F}_{\text{BB},p}^{\text{T}}\boldsymbol{F}_{\text{BB},p}^{*} - \boldsymbol{I}_{N_{\text{T}}^{\text{P}}/N_{\text{T}}^{\text{G}}} \right\|_F^2 + G_{\text{h}}G_{\text{v}} - N_{\text{T}}^{\text{P}}
\end{aligned} \tag{3.31}$$

where the approximation is based on an empirical conclusion $\boldsymbol{A}_{\text{T}}^{\text{H}}\boldsymbol{A}_{\text{T}} \approx c\boldsymbol{I}_{N_{\text{T}}}$ (c is a constant). It can be observed that the columns of $\boldsymbol{F}_{\text{BB},p}$, $\forall p$, should be orthogonal to minimize the objective function in (3.31). Therefore, we obtain the optimized solution for baseband precoder $\boldsymbol{F}_{\text{BB}}^{\text{opt}}$, which is given by

$$\boldsymbol{F}_{\text{BB}}^{\text{opt}} = \text{Bdiag}\left(\boldsymbol{F}_{\text{BB},1}^{\text{opt}}, \boldsymbol{F}_{\text{BB},2}^{\text{opt}}, \ldots, \boldsymbol{F}_{\text{BB},N_{\text{T}}^{\text{G}}}^{\text{opt}} \right) \tag{3.32}$$

where $\boldsymbol{F}_{\text{BB},p}^{\text{opt}}$ is the matrix satisfying $\left(\boldsymbol{F}_{\text{BB},p}^{\text{opt}}\right)^{\text{H}}\boldsymbol{F}_{\text{BB},p}^{\text{opt}} = \boldsymbol{I}_{N_{\text{T}}^{\text{P}}/N_{\text{T}}^{\text{G}}}$, $\forall p$. In our simulations, we set all $\boldsymbol{F}_{\text{BB},p}^{\text{opt}}$, $\forall p$, to the partial normalized DFT matrix, which is a sub-matrix consisting of the first $N_{\text{T}}^{\text{P}}/N_{\text{T}}^{\text{G}}$ columns of an $N_{\text{T}}^{\text{RF}} \times N_{\text{T}}^{\text{RF}}$ normalized DFT matrix. Similarly, the optimal baseband combiner can be obtained as

$$\boldsymbol{W}_{\text{BB}}^{\text{opt}} = \text{Bdiag}\left(\boldsymbol{W}_{\text{BB},1}^{\text{opt}}, \boldsymbol{W}_{\text{BB},2}^{\text{opt}}, \ldots, \boldsymbol{W}_{\text{BB},N_{\text{T}}^{\text{G}}}^{\text{opt}} \right) \tag{3.33}$$

where $\boldsymbol{W}_{\text{BB},n}^{\text{opt}}$, $\forall n$, are arbitrary unitary matrices. In our simulations, we set all $\boldsymbol{W}_{\text{BB},n}^{\text{opt}}$, $\forall n$, to an $N_{\text{R}}^{\text{RF}} \times N_{\text{R}}^{\text{RF}}$ normalized DFT matrix.

3.4.5 Computational Complexity Analysis

Since the CE problem has been formulated as the sparse signal recovery problem in Sects. 3.4.2 and 3.4.3, various off-the-shelf CS algorithms can be used to estimate the channels for the proposed scheme. Clearly, the computational complexity of the proposed scheme heavily depends on the adopted sparse signal recovery algorithms. In the simulations, we employ the well-known orthogonal matching pursuit (OMP) algorithm [37] to solve the CS problem. The computational complexity of our scheme

Table 3.1 Computational complexity of two schemes in each Iiteration [121]

Operations	Computational complexity
Proposed CE scheme	
Correlation	$\mathcal{O}(N_\mathrm{R} N_\mathrm{T}^\mathrm{P} G_\mathrm{v} G_\mathrm{h})$
Project subspace	$\mathcal{O}(i^3 + 2N_\mathrm{R} N_\mathrm{T}^\mathrm{P} i^2 + N_\mathrm{R} N_\mathrm{T}^\mathrm{P} i)$
Update residual	$\mathcal{O}(N_\mathrm{R} N_\mathrm{T}^\mathrm{P} i)$
Stop criteria	$\mathcal{O}(N_\mathrm{R} N_\mathrm{T}^\mathrm{P})$
DC-based support detection scheme [48]	
Correlation	$\mathcal{O}((N_\mathrm{h}-1)(8N_\mathrm{h}^3 + 8N_\mathrm{h}^2 + 2N_\mathrm{h} N_\mathrm{T}^\mathrm{P} + 2N_\mathrm{h}) + (N_\mathrm{v}-1)(8N_\mathrm{v}^3 + 8N_\mathrm{v}^2 + 2N_\mathrm{v} N_\mathrm{T}^\mathrm{P} + 2N_\mathrm{v}))$
Project subspace	$\mathcal{O}(J^3 + 2J^2 N_\mathrm{T}^\mathrm{P} + J N_\mathrm{T}^\mathrm{P})$
Update residual	$\mathcal{O}(N_\mathrm{T}^\mathrm{P} J)$
Stop criteria	N/A

is summarized in Table 3.1. We consider four major steps of the OMP algorithm, namely correlation, project subspace, residual update, and stop criteria identification, and the computational complexity of each step in the i-th iteration is summarized in the table. To compare the proposed scheme with existing techniques, we introduce the dual crossing (DC)-based support detection algorithm proposed in [48], which is an extension of OMP algorithm to estimate the channel associated with a lens antenna array with $N_\mathrm{h} \times N_\mathrm{v}$ antenna elements. The DC-based algorithm models the pattern of power leakage as "dual crossing", and accordingly selects $J \gg 1$ atoms (while the OMP algorithm only selects 1 atom) in each iteration. The large value of J (e.g., $J = 64$ in [48]) causes unaffordable computational complexity, as listed in Table 3.1. It is also necessary to stress that the DC-based CE scheme requires the bulky PSN to assist CE, while the proposed scheme only needs simple ASN. In the following simulations, we will compare the performance of two schemes.

3.4.6 Simulation Results

In the simulations, we set the carrier frequency to 30 GHz, $g_l \sim \mathcal{CN}(0, 1)$, θ_T^l, ϕ_T^l, θ_R^l, and ϕ_R^l all follow the uniform distribution $\mathcal{U}(-\pi/2, \pi/2)$. We define the signal-to-noise ratio (SNR) as $-10\log_{10}\sigma^2$ dB. The normalized mean square error (NMSE) $\mathbb{E}\left(\left\|\boldsymbol{H} - \widehat{\boldsymbol{H}}\right\|_F^2 / \|\boldsymbol{H}\|_F^2\right)$ is adopted as the metric for CE performance, where $\widehat{\boldsymbol{H}}$ is the estimate of the channel $\boldsymbol{H}$.

In Fig. 3.6a, we plot the NMSE of the proposed scheme and the DC-based scheme with different $\{G_\mathrm{v}, G_\mathrm{h}\}$. It is observed that even with a simple ASN, the proposed scheme outperforms the DC-based scheme when the appropriate parameters $\{G_\mathrm{v}, G_\mathrm{h}\}$ are selected. More importantly, we can see that increasing $\{G_\mathrm{v}, G_\mathrm{h}\}$ can

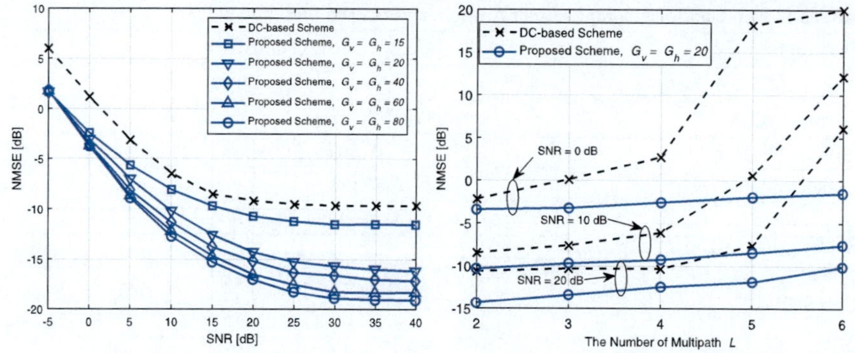

(a) NMSE performance with $L = 3$ (b) NMSE performance with different L

Fig. 3.6 Performance comparison between the proposed scheme and the DC-based scheme $N_{\mathrm{R}} = N_{\mathrm{R}}^{\mathrm{RF}} = 1$, $N_{\mathrm{T}} = 128$ (i.e., $\tilde{D}_{\mathrm{T}}^{\mathrm{v}} = \tilde{D}_{\mathrm{T}}^{\mathrm{h}} = 6.4$), $N_{\mathrm{T}}^{\mathrm{RF}} = 4$ [121]

significantly improve the CE performance of the proposed scheme, which verifies the proposed redundant dictionary design. However, one must remember that larger $\{G_{\mathrm{v}}, G_{\mathrm{h}}\}$ also brings higher computational complexity, and, according to Fig. 3.6a, when $\{G_{\mathrm{v}}, G_{\mathrm{h}}\}$ is larger than some threshold (e.g., 20), the performance improvement with $\{G_{\mathrm{v}}, G_{\mathrm{h}}\}$ becomes minor. Therefore, in practice, we should judiciously choose the value of $\{G_{\mathrm{v}}, G_{\mathrm{h}}\}$ to trade off the CE performance and the computational complexity. We set $G_{\mathrm{v}} = G_{\mathrm{h}} = 20$ in the following experiments.

Figure 3.6b investigates the robustness of different schemes as a function of the number of multipath L. When L increases, the power leakage becomes severe and the structured sparsity pattern of channels may be destroyed. Consequently, the DC-based scheme which exploits the structured sparsity suffers from the increase of L and the corresponding NMSE performance degrades, as indicated in Fig. 3.6b. By contrast, the propose scheme with redundant dictionary has enhanced angular resolution and thus can keep the channel sparsity stable when L increases. Therefore, as shown in Fig. 3.6b, the proposed scheme exhibits better robustness against increasing L. In other words, the proposed scheme can effectively estimate the channels with more multipath components.

Another drawback of the DC-based scheme in [48] is that it cannot be directly applied to the system with lens antenna arrays at both the transmitter and receiver. In the following text, we consider such a more general scenario and investigate the performance of the proposed scheme. We set $\tilde{D}_{\mathrm{T}}^{\mathrm{v}} \times \tilde{D}_{\mathrm{T}}^{\mathrm{h}} = \tilde{D}_{\mathrm{R}}^{\mathrm{v}} \times \tilde{D}_{\mathrm{R}}^{\mathrm{h}} = 4.7 \times 4.7$ which yields $N_{\mathrm{T}} = N_{\mathrm{R}} = 64$. We assume that the numbers of RF chains are $N_{\mathrm{T}}^{\mathrm{RF}} = N_{\mathrm{R}}^{\mathrm{RF}} = 4$ (i.e., $N_{\mathrm{T}}^{\mathrm{G}} = N_{\mathrm{R}}^{\mathrm{G}} = 16$), and $G_{\mathrm{v}} = G_{\mathrm{h}} = 20$. For comparison, we also weigh in the following schemes: (1) the proposed pilot training and redundant dictionary with random baseband pilot signals and random baseband combiner [40]; (2) the proposed pilot training without redundant dictionary, where the channel $\boldsymbol{H}$ is directly estimated from (3.20); and (3) traditional well-determined least square (LS) algorithm, which requires $N_{\mathrm{T}}^{\mathrm{P}} = N_{\mathrm{T}}$ and $\boldsymbol{F}_{\mathrm{BB}} = \boldsymbol{I}_{N_{\mathrm{T}}}$, $\boldsymbol{W}_{\mathrm{BB}} = \boldsymbol{I}_{N_{\mathrm{R}}}$ to transform (3.20)

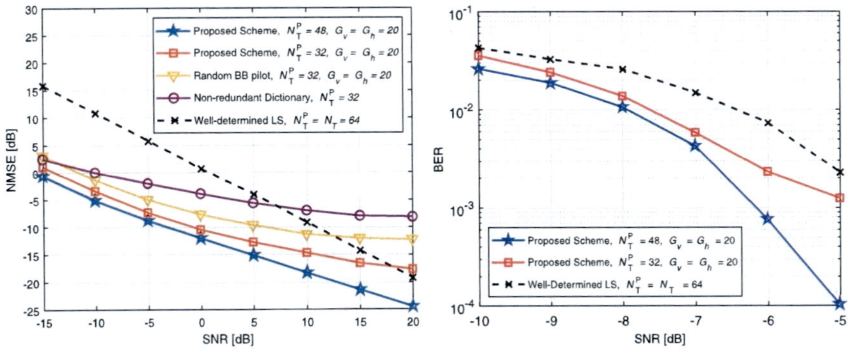

(a) NMSE performance (b) BER performance

Fig. 3.7 Simulation results of different schemes $N_T = N_R = 64$, $N_T^{RF} = N_r^{RF} = 4$ [121]

into well-determined equations. Figure 3.7a demonstrates the NMSE performance of different schemes as a function of SNR. We observe that the gain of the proposed baseband pilot signal design and the redundant dictionary design is considerable within a wide range of SNR. Moreover, the proposed CS-based scheme can outperform the traditional well-determined LS algorithm, especially in the low SNR regime. For example, when $N_T^P = 32$ and $\mathrm{SNR} = 0\,\mathrm{dB}$, the proposed scheme has more than 10 dB gain over the well-determined LS scheme with only half the pilot overhead. Note that in absence of the complete channel state information, the SNR during the mmWave CE stage is usually low in practice. Therefore, the proposed scheme is a more appealing CE solution for the practical mmWave massive MIMO with lens antenna arrays.

We further compare the bit error rate (BER) performance of the proposed scheme and the well-determined LS scheme, in order to show the direct improvement on communication performance brought by the proposed CE scheme. We adopt 64-Quadrature Amplitude Modulation (QAM) and turbo channel coding with code rate 1/3. When investing BER, we let the ASNs at the transceiver select those antenna elements with most significant energy, as pointed out by the estimated channel matrix, and the baseband precoder/combiner are designed based on the singular value decomposition of the low-dimensional effective channel after RF precoding/combining. Figure 3.7b shows that the proposed scheme can achieve better BER performance with a reduced pilot overhead. When BER is 10^{-2}, for example, the proposed scheme with pilot overhead $N_T^P = 32$ has about 1 dB gain over the well-determined LS scheme with $N_T^P = 64$.

3.5 Summary

In this chapter, we investigate the CE scheme for mmWave massive MIMO with FD lens antenna arrays based on CS theory. As a promising candidate for realizing mmWave massive MIMO, the lens antenna array adopts passive EM lens, rather than traditional bulky PSN, in order to achieve high energy efficiency and low hardware complexity in mmWave massive MIMO systems. To accurately estimate the channels associated with lens antenna arrays with reduced pilot overhead, we introduce advanced CS techniques. First, we propose a pilot training scheme under the simple ASN, which ensures that the FD spatial range can be sounded with compressed observation. On this basis, we formulate the CE problem as the sparse signal recovery problem which can be solved by CS techniques. Subsequently, we design a redundant dictionary to re-represent the original sparse channel as a sparser form, which is expected to benefit the applied CS-based algorithms and enhance the robustness against the number of multipath components. Finally, relying on the CS theory, we propose to optimize the pilot signals (including baseband transmitted signal and the baseband combiner) by minimizing the total mutual coherence of the sensing matrix, and we obtain the closed-form optimization results. Simulation results demonstrate the superiority of the proposed CS-based CE scheme over its counterparts.

Chapter 4
Hybrid Beamforming Design for MmWave Massive MIMO Systems

Abstract In this chapter, we propose an efficient hybrid beamforming design framework for downlink multi-user multi-stream transmission in mmWave massive MIMO systems. First, the analytic solutions for digital precoder are derived to minimize the sum mean square error between the transmitted and received signals. Then, we propose a tractable criterion for the codebook-based fully-connected analog beamforming design, and further extend it to the case where the dynamic partially-connected structure is considered. A low-complexity energy-based greedy antenna grouping scheme is proposed for the dynamic hybrid structure. Simulation results demonstrate the effectiveness and the superiority of the proposed scheme in terms of spectral efficiency and bit error rate over its counterparts.

4.1 Introduction

In order to reduce the hardware cost and power consumption, the mmWave massive MIMO system is usually accomplished with hybrid beamforming architecture. Compared to the traditional fully-digital architecture, hybrid beamforming architecture greatly reduces the number of required radio frequency (RF) chains, yet it also reduces the degree-of-freedom (DoF) for beamforming design and thus causes performance loss. Designing an effective hybrid beamformer (including both analog part and digital part) under the constraint imposed by the analog network is a non-trivial task. This chapter studies the hybrid beamforming in the mmWave massive MIMO system. Unlike most of the existing works introduced in Sect. 1.3.2 whose objective is maximizing spectral efficiency, the proposed beamforming scheme in this chapter is designed to minimize the sum mean square error between the transmitted and received signals.

Z. Gao et al., *Millimeter-Wave/Sub-Terahertz Ultra-Massive MIMO Transmission Technology*, https://doi.org/10.1007/978-981-99-2388-5_4

4.2 Narrowband Hybrid Beamforming Design with Fully-Connected Architecture

In this section, we present a narrowband hybrid beamforming design for a multi-user multi-stream massive MIMO system, as shown in Fig. 4.1.

4.2.1 System Model

In the considered system in Fig. 4.1, the base station (BS) is equipped with N_t antennas and M_t RF chains, where $M_t \ll N_t$. U users are simultaneously served by the BS, and each user is equipped with N_r antennas and M_r RF chains, where $M_r \ll N_r$. We assume that the number of data streams for each user is $N_S \geq 1$, so the BS should be able to support at least UN_s data steams. To make full use of the hardware resources, we assume $N_s = M_r$ and $UN_s = M_t$ in the rest of this chapter.

In the downlink transmission, the received signal $\boldsymbol{y}_u$ at the uth ($1 \leq u \leq U$) user can be expressed as

$$\boldsymbol{y}_u = \boldsymbol{V}_u^{\mathrm{H}} \boldsymbol{M}_u^{\mathrm{H}} \left(\gamma \boldsymbol{H}_u \boldsymbol{F} \boldsymbol{W} \boldsymbol{x} + \boldsymbol{n}_u\right) \tag{4.1}$$

where $\boldsymbol{V}_u \in \mathbb{C}^{M_r \times N_s}$ is the digital combiner of the uth user, $\boldsymbol{M}_u \in \mathbb{C}^{N_r \times M_r}$ is the analog combiner of the uth user, and $\boldsymbol{H}_u \in \mathbb{C}^{N_r \times N_t}$ is the downlink channel between the BS and the uth user, while $\boldsymbol{F} = [\boldsymbol{F}_1, \boldsymbol{F}_2, \cdots, \boldsymbol{F}_U] \in \mathbb{C}^{N_t \times M_t}$ and $\boldsymbol{W} = [\boldsymbol{W}_1, \boldsymbol{W}_2, \cdots, \boldsymbol{W}_U] \in \mathbb{C}^{M_t \times UN_s}$ represent the analog precoder and the digital precoder, respectively, at the BS, where $\boldsymbol{F}_u \in \mathbb{C}^{N_t \times M_r}$ and $\boldsymbol{W}_u \in \mathbb{C}^{M_t \times N_s}$ is the analog precoder and the digital precoder dedicated for the uth user. Also, γ is a normalization factor to ensure the power constraint $\gamma^2 \mathrm{tr}\left(\boldsymbol{F}\boldsymbol{W}\boldsymbol{W}^{\mathrm{H}}\boldsymbol{F}^{\mathrm{H}}\right) = UN_s$ at the

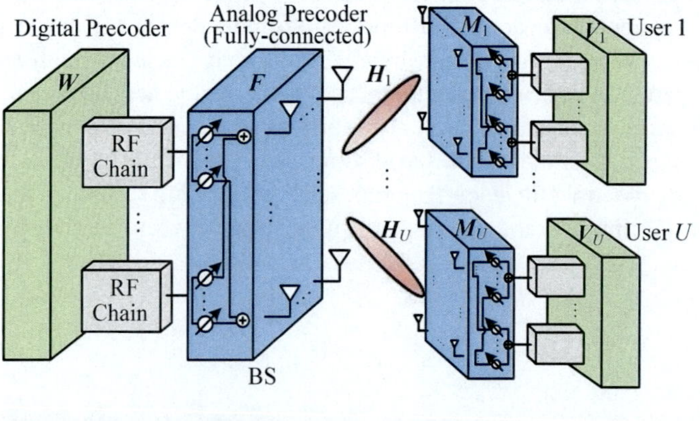

Fig. 4.1 A block diagram of the considered mmWave massive MIMO system with hybrid beamforming architecture. The BS supports multiple users and multi-stream transmission for each user

BS, $\boldsymbol{x} = \left[\boldsymbol{x}_1^{\mathrm{T}}, \boldsymbol{x}_2^{\mathrm{T}}, \cdots, \boldsymbol{x}_U^{\mathrm{T}}\right]^{\mathrm{T}} \in \mathbb{C}^{UN_s}$, where $\boldsymbol{x}_u \in \mathbb{C}^{N_s}$ is the transmitted signal for the uth user, satisfying $\mathbb{E}\left(\boldsymbol{x}\boldsymbol{x}^{\mathrm{H}}\right) = \boldsymbol{I}_{UN_s}, \boldsymbol{n}_u \sim CN\left(\boldsymbol{0}_{N_r}, \sigma^2 \boldsymbol{I}_{N_r}\right)$ is the additive white Guassian noise (AWGN) at the uth user. Note the the constant modulus constraint $\left|[\boldsymbol{F}]_{i,j}\right| = \frac{1}{\sqrt{N_t}}, \forall i, j$ and $\left|[\boldsymbol{M}_u]_{i,j}\right| = \frac{1}{\sqrt{N_r}}, \forall u, i, j$, should be considered.

The mmWave channel $\boldsymbol{H}_u$ is assumed to contain N_c scattering clusters with N_p effective paths for each cluster, that is

$$\boldsymbol{H}_u = \sqrt{\frac{N_t N_r}{N_c N_p}} \sum_{i=1}^{N_c} \sum_{l=1}^{N_p} \alpha_{il}^u \boldsymbol{a}_r^u(\theta_{il}^{ru}, \phi_{il}^{ru}) \boldsymbol{a}_t^u(\theta_{il}^{tu}, \phi_{il}^{tu})^{\mathrm{H}} \tag{4.2}$$

where α_{il}^u is the complex gain of the lth path in the ith cluster, θ_{il}^{ru} and ϕ_{il}^{ru} are the azimuth angle of arrival (AoA) and elevation AoA, respectively, of the lth path in the ith cluster, θ_{il}^{tu} and ϕ_{il}^{tu} are the azimuth angle of departure (AoD) and elevation AoD, respectively, of the lth path in the ith cluster, and $\boldsymbol{a}_r^u(\theta_{il}^{ru}, \phi_{il}^{ru}) \in \mathbb{C}^{N_r}$ and $\boldsymbol{a}_t^u(\theta_{il}^{tu}, \phi_{il}^{tu}) \in \mathbb{C}^{N_t}$ are the steering vectors at the user and the BS, respectively. In this chapter, we consider the half-wavelength spaced uniform planar array (UPA) and the corresponding steering vector, taking $\boldsymbol{a}_t^u(\theta_{il}^{tu}, \phi_{il}^{tu})$ as an example, can be written as

$$\boldsymbol{a}_t^u(\theta_{il}^{tu}, \phi_{il}^{tu}) = \boldsymbol{a}_t^{\mathrm{h}}\left(\mu_{il}^{tu}\right) \otimes \boldsymbol{a}_t^{\mathrm{v}}\left(\nu_{il}^{tu}\right) \tag{4.3}$$

where $\mu_{il}^{tu} = \pi \sin\phi_{il}^{tu} \cos\theta_{il}^{tu}$, $\nu_{il}^{tu} = \pi \sin\theta_{il}^{tu}$,

$$\boldsymbol{a}_t^{\mathrm{h}}\left(\mu_{il}^{tu}\right) = \frac{1}{\sqrt{N_t^{\mathrm{h}}}}\left[1, e^{\mathrm{j}\mu_{il}^{tu}}, \cdots, e^{\mathrm{j}(N_t^{\mathrm{h}}-1)\mu_{il}^{tu}}\right]^{\mathrm{T}} \in \mathbb{C}^{N_t^{\mathrm{h}}}$$

$$\boldsymbol{a}_t^{\mathrm{v}}\left(\nu_{il}^{tu}\right) = \frac{1}{\sqrt{N_t^{\mathrm{v}}}}\left[1, e^{\mathrm{j}\nu_{il}^{tu}}, \cdots, e^{\mathrm{j}(N_t^{\mathrm{v}}-1)\nu_{il}^{tu}}\right]^{\mathrm{T}} \in \mathbb{C}^{N_t^{\mathrm{v}}}$$

N_t^{h} and N_t^{v} are the numbers of antennas along the azimuth direction and the elevation direction, respectively, and $N_t = N_t^{\mathrm{h}} \times N_t^{\mathrm{v}}$. The steering vector at the user $\boldsymbol{a}_r^u(\theta_{il}^{ru}, \phi_{il}^{ru})$ can be formulated similarly to (4.3), and we omit the details for brevity.

Based on (4.1), the mean square error (MSE) between the received signal $\hat{\boldsymbol{x}}_u = \boldsymbol{y}_u/\gamma$ at the uth user and the transmitted signal $\boldsymbol{x}_u$ at the BS can be denoted as $\xi_u = \mathbb{E}\left(\left\|\hat{\boldsymbol{x}}_u - \boldsymbol{x}_u\right\|_2^2\right)$, and further, the sum MSE (SMSE) can be denoted as $\xi = \sum_{u=1}^{U} \xi_u$. The objective is jointly designing the analog/digital precoder/combiner to minimize SMSE ξ, which is formulated as

$$\min_{\boldsymbol{W},\boldsymbol{F},\{\boldsymbol{V}_u\}_{u=1}^U,\{\boldsymbol{M}_u\}_{u=1}^U} \quad \xi \tag{4.4a}$$

$$\text{s.t.}\gamma^2\text{tr}\left(\boldsymbol{F}\boldsymbol{W}\boldsymbol{W}^{\text{H}}\boldsymbol{F}^{\text{H}}\right) = UN_s \tag{4.4b}$$

$$\left|[\boldsymbol{F}]_{i,j}\right| = \frac{1}{\sqrt{N_t}}, \forall i, j \tag{4.4c}$$

$$\left|[\boldsymbol{M}_u]_{i,j}\right| = \frac{1}{\sqrt{N_r}}, \forall u, i, j \tag{4.4d}$$

Due to the non-convex constant modulus constraint imposed by the analog beamformer, it is difficult to solve the above problem directly. Later in this chapter, we will simplify the optimization problem above by decoupling the design of digital part and analog part.

4.2.2 Digital Precoder/Combiner Design

With the assumption that the analog precoder/combiner have been well designed, we first consider how to design the digital precoder/combiner. Let $\boldsymbol{H}_{\text{eff}}^u = \boldsymbol{M}_u^{\text{H}}\boldsymbol{H}_u\boldsymbol{F} \in {}^{M_r \times M_t}$ be the *effective baseband channel* between the BS and the uth user, and thus the received signal $\hat{\boldsymbol{x}}_u$ can be written as

$$\hat{\boldsymbol{x}}_u = \boldsymbol{V}_u^{\text{H}}\boldsymbol{H}_{\text{eff}}^u\boldsymbol{W}\boldsymbol{x} + \boldsymbol{V}_u^{\text{H}}\boldsymbol{M}_u^{\text{H}}\boldsymbol{n}_u/\gamma \tag{4.5}$$

At the same time, we assume that the inter-user interference (IUI) can be well eliminated with the aid of analog precoder/combiner, that is $\sum_{i=1,i\neq u}^{U} \boldsymbol{V}_u^{\text{H}}\boldsymbol{H}_{\text{eff}}^u\boldsymbol{W}_i\boldsymbol{x}_i \approx \boldsymbol{0}_{N_s}$. Accordingly, the MSE of the uth user can be expressed as

$$\begin{aligned}\xi_u \approx &\text{tr}\left(\boldsymbol{V}_u^{\text{H}}\boldsymbol{H}_{\text{eff}}^u\boldsymbol{W}\boldsymbol{W}^{\text{H}}(\boldsymbol{H}_{\text{eff}}^u)^{\text{H}}\boldsymbol{V}_u\right) + \frac{\sigma^2}{\gamma^2}\text{tr}\left(\boldsymbol{V}_u^{\text{H}}\boldsymbol{M}_u^{\text{H}}\boldsymbol{M}_u\boldsymbol{V}_u\right) \\ &- \text{tr}\left(\boldsymbol{V}_u^{\text{H}}\boldsymbol{H}_{\text{eff}}^u\boldsymbol{W}_u\right) - \text{tr}\left(\boldsymbol{W}_u^{\text{H}}(\boldsymbol{H}_{\text{eff}}^u)^{\text{H}}\boldsymbol{V}_u\right) + N_s\end{aligned} \tag{4.6}$$

Note that the dependence between the transmitted signal $\boldsymbol{x}_u$ and the noise $\boldsymbol{n}_u$ is utilized when deriving (4.6).

We discuss the design of digital precoder and digital combiner separately.

- **Digital Combiner $\boldsymbol{V}_{\mathbf{u}}$.** Taking the uth user as the example, we calculate the partial derivative of $\boldsymbol{V}_u^{\text{H}}$ with respect to ξ_u as

$$\frac{\partial \xi_u}{\partial \boldsymbol{V}_u^{\text{H}}} = \left(\boldsymbol{H}_{\text{eff}}^u\boldsymbol{W}\boldsymbol{W}^{\text{H}}(\boldsymbol{H}_{\text{eff}}^u)^{\text{H}}\boldsymbol{V}_k\right)^{\text{T}} + \frac{\sigma^2}{\gamma^2}\left(\boldsymbol{M}_u^{\text{H}}\boldsymbol{M}_u\boldsymbol{V}_u\right)^{\text{T}} - \left(\boldsymbol{H}_{\text{eff}}^u\boldsymbol{W}_u\right)^{\text{T}} \tag{4.7}$$

Let the derivative above equal to zero, we obtain

$$V_u^{\mathrm{H}} = W_u^{\mathrm{H}}(H_{\mathrm{eff}}^u)^{\mathrm{H}}\left(H_{\mathrm{eff}}^u W W^{\mathrm{H}}(H_{\mathrm{eff}}^u)^{\mathrm{H}} + \frac{\sigma^2}{\gamma^2} M_u^{\mathrm{H}} M_u\right)^{-1} \tag{4.8}$$

- **Digital Precoder** $\boldsymbol{W}$. Rather than the single-user processing when designing the digital cominer, the BS should take all users' signals into account, so the SMSE should be chosen as the objective. Substituting (4.6) into $\xi = \sum_{u=1}^{U} \xi_u$, we have

$$\begin{aligned} \xi \approx & \operatorname{tr}\left(V^{\mathrm{H}} H_{\mathrm{eff}} W W^{\mathrm{H}} H_{\mathrm{eff}}^{\mathrm{H}} V\right) + \frac{\sigma^2}{\gamma^2} \operatorname{tr}\left(V^{\mathrm{H}} M^{\mathrm{H}} M V\right) \\ & - \operatorname{tr}\left(V^{\mathrm{H}} H_{\mathrm{eff}} W\right) - \operatorname{tr}\left(W^{\mathrm{H}} H_{\mathrm{eff}}^{\mathrm{H}} V\right) + U N_s \end{aligned} \tag{4.9}$$

where $H_{\mathrm{eff}} = \left[(H_{\mathrm{eff}}^1)^{\mathrm{T}}, (H_{\mathrm{eff}}^2)^{\mathrm{T}}, \ldots, (H_{\mathrm{eff}}^U)^{\mathrm{T}}\right]^{\mathrm{T}}$, $V = \mathrm{Bdiag}\,(V_1, V_2, \cdots, V_U)$, $M = \mathrm{Bdiag}\,(M_1, M_2, \cdots, M_U)$. Similarly, we calculate the partial derivative of W with respect to ξ as

$$\frac{\partial \xi}{\partial W} = \left(W^{\mathrm{H}} H_{\mathrm{eff}}^{\mathrm{H}} V V^{\mathrm{H}} H_{\mathrm{eff}}\right)^{\mathrm{T}} - \left(V^{\mathrm{H}} H_{\mathrm{eff}}\right)^{\mathrm{T}} \tag{4.10}$$

Let the derivative above equal to zero, we have

$$W = \left(H_{\mathrm{eff}}^{\mathrm{H}} V V^{\mathrm{H}} H_{\mathrm{eff}}\right)^{-1} H_{\mathrm{eff}}^{\mathrm{H}} V \tag{4.11}$$

The normalization factor γ can be simply obtained as $\gamma = \sqrt{U N_s / \operatorname{tr}\left(F W W^{\mathrm{H}} F^{\mathrm{H}}\right)}$.

Now we have obtained the closed-form solutions for the digital combiner and precoder as shown in (4.8) and (4.11), respectively. However, there are still the following two unsolved problems: (1) From (4.8) and (4.11), it can be seen that there is a coupling between the two solutions, which motivates us to find a suitable initial solution to decouple them so as to avoid complex alternating optimization; (2) The obtained closed-form solutions are based on the assumption that the analog precoder/combiner can eliminate IUI well, and thus the design of analog precoder/combiner is essential while it has yet been discussed. In the next section, we will propose an efficient and low-complexity analog beamforming design to ensure good performance of IUI elimination. We ocus on the first problem in the following text of this section.

Consider a matrix $V_{\mathrm{ini}} = \mathrm{Bdiag}\left(V_{\mathrm{ini}}^1, V_{\mathrm{ini}}^2, \cdots, V_{\mathrm{ini}}^U\right)$, where $V_{\mathrm{ini}}^u \in \mathbb{C}^{M_r \times N_s}, \forall u$ are unitary matrices. We propose the following procedure to design the digital precoder/combiner. Firstly, V_{ini} is used as the initial solution of the digital combiners of all users, and we substitute it into (4.11) to obtain the digital precoder at the BS. Then, the obtained digital precoder solution is substituted into (4.8) to update the digital combiner solution at the users. Let us elaborate on the rationality of this procedure. We first introduce Lemma 1 as follows, which is important to the further discussion.

Lemma 1 *Assuming that the square matrix* $\boldsymbol{A} = \left[\boldsymbol{A}_1^{\mathrm{T}}, \cdots, \boldsymbol{A}_N^{\mathrm{T}}\right]^{\mathrm{T}}$ *is invertible, we have* $\boldsymbol{A}_i\left(\boldsymbol{A}^{\mathrm{H}}\boldsymbol{A}\right)^{-1}\boldsymbol{A}_j^{\mathrm{H}} = \begin{cases} \boldsymbol{I}, & i = j \\ \boldsymbol{O}, & i \neq j \end{cases}$, *where* $\boldsymbol{I}$ *and* $\boldsymbol{O}$ *are the identity matrix and all-zeros matrix, respectively,* $1 \le i \le N$, $1 \le j \le N$.

Proof Since $\boldsymbol{A}$ is invertible,

$$\boldsymbol{A}\left(\boldsymbol{A}^{\mathrm{H}}\boldsymbol{A}\right)^{-1}\boldsymbol{A}^{\mathrm{H}} = \boldsymbol{A}\boldsymbol{A}^{-1}\left(\boldsymbol{A}^{\mathrm{H}}\right)^{-1}\boldsymbol{A}^{\mathrm{H}} = \boldsymbol{I}$$

On the other hand,

$$\boldsymbol{A}\left(\boldsymbol{A}^{\mathrm{H}}\boldsymbol{A}\right)^{-1}\boldsymbol{A}^{\mathrm{H}} = \begin{bmatrix} \boldsymbol{A}_1\left(\boldsymbol{A}^{\mathrm{H}}\boldsymbol{A}\right)^{-1}\boldsymbol{A}_1^{\mathrm{H}} & \boldsymbol{A}_1\left(\boldsymbol{A}^{\mathrm{H}}\boldsymbol{A}\right)^{-1}\boldsymbol{A}_2^{\mathrm{H}} & \cdots & \boldsymbol{A}_1\left(\boldsymbol{A}^{\mathrm{H}}\boldsymbol{A}\right)^{-1}\boldsymbol{A}_N^{\mathrm{H}} \\ \boldsymbol{A}_2\left(\boldsymbol{A}^{\mathrm{H}}\boldsymbol{A}\right)^{-1}\boldsymbol{A}_1^{\mathrm{H}} & \boldsymbol{A}_2\left(\boldsymbol{A}^{\mathrm{H}}\boldsymbol{A}\right)^{-1}\boldsymbol{A}_2^{\mathrm{H}} & \cdots & \boldsymbol{A}_2\left(\boldsymbol{A}^{\mathrm{H}}\boldsymbol{A}\right)^{-1}\boldsymbol{A}_N^{\mathrm{H}} \\ \vdots & \vdots & \ddots & \vdots \\ \boldsymbol{A}_N\left(\boldsymbol{A}^{\mathrm{H}}\boldsymbol{A}\right)^{-1}\boldsymbol{A}_1^{\mathrm{H}} & \boldsymbol{A}_N\left(\boldsymbol{A}^{\mathrm{H}}\boldsymbol{A}\right)^{-1}\boldsymbol{A}_2^{\mathrm{H}} & \cdots & \boldsymbol{A}_N\left(\boldsymbol{A}^{\mathrm{H}}\boldsymbol{A}\right)^{-1}\boldsymbol{A}_N^{\mathrm{H}} \end{bmatrix}$$

Comparing the above two equations block-by-block, we can see that Lemma 1 holds. □

Substituting the unitary matrices $\{\boldsymbol{V}_{\mathrm{ini}}^{u}\}_{u=1}^{U}$ into (4.11), the digital precoder can be obtained as $\boldsymbol{W} = \left(\boldsymbol{H}_{\mathrm{eff}}^{\mathrm{H}}\boldsymbol{H}_{\mathrm{eff}}\right)^{-1}\boldsymbol{H}_{\mathrm{eff}}^{\mathrm{H}}\boldsymbol{V}_{\mathrm{ini}}$, therefore

$$\boldsymbol{W}_u = \left(\boldsymbol{H}_{\mathrm{eff}}^{\mathrm{H}}\boldsymbol{H}_{\mathrm{eff}}\right)^{-1}\left(\boldsymbol{H}_{\mathrm{eff}}^{u}\right)^{\mathrm{H}}\boldsymbol{V}_{\mathrm{ini}}^{u} \tag{4.12}$$

We assume that $\boldsymbol{H}_{\mathrm{eff}}$ is of full rank and thus invertible (which is a commonly-adopted assumption), so $\boldsymbol{W}\boldsymbol{W}^{\mathrm{H}} = \left(\boldsymbol{H}_{\mathrm{eff}}^{\mathrm{H}}\boldsymbol{H}_{\mathrm{eff}}\right)^{-1}$. Then, we substitute $\boldsymbol{W}$ in (4.12) into (4.8) to obtain the final digital combiner solution for the uth user, that is

$$\begin{aligned} \boldsymbol{V}_u &= \left(\boldsymbol{H}_{\mathrm{eff}}^{u}\left(\boldsymbol{H}_{\mathrm{eff}}^{\mathrm{H}}\boldsymbol{H}_{\mathrm{eff}}\right)^{-1}\left(\boldsymbol{H}_{\mathrm{eff}}^{u}\right)^{\mathrm{H}} + \frac{\sigma^2}{\gamma^2}\boldsymbol{M}_u^{\mathrm{H}}\boldsymbol{M}_u\right)^{-1}\boldsymbol{H}_{\mathrm{eff}}^{u}\left(\boldsymbol{H}_{\mathrm{eff}}^{\mathrm{H}}\boldsymbol{H}_{\mathrm{eff}}\right)^{-1}\left(\boldsymbol{H}_{\mathrm{eff}}^{u}\right)^{\mathrm{H}}\boldsymbol{V}_{\mathrm{ini}}^{u} \\ &\approx \left(\boldsymbol{I}_{M_r} + \frac{\sigma^2}{\gamma^2}\boldsymbol{I}_{M_r}\right)^{-1}\boldsymbol{I}_{M_r}\boldsymbol{V}_{\mathrm{ini}}^{u} \\ &= \eta\boldsymbol{V}_{\mathrm{ini}}^{u} \end{aligned} \tag{4.13}$$

where $\eta = \frac{\gamma^2}{\gamma^2+\sigma^2}$. Note that when deriving (4.13), Lemma 1 has been utilized for several times, and the approximation is due to $\boldsymbol{M}_u^{\mathrm{H}}\boldsymbol{M}_u \approx \boldsymbol{I}_{M_r}$ in the massive MIMO regime [6].

Based on the final solutions in (4.12) and (4.13), we have

$$\begin{aligned} \boldsymbol{V}_u^{\mathrm{H}} \boldsymbol{H}_{\mathrm{eff}}^u \boldsymbol{W}_{u'} &\approx \eta (\boldsymbol{V}_{\mathrm{ini}}^u)^{\mathrm{H}} \boldsymbol{H}_{\mathrm{eff}}^u (\boldsymbol{H}_{\mathrm{eff}}^{\mathrm{H}} \boldsymbol{H}_{\mathrm{eff}})^{-1} \left(\boldsymbol{H}_{\mathrm{eff}}^{u'}\right)^{\mathrm{H}} \boldsymbol{V}_{\mathrm{ini}}^{u'} \\ &= \begin{cases} \eta \boldsymbol{I}_{N_s}, & u = u' \\ \boldsymbol{O}_{N_s \times N_s}, & u \neq u' \end{cases} \end{aligned} \tag{4.14}$$

Note that Lemma 1 has also been utilized when deriving (4.14). (4.14) indicates that the proposed digital beamforming design can well eliminate both IUI and the inter-stream interference, that is $\boldsymbol{V}^{\mathrm{H}} \boldsymbol{H}_{\mathrm{eff}} \boldsymbol{W} \approx \eta \boldsymbol{I}_{K N_s}$. Therefore, the proposed digital beamforming design is reasonable.

4.2.3 Analog Precoder/Combiner Design

In this section, we consider designing analog beamforming, including analog precoder and combiner. Due to the non-convex constant modulus constraint, the analog beamforming is more challenging than the previous digital beamforming design. Therefore, we propose to further simplify the optimization problem. Specifically, the key idea is two-fold: (1) Construct the analog beamforming codebook, that is to say, the analog precoder/combiner can only be selected from the pre-defined set with limited number of codewords; and (2) Find a more feasible optimization objective instead of the abstract min-SMSE criterion. We will discuss these two points in the following text.

Algorithm 4.1: Oversampled Codebook Design

Input: The number of antennas $N = N^{\mathrm{h}} \times N^{\mathrm{v}}$, the quantization level of phase shifters q (in bit), and the oversampling factor ρ

Output: Oversampled codebook $\mathcal{D}$

1 **Initialization:** Generate the set $Q = \{0, 2\pi/2^q, \cdots, 2\pi(2^q - 1)/2^q\}$ and the normalized DFT matrices $\boldsymbol{F}^{\mathrm{h}}$ and $\boldsymbol{F}^{\mathrm{h}}$ of order ρN^{h} and ρN^{v}, respectively ;

2 $\boldsymbol{F}^{\mathrm{h}}_{\mathrm{oversampled}} = \boldsymbol{F}^{\mathrm{h}}_{\{1:N^{\mathrm{h}},:\}}$, $\boldsymbol{F}^{\mathrm{v}}_{\mathrm{oversampled}} = \boldsymbol{F}^{\mathrm{v}}_{\{1:N^{\mathrm{h}},:\}}$;

3 $\boldsymbol{F}_{\mathrm{oversampled}} = \boldsymbol{F}^{\mathrm{h}}_{\mathrm{oversampled}} \otimes \boldsymbol{F}^{\mathrm{v}}_{\mathrm{oversampled}}$;

4 Quantization: $\angle\left(\left[\boldsymbol{F}_{\mathrm{oversampled}}\right]_{i,j}\right) = \arg\min\limits_{\alpha \in Q} \left|\angle\left(\left[\boldsymbol{F}_{\mathrm{oversampled}}\right]_{i,j}\right) - \alpha\right|, \forall i, j$;

5 Deduplication: make each column vector of $\boldsymbol{F}_{\mathrm{oversampled}}$ be unique;

6 **Return:** $\mathcal{D} = \{$all the column vectors of $\boldsymbol{F}_{\mathrm{oversampled}}\}$

In terms of codebook design, most of the traditional methods consider discrete Fourier transformation (DFT) codebook, which is equivalent to dividing the whole space equally according to the (virtual) angle. The DFT codebook has several advantages, for example, it is simple to implement, the required quantization level of phase

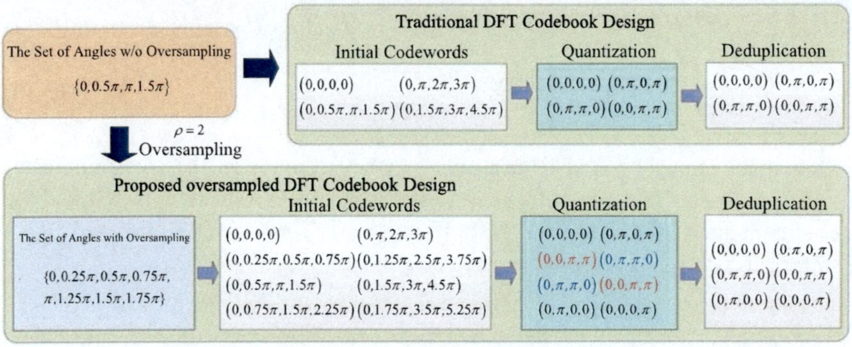

Fig. 4.2 Comparison between the traditional DFT codebook and the proposed oversampled DFT codebook. We consider a half-wavelenth spaced uniform linear array with 4 antennas, and $\rho = 2$

shifter is low, and the codewords in it are orthogonal to each other, which can avoid inter-beam interference. However, the resolution of the DFT codebook is strictly limited by the number of antennas, that is to say, if the number of antennas is small (especially at the user side), the number of available codewords in the DFT codebook will also be small, which may not provide the efficient spatial resolution to meet the requirement of MIMO communications. In order to overcome the drawback of traditional DFT codebook, we propose a codebook design based on oversampled DFT. The key feature of the proposed design is that the angular range in the original DFT codebook is oversampled with an oversampling factor $\rho > 1$, and then the oversampled codebook is generated similarly to the construction of a DFT codebook. After quantization and deduplication operations, the final codebook is obtained. We summarize the proposed codebook design as Algorithm 4.1, and illustrate a simple example in Fig. 4.2. As seen from Fig. 4.2, if only the traditional DFT codebook design is used, we can only get 4 available codewords, while we can obtain 6 available codewords after conducting Algorithm 4.1 with $\rho = 2$. Richer codewords improve the DoF for codebook-based analog beamforming design, making beamforming scheme better adapt to various channel conditions.

Next, we focus on how to find a new optimization objective to replace the abstract min-SMSE criterion. Note that we have argue that the proposed digital precoder/combiner design yields $\boldsymbol{V}^{\mathrm{H}}\boldsymbol{H}_{\mathrm{eff}}\boldsymbol{W} \approx \mu \boldsymbol{I}_{KN_s}$, and we substitute it into (4.9) to re-write the SMSE as

$$\begin{aligned}
\xi &\approx \left(\mu^2 - 2\mu + 1\right) U N_s + \frac{\sigma^2}{\gamma^2} \mathrm{tr}\left(\boldsymbol{V}^{\mathrm{H}}\boldsymbol{M}^{\mathrm{H}}\boldsymbol{M}\boldsymbol{V}\right) \\
&\approx \left(\mu^2 - 2\mu + 1\right) U N_s + \frac{\sigma^2\mu^2}{\gamma^2} U N_s \\
&= \frac{\sigma^2 U N_s}{\gamma^2 + \sigma^2}
\end{aligned} \tag{4.15}$$

where the second approximation is due to $\boldsymbol{M}_u^{\mathrm{H}}\boldsymbol{M}_u \approx \boldsymbol{I}_{M_r}, \forall u$, in the massive MIMO regime. Equation (4.15) reveals that the SMSE has a negative correlation with γ^2, which indicates that maximizing γ^2 is a feasible criterion for designing the analog beamforming scheme. In light of this, we further investigate γ^2. Apply the singular value decomposition (SVD) to the effective baseband channel $\boldsymbol{H}_{\text{eff}}$, and we have $\boldsymbol{H}_{\text{eff}} = \boldsymbol{L}\boldsymbol{\Sigma}\boldsymbol{R}^{\mathrm{H}}$, where $\boldsymbol{L}$ and $\boldsymbol{R}$ are the left singular matrix and right singular matrix, respectively, and $\boldsymbol{\Sigma}$ is a positive definite diagonal matrix (since we have assumed that $\boldsymbol{H}_{\text{eff}}$ is of full rank) consisting of the singular values of $\boldsymbol{H}_{\text{eff}}$. On this basis, γ^2 can be re-written as

$$
\begin{aligned}
\gamma^2 &= \frac{UN_s}{\operatorname{tr}\left(\boldsymbol{F}\boldsymbol{W}\boldsymbol{W}^{\mathrm{H}}\boldsymbol{F}^{\mathrm{H}}\right)} \\
&= \frac{UN_s}{\operatorname{tr}\left(\boldsymbol{F}\left(\boldsymbol{H}_{\text{eff}}^{\mathrm{H}}\boldsymbol{H}_{\text{eff}}\right)^{-1}\boldsymbol{F}^{\mathrm{H}}\right)} \\
&= \frac{UN_s}{\operatorname{tr}\left(\boldsymbol{F}^{\mathrm{H}}\boldsymbol{F}\left(\boldsymbol{H}_{\text{eff}}^{\mathrm{H}}\boldsymbol{H}_{\text{eff}}\right)^{-1}\right)} \\
&\approx \frac{UN_s}{\operatorname{tr}\left(\left(\boldsymbol{H}_{\text{eff}}^{\mathrm{H}}\boldsymbol{H}_{\text{eff}}\right)^{-1}\right)} \\
&= \frac{UN_s}{\operatorname{tr}\left(\boldsymbol{\Sigma}^{-2}\right)} \\
&\leq \frac{\operatorname{tr}\left(\boldsymbol{\Sigma}^{2}\right)}{UN_s} \\
&= \frac{\|\boldsymbol{H}_{\text{eff}}\|_2^2}{UN_s}
\end{aligned} \tag{4.16}
$$

The approximation above is due to $\boldsymbol{F}^{\mathrm{H}}\boldsymbol{F} \approx \boldsymbol{I}_{M_t}$ [6], and the inequality is based on the famous harmonic mean-arithmetic mean (HM-AM) inequality, where "=" holds if and only if all the singular values of $\boldsymbol{H}_{\text{eff}}$ are equal. (4.16) provides an insight that in order to maximize γ^2 (i.e., minimize SMSE), we can maximize the total energy of $\boldsymbol{H}_{\text{eff}}$ (i.e., $\|\boldsymbol{H}_{\text{eff}}\|_2^2$) while keeping all of its singular values as equal as possible. Note that $\boldsymbol{H}_{\text{eff}}$ is expected to be a diagonally-dominant matrix so that the IUI and inter-stream interference can be suppressed. Therefore we have $\|\boldsymbol{H}_{\text{eff}}\|_2^2 \approx \sum_{i=1}^{UN_s}\left|[\boldsymbol{H}_{\text{eff}}]_{i,i}\right|^2$, that is to say, we can replace $\|\boldsymbol{H}_{\text{eff}}\|_2^2$ by $\sum_{i=1}^{UN_s}\left|[\boldsymbol{H}_{\text{eff}}]_{i,i}\right|^2$ in (4.16) to further simplify the optimization objective.

Based on the analysis above, we can formulate the analog beamforming design as the following optimization problem

$$\max_{\{M_u\}_{u=1}^{U}, F} \sum_{i=1}^{UN_s} \left|[H_{\text{eff}}]_{i,i}\right|^2 - \text{cond}\left(H_{\text{eff}}\right) \tag{4.17a}$$

$$\text{s.t.} F_{\{:,m\}} \in \mathcal{D}_t, \forall m \tag{4.17b}$$

$$\left[M_u\right]_{\{:,n\}} \in \mathcal{D}_r^u, \forall u, n \tag{4.17c}$$

where $\mathcal{D}_t$ and $\mathcal{D}_r^u$ are the codebooks obtained by Algorithm 4.1 at the BS and the uth user, respectively, while $\text{cond}\left(H_{\text{eff}}\right) = \sigma_{\max}\left(H_{\text{eff}}\right)/\sigma_{\min}\left(H_{\text{eff}}\right)$ is the condition number of H_{eff} where $\sigma_{\max}\left(H_{\text{eff}}\right)$ and $\sigma_{\min}\left(H_{\text{eff}}\right)$ are the maximum singular value and the minimum singular value of H_{eff}, respectively. In a nutshell, the objective function in (4.17a) tries to enlarge the total energy of the diagonal elements in H_{eff} while keeping $\text{cond}\left(H_{\text{eff}}\right)$ as small as possible (i.e., keeping all singular values of H_{eff} as equal as possible). A greedy algorithm shown in Algorithm 4.2 is proposed to solve the problem. To be specific, in each iteration of Algorithm 4.2, the codewords which maximize the total energy of the diagonal elements in H_{eff} are selected as the analog precoder/combiner. Meanwhile, in order to guarantee a good $\text{cond}\left(H_{\text{eff}}\right)$, a correlation factor β is introduced in Algorithm 4.2. The candidates in the codebooks having strong correlation (lager than β) with the selected codewords are removed.

Algorithm 4.2: Codebook-based Narrowband Analog Beamforming Design

Input: Correlation factor β, BS codebook $\mathcal{D}_t$, user codebooks $\mathcal{D}_r^u$, $\forall u$, and the channels H_u, $\forall u$

Output: Analog precoder $F = [F_1, F_2, \cdots, F_U]$ and combiners M_u, $\forall u$

Initialization: $\mathcal{U} = \{1, 2, ..., U\}$, and let F and M_u, $\forall u$, be empty matrices ;

while *the number of columns in* F *is less than* UN_s **do**

- $\{u^{\text{opt}}, d_r^{\text{opt}}, d_t^{\text{opt}}\} = \underset{u\in\mathcal{U}, d_r\in\mathcal{D}_r^u, d_t\in\mathcal{D}_t}{\arg\max} \left|d_r^{\text{H}} H_u d_t\right|^2$;
- $M_{u^{\text{opt}}} = \left[M_{u^{\text{opt}}} | d_r^{\text{opt}}\right]$, $F_{u^{\text{opt}}} = \left[F_{u^{\text{opt}}} | d_t^{\text{opt}}\right]$;
- $\mathcal{D}_r^{u^{\text{opt}}} = \mathcal{D}_r^{u^{\text{opt}}} - \left\{d_r | d_r \in \mathcal{D}_r^{u^{\text{opt}}}, \left|d_r^{\text{H}} d_r^{\text{opt}}\right| \geq \beta\right\}$;
- $\mathcal{D}_t = \mathcal{D}_t - \left\{d_t | d_t \in \mathcal{D}_t, \left|d_t^{\text{H}} d_t^{\text{opt}}\right| \geq \beta\right\}$;
- **if** *the number of columns in* $M_{u^{\text{opt}}}$ *is* N_s **then**
  - $\mathcal{U} = \mathcal{U} - u^{\text{opt}}$;
- **end**

end

Return: $F = [F_1, F_2, \cdots, F_U]$ and M_u, $\forall u$

4.2.4 Simulation Results

In the simulations, we adopt average spectrum efficiency (ASE) and bit error rate (BER) as the metric for evaluating the beamforming performance. Note that the proposed analog beamforming design and digital beamforming design are independent to each other. We consider the following two schemes as comparison: (1) two-stage hybrid beamforming scheme [52] and (2) hybrid block diagonalization (Hy-BD) scheme [54]. It is also noteworthy that the analog beamforming design and digital beamforming design in Hy-BD scheme are also independent to each other, so we can apply the analog beamforming design digital beamforming design from the proposed scheme and/or the hybrid BD scheme to verify the respective superiority of our scheme. At the BS, $N_t = 8 \times 8$, and the quantization level of the phase shifters is 3 bits, while at each user, $N_r = 4 \times 4$ and the quantization level of the phase shifters is 2 bits. For the channel model, $N_c = 8$, $N_p = 10$, $\alpha_{il}^{u} \sim \mathcal{CN}(0, 1)$. The central AoA/AoD in each cluster follow the uniform distribution $\mathcal{U}(-\pi/2, \pi/2)$, while the angular spread is set to 7.5 deg. When investing BER performance, we adopt 4-quadrature amplitude modulation (QAM). The signal-to-noise ratio (SNR) is defined as $-10\log_{10}\sigma^2$.

In Figs. 4.3 and 4.4, we respectively illustrate the ASE and BER performance of the proposed scheme and its counterparts. For Fig. 4.3, we can readily see that the proposed beamforming scheme outperforms the two-stage scheme and the Hy-BD scheme. Moreover, combining the proposed digital/analog scheme with the Hy-BD analog/digital scheme also results in the better ASE performance compared to the pure Hy-BD scheme. This further indicates that the proposed digital beamforming design and analog beamforming design are able to improve the system performance independently. The similar conclusion can also be drawn under the BER metric, as shown in Fig. 4.4.

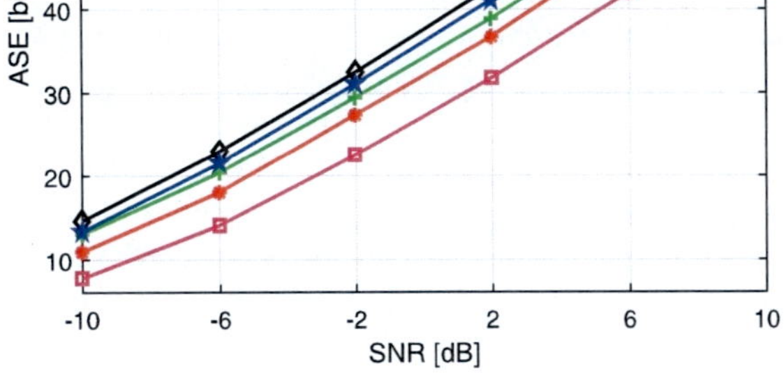

Fig. 4.3 ASE performance verse SNR. $U = 8$, $N_s = 1$, and $\rho = 8$, $\beta = 0.15$ for the proposed scheme

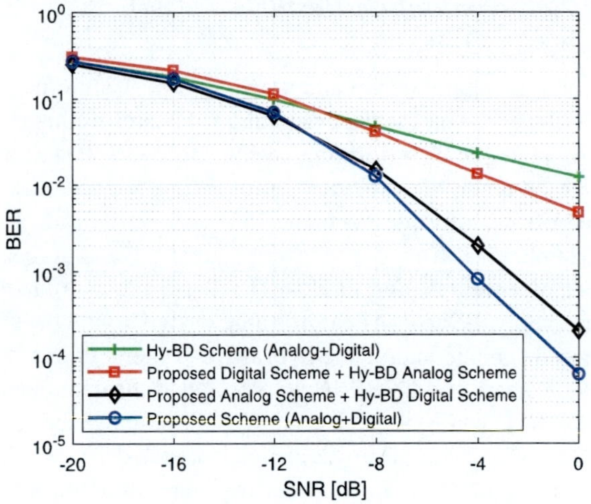

Fig. 4.4 BER performance verse SNR. $U = 2$, $N_s = 2$, and $\rho = 8$, $\beta = 0.15$ for the proposed scheme

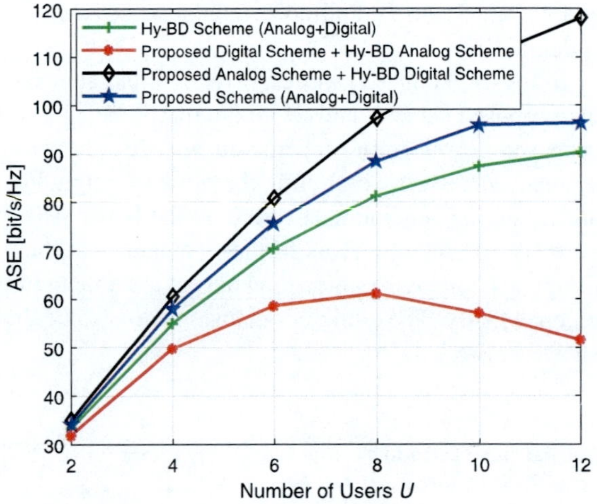

Fig. 4.5 ASE performance verse the number of users. SNR = 10 dB, $N_s = 2$, and $\rho = 8$, $\beta = 0.15$ for the proposed scheme

In Fig. 4.5, the ASE performance as a function of the number of users U is investigated, where it is observed that when increasing U within a certain range (e.g., $U \leq 6$), the ASE performance can be significantly improved. However, when U exceeds a threshold (e.g., $U \geq 8$, as shown in Fig. 4.5), increasing U results in the decrease of ASE performance, since in such case the IUI is the dominating factor.

Last, we investigate the performance of the proposed beamforming design under different ρ and β, and the results are presented in Figs. 4.6 and 4.7. It is readily seen from Fig. 4.6 that increasing ρ significantly improves the ASE performance of the proposed scheme, which verifies the effectiveness of the proposed over-sampled codebook design. On the other hand, increasing correlation factor β fails to guarantee a good condition number of the effective baseband channel matrix, and thus the ASE

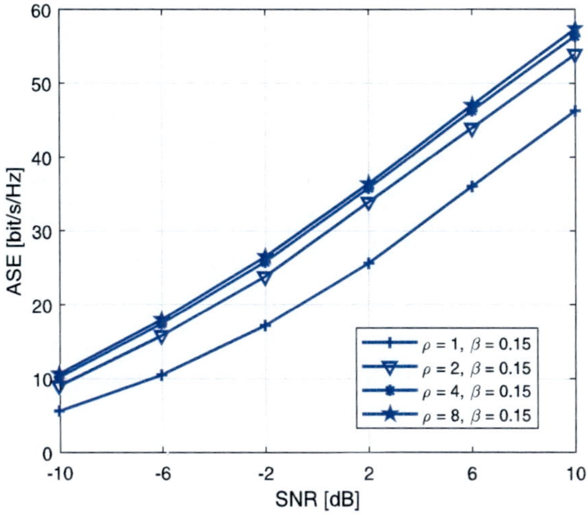

Fig. 4.6 ASE performance under different ρ. $U = 4$, $N_s = 2$

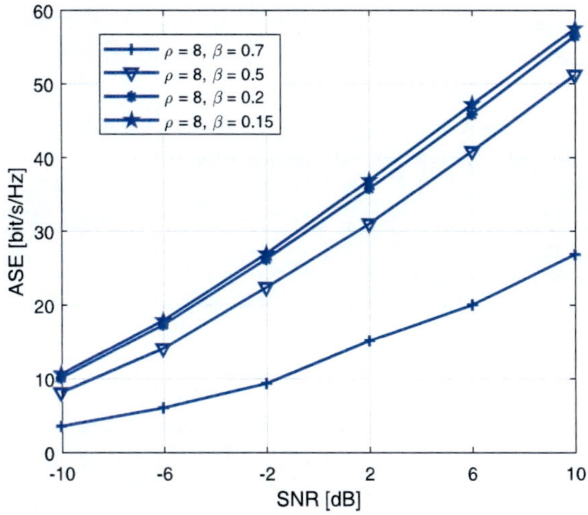

Fig. 4.7 ASE performance under different β. $U = 4$, $N_s = 2$

performance deteriorates rapidly, as shown in Fig. 4.7. In practice, the coefficients ρ and β should be carefully chosen to strike a balance between the system performance and computational complexity.

4.3 Wideband Hybrid Beamforming Design with Fully-Connected Architecture

In practical mmWave channels, the transmitted signal is likely to arrive at the receiver via different paths with different delays (i.e., delay spread). When the bandwidth of

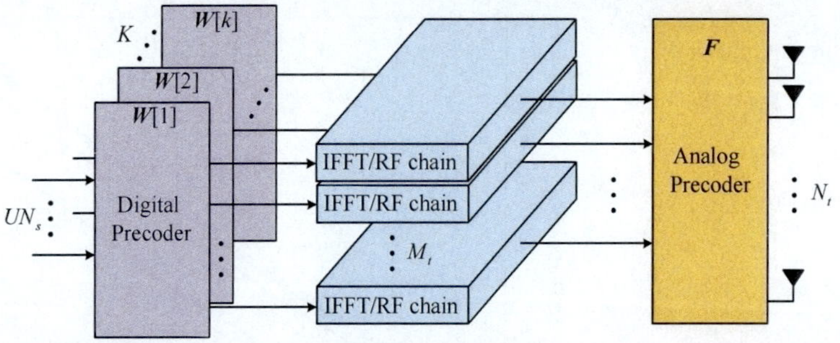

Fig. 4.8 The hardware architecture at the BS of a hybrid massive MIMO system adopting OFDM. If the user side is considered, the inverse fast Fourier transformation (inverse FFT, IFFT) module should be replaced by the FFT module

the transmitted signal is larger than the coherence bandwidth of the channel, the signal will experience different channel responses at different frequencies, which is called the *frequency selective fading* effect in wideband mmWave communications. In this section, in order to overcome the frequency selective fading effect, we adopt an orthogonal frequency division multiplexing (OFDM) transmission scheme in mmWave massive MIMO with hybrid beamforming architecture, and design a tailored wideband hybrid beamforming scheme.

4.3.1 System Model

In general, we follow the symbol definition in the previous section. The hardware architecture of the BS in a MIMO-OFDM system is illustrated in Fig. 4.8 (the architecture at each user is similar so the sketch map is omitted for brevity), where K is the number of sub-carriers in the OFDM system. Note that we add a suffix "$[k]$" for a symbol to represent that it is related to the kth sub-carrier.

In wideband OFDM system, the received signal needs to be re-written from (4.1) to

$$\boldsymbol{y}_u[k] = \boldsymbol{V}_u^{\mathrm{H}}[k]\,\boldsymbol{M}_u^{\mathrm{H}}\left(\boldsymbol{H}_u[k]\,\boldsymbol{F}\boldsymbol{W}[k]\,\boldsymbol{x}[k] + \boldsymbol{n}_u[k]\right) \tag{4.18}$$

It should be stressed again that the definition of the symbols above is totally the same as that in (4.1) but only the suffix "$[k]$" is attached. Also note that the phase shifter in the analog network are deemed to be frequency-flat, so "$[k]$" is omitted for the analog precoder $\boldsymbol{F}$ and analog combiner $\boldsymbol{M}_u$. The channel at the kth sub-carrier $\boldsymbol{H}_u[k]$ in (4.18) can be expressed as

$$\boldsymbol{H}_u[k] = \sqrt{\frac{N_t N_r}{N_c N_p}} \sum_{i=1}^{N_c}\sum_{l=1}^{N_p} \alpha_{il}^u e^{-j\frac{2\pi\tau_{il}^u k}{KT_s}} \boldsymbol{a}_r^u(\theta_{il}^{ru}, \phi_{il}^{ru})\boldsymbol{a}_t^u(\theta_{il}^{tu}, \phi_{il}^{tu})^{\mathrm{H}} \tag{4.19}$$

where τ_{il}^{u} is the delay offset of each path, and T_s is the sampling period of the system. The definition of other symbols in (4.19) is the same as that in (4.2).

Similarly, the MSE of the uth user at the kth sub-carrier can be written as $\xi_u[k] = \mathbb{E}\left(\left\|\gamma^{-1}[k]\,\boldsymbol{y}_u[k] - \boldsymbol{x}_u[k]\right\|_2^2\right)$, where $\gamma[k]$ is the normalization factor at the BS. The wideband beamforming problem can be cast to the following optimization problem

$$\min_{\{\boldsymbol{W}[k]\}_{k=1}^{K},\{\boldsymbol{V}[k]\}_{k=1}^{K},\{\gamma[k]\}_{k=1}^{K},\boldsymbol{F},\boldsymbol{M}} \sum_{k=1}^{K}\sum_{u=1}^{U}\xi_u[k] \tag{4.20a}$$

$$\text{s.t.}\operatorname{tr}\left(\boldsymbol{F}\boldsymbol{W}[k]\,\boldsymbol{W}^{\mathrm{H}}[k]\,\boldsymbol{F}^{\mathrm{H}}\right) = P, \forall k \tag{4.20b}$$

$$\left|[\boldsymbol{F}]_{i,j}\right| = \frac{1}{\sqrt{N_t}}, \forall i, j \tag{4.20c}$$

$$\left|[\boldsymbol{M}_u]_{i,j}\right| = \frac{1}{\sqrt{N_r}}, \forall u, i, j \tag{4.20d}$$

where $\boldsymbol{V}[k] = \operatorname{Bdiag}(\boldsymbol{V}_1[k], \boldsymbol{V}_2[k], \cdots, \boldsymbol{V}_U[k])$, $\boldsymbol{M} = \operatorname{Bdiag}(\boldsymbol{M}_1, \boldsymbol{M}_2, \cdots, \boldsymbol{M}_U)$, and P is the transmit power at each sub-carrier (the uniform power allocation among all sub-carriers is assumed).

4.3.2 Wideband Digital Precoder/Combiner Design

Similarly to Sect. 4.2.2, we assume that the analog beamforming has been designed well so that the IUI is almost eliminated. $\xi_u[k]$ can be written as

$$\begin{aligned}\xi_u[k] =& \operatorname{tr}\left(\boldsymbol{V}_u^{\mathrm{H}}[k]\,\boldsymbol{H}_{\mathrm{eff}}^{u}[k]\,\tilde{\boldsymbol{W}}[k]\,\tilde{\boldsymbol{W}}^{\mathrm{H}}[k]\left(\boldsymbol{H}_{\mathrm{eff}}^{u}[k]\right)^{\mathrm{H}}\boldsymbol{V}_u[k]\right)\\ &- 2\operatorname{Re}\left(\operatorname{tr}\left(\boldsymbol{V}_u^{\mathrm{H}}[k]\,\boldsymbol{H}_{\mathrm{eff}}^{u}[k]\,\tilde{\boldsymbol{W}}_u[k]\right)\right)\\ &+ \sigma^2\gamma^{-2}[k]\operatorname{tr}\left(\boldsymbol{V}_u^{\mathrm{H}}[k]\,\boldsymbol{M}_u^{\mathrm{H}}\boldsymbol{M}_u\boldsymbol{V}_u[k]\right) + N_s\end{aligned} \tag{4.21}$$

where $\boldsymbol{H}_{\mathrm{eff}}^{u}[k] = \boldsymbol{M}_u^{\mathrm{H}}\boldsymbol{H}_u[k]\,\boldsymbol{F}$, and $\tilde{\boldsymbol{W}}[k] = \gamma^{-1}[k]\,\boldsymbol{W}[k]$ stands for the digital precoder without power constraint. Given the independence of different sub-carriers, we focus only on the kth sub-carrier without loss of generality. Calculating the partial derivative of $\boldsymbol{V}_u^{\mathrm{H}}[k]$ with respect to $\xi_u[k]$ and letting it be zero, we obtain the solution for digital combiner

$$\boldsymbol{V}_u^{\mathrm{H}}[k] = \tilde{\boldsymbol{W}}_u^{\mathrm{H}}[k]\left(\boldsymbol{H}_{\mathrm{eff}}^{u}[k]\right)^{\mathrm{H}}\left(\boldsymbol{H}_{\mathrm{eff}}^{u}[k]\,\tilde{\boldsymbol{W}}[k]\,\tilde{\boldsymbol{W}}^{\mathrm{H}}[k]\left(\boldsymbol{H}_{\mathrm{eff}}^{u}[k]\right)^{\mathrm{H}} + \frac{\sigma^2}{\gamma^2[k]}\boldsymbol{M}_u^{\mathrm{H}}\boldsymbol{M}_u\right)^{-1} \tag{4.22}$$

For the digital precoder design at the BS, we consider the SMSE $\xi[k]$

$$\begin{aligned}\xi[k] = &\sum_{u=1}^{U} \xi_u[k]\\ =&\gamma^{-2}[k]\,\mathrm{tr}\left(\boldsymbol{V}^{\mathrm{H}}[k]\,\boldsymbol{H}_{\mathrm{eff}}[k]\,\boldsymbol{W}[k]\,\boldsymbol{W}^{\mathrm{H}}[k]\,\boldsymbol{H}_{\mathrm{eff}}^{\mathrm{H}}[k]\,\boldsymbol{V}[k]\right)\\ &-2\gamma^{-1}[k]\,\mathrm{Re}\left(\mathrm{tr}\left(\boldsymbol{V}^{\mathrm{H}}[k]\,\boldsymbol{H}_{\mathrm{eff}}[k]\,\boldsymbol{W}[k]\right)\right)\\ &+\sigma^2\gamma^{-2}[k]\,\mathrm{tr}\left(\boldsymbol{V}^{\mathrm{H}}[k]\,\boldsymbol{M}^{\mathrm{H}}\boldsymbol{M}\boldsymbol{V}[k]\right)+UN_s\end{aligned} \tag{4.23}$$

where $\boldsymbol{H}_{\mathrm{eff}}[k] = \left[\left(\boldsymbol{H}_{\mathrm{eff}}^{1}[k]\right)^{\mathrm{T}}, \left(\boldsymbol{H}_{\mathrm{eff}}^{2}[k]\right)^{\mathrm{T}}, \cdots, \left(\boldsymbol{H}_{\mathrm{eff}}^{U}[k]^{\mathrm{T}}\right)\right]^{\mathrm{T}}$. Instead of directly calculating the partial derivative of $\boldsymbol{W}[k]$, we introduce the Lagrange multiplier $\mu[k]$ and construct the following Lagrange function

$$L = \xi[k] + \mu[k]\left(\mathrm{tr}\left(\boldsymbol{F}\boldsymbol{W}[k]\,\boldsymbol{W}^{\mathrm{H}}[k]\,\boldsymbol{F}^{\mathrm{H}}\right) - P\right) \tag{4.24}$$

Then we can calculate the partial derivative of $\boldsymbol{W}[k]$ with respect to L as

$$\begin{aligned}\frac{\partial L}{\partial \boldsymbol{W}[k]} =&\gamma^{-2}[k]\left(\boldsymbol{W}^{\mathrm{H}}[k]\,\boldsymbol{H}_{\mathrm{eff}}^{\mathrm{H}}[k]\,\boldsymbol{V}[k]\,\boldsymbol{V}^{\mathrm{H}}[k]\,\boldsymbol{H}_{\mathrm{eff}}[k]\right)^{\mathrm{T}}\\ &-\gamma^{-1}[k]\left(\boldsymbol{V}^{\mathrm{H}}[k]\,\boldsymbol{H}_{\mathrm{eff}}[k]\right)^{\mathrm{T}} + \mu[k]\left(\boldsymbol{W}^{\mathrm{H}}[k]\,\boldsymbol{F}^{\mathrm{H}}\boldsymbol{F}\right)^{\mathrm{T}}\end{aligned} \tag{4.25}$$

and let it equal to zero to get

$$\boldsymbol{W}[k] = \gamma[k]\left(\boldsymbol{H}_{\mathrm{eff}}^{\mathrm{H}}[k]\,\boldsymbol{V}[k]\,\boldsymbol{V}^{\mathrm{H}}[k]\,\boldsymbol{H}_{\mathrm{eff}}[k] + \mu[k]\,\gamma^2[k]\,\boldsymbol{F}^{\mathrm{H}}\boldsymbol{F}\right)^{-1}\boldsymbol{H}_{\mathrm{eff}}^{\mathrm{H}}[k]\,\boldsymbol{V}[k] \tag{4.26}$$

Furthermore, we consider the derivative of normalization factor $\gamma[k]$ with respect to L

$$\begin{aligned}\frac{\partial L}{\partial \gamma[k]} =&2\gamma^{-2}[k]\,\mathrm{Re}\left(\mathrm{tr}\left(\boldsymbol{V}^{\mathrm{H}}[k]\,\boldsymbol{H}_{\mathrm{eff}}[k]\,\boldsymbol{W}[k]\right)\right)\\ &-2\gamma^{-3}[k]\,\mathrm{tr}\left(\boldsymbol{V}^{\mathrm{H}}[k]\,\boldsymbol{H}_{\mathrm{eff}}[k]\,\boldsymbol{W}[k]\,\boldsymbol{W}^{\mathrm{H}}[k]\,\boldsymbol{H}_{\mathrm{eff}}^{\mathrm{H}}[k]\,\boldsymbol{V}[k]\right)\\ &-2\sigma^2\gamma^{-3}[k]\,\mathrm{tr}\left(\boldsymbol{V}^{\mathrm{H}}[k]\,\boldsymbol{M}^{\mathrm{H}}\boldsymbol{M}\boldsymbol{V}[k]\right)\end{aligned} \tag{4.27}$$

Let the derivative be 0 and after complex math we have

$$\mu[k]\,\gamma^2[k] = \sigma^2\mathrm{tr}\left(\boldsymbol{V}^{\mathrm{H}}[k]\,\boldsymbol{M}^{\mathrm{H}}\boldsymbol{M}\boldsymbol{V}[k]\right)/P \tag{4.28}$$

Observing (4.26) and (4.28), we can see that although the value of $\mu[k]$ cannot be specifically obtained, $\mu[k]\,\gamma^2[k]$ in (4.28) can be regarded as a whole and substituted into (4.26) to get the solution for unconstrained digital precoder $\tilde{\boldsymbol{W}}[k] = \gamma^{-1}[k]\,\boldsymbol{W}[k]$. After that, the normalization factor $\gamma[k]$ is given by

$$\gamma\left[k\right]=\sqrt{P/\mathrm{tr}\left(\boldsymbol{F}\tilde{\boldsymbol{W}}\left[k\right]\tilde{\boldsymbol{W}}^{\mathrm{H}}\left[k\right]\boldsymbol{F}^{\mathrm{H}}\right)} \tag{4.29}$$

At this point, the closed-form solutions for $\boldsymbol{W}\left[k\right]$ and $\boldsymbol{V}_u\left[k\right]$ have been obtained, as shown in (4.22) and (4.26), respectively. Note that the terms $\mu\left[k\right]\gamma^2\left[k\right]$ and $\gamma\left[k\right]$ in (4.26) are presented in (4.28) and (4.29), respectively. Although there is coupling between $\boldsymbol{W}\left[k\right]$ and $\boldsymbol{V}_u\left[k\right]$, we can resort to the method mentioned in the previous section, that is to say, we use a matrix $\boldsymbol{V}_{\mathrm{ini}}=\mathrm{Bdiag}\left(\boldsymbol{V}_{\mathrm{ini}}^1,\boldsymbol{V}_{\mathrm{ini}}^2,\cdots,\boldsymbol{V}_{\mathrm{ini}}^U\right)$, where each $\boldsymbol{V}_{\mathrm{ini}}^u$ is a unitary matrix, as an initial solution. This procedure has been discussed in detail in the previous section so we do not repeat it here.

4.3.3 Wideband Analog Precoder/Combiner Design

In wideband OFDM system, the codebook-based analog beamforming design proposed in Sect. 4.2.3 is still valid. One thing that should be modified is that we should take the channels at all the sub-carriers into account, as detailed in *line 3* of Algorithm 4.3. Algorithm 4.3 gives the procedure of codebook-based wideband analog beamforming design, which is very similar to Algorithm 4.2.

Algorithm 4.3: Codebook-based Wideband Analog Beamforming Design

---

Input: Correlation factor β, BS codebook $\mathcal{D}_t$, user codebook $\mathcal{D}_r^u$, $\forall u$, and channels $\boldsymbol{H}_u[k]$, $\forall u, k$

Output: Analog precoder $\boldsymbol{F}=[\boldsymbol{F}_1,\boldsymbol{F}_2,\cdots,\boldsymbol{F}_U]$ and combiners $\boldsymbol{M}_u$, $\forall u$

1 **Initialization:** $\mathcal{U}=\{1,2,...,U\}$, and let $\boldsymbol{F}$ and $\boldsymbol{M}_u$, $\forall u$, be empty matrices ;

2 **while** *the number of columns in* $\boldsymbol{F}$ *is less than* UN_s **do**

3 $\quad\{u^{\mathrm{opt}},\boldsymbol{d}_r^{\mathrm{opt}},\boldsymbol{d}_t^{\mathrm{opt}}\}=\underset{u\in\mathcal{U},\boldsymbol{d}_r\in\mathcal{D}_r^u,\boldsymbol{d}_t\in\mathcal{D}_t}{\arg\max}\sum_{k=1}^{K}\left|\boldsymbol{d}_r^{\mathrm{H}}\boldsymbol{H}_u[k]\boldsymbol{d}_t\right|^2$;

4 $\quad\boldsymbol{M}_{u^{\mathrm{opt}}}=\left[\boldsymbol{M}_{u^{\mathrm{opt}}}|\boldsymbol{d}_r^{\mathrm{opt}}\right]$, $\boldsymbol{F}_{u^{\mathrm{opt}}}=\left[\boldsymbol{F}_{u^{\mathrm{opt}}}|\boldsymbol{d}_t^{\mathrm{opt}}\right]$;

5 $\quad\mathcal{D}_r^{u^{\mathrm{opt}}}=\mathcal{D}_r^{u^{\mathrm{opt}}}-\left\{\boldsymbol{d}_r|\boldsymbol{d}_r\in\mathcal{D}_r^{u^{\mathrm{opt}}},\left|\boldsymbol{d}_r^{\mathrm{H}}\boldsymbol{d}_r^{\mathrm{opt}}\right|\geq\beta\right\}$;

6 $\quad\mathcal{D}_t=\mathcal{D}_t-\left\{\boldsymbol{d}_t|\boldsymbol{d}_t\in\mathcal{D}_t,\left|\boldsymbol{d}_t^{\mathrm{H}}\boldsymbol{d}_t^{\mathrm{opt}}\right|\geq\beta\right\}$;

7 $\quad$**if** *the number of columns in* $\boldsymbol{M}_{u^{\mathrm{opt}}}$ *is* N_s **then**

8 $\quad\quad\mathcal{U}=\mathcal{U}-u^{\mathrm{opt}}$;

9 $\quad$**end**

10 **end**

11 **Return:** $\boldsymbol{F}=[\boldsymbol{F}_1,\boldsymbol{F}_2,\cdots,\boldsymbol{F}_U]$ and $\boldsymbol{M}_u$, $\forall u$

4.3.4 Simulation Results

Following the system setting in Sect. 4.2.4, we further assume that the system bandwidth is 100 MHz (i.e., $T_s = 10$ ns), $K = 64$, and the delay offset follows the uniform distribution $\tau_{il}^{u} \sim \mathcal{U}(0, \tau_{\max})$, where the maximum delay spread $\tau_{\max} = 160$ ns. The number of antennas at each user is fixed as 2×2 in this section. We consider an existing wideband beamforming scheme, simultaneous greedy hybrid precoding (S-GHP) scheme shown in [59], as comparison. Also, we weigh in the analog beamforming based on traditional DFT codebook, which corresponds to the codebook obtained by Algorithm 4.1 with $\rho = 1$ and $q = \infty$. We denote the digital beamforming schemes in 4.2.2 and 4.3.2 as "Old Digital Scheme" and "New Digital Scheme", respectively.

From Fig. 4.9, we can see that the proposed scheme can significantly outperform its counterparts in terms of ASE under the wideband scenario. Meanwhile, the "New Digital Scheme" proposed in this section has the superiority over the "Old Digital Scheme", since the former considers the normalization factor as one of the optimization objective. The similar conclusion can be drawn under the BER metric, as shown in Fig. 4.10.

Figure 4.11 compares the robustness of different wideband beamforming schemes to the imperfect channel state information (CSI), which results from the channel perturbation including channel estimation error, the CSI quantization in channel feedback, and/or outdated CSI [111]. For simplicity, we consider using the channel matrix perturbed by the AWGN whose power is called channel perturbation error (CPE). We observe from Fig. 4.11 that except for S-GHP, all other approaches have the similar robustness to channel perturbation. However, the proposed scheme outperforms S-GHP within a wide range of CPE.

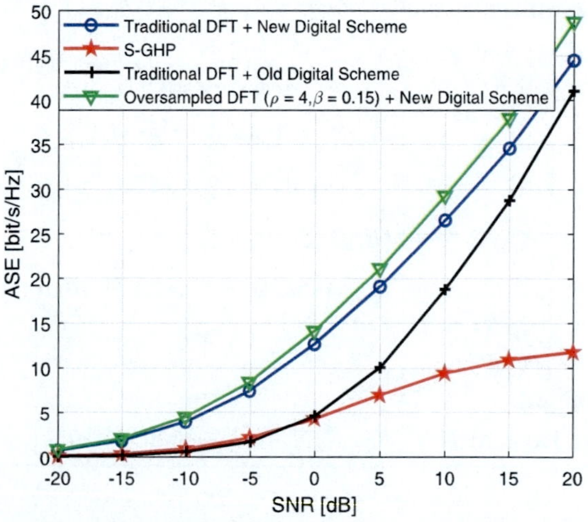

Fig. 4.9 ASE performance verse SNR. $U = 4$, $N_s = 2$, and $\rho = 4$, $\beta = 0.15$ for the proposed scheme

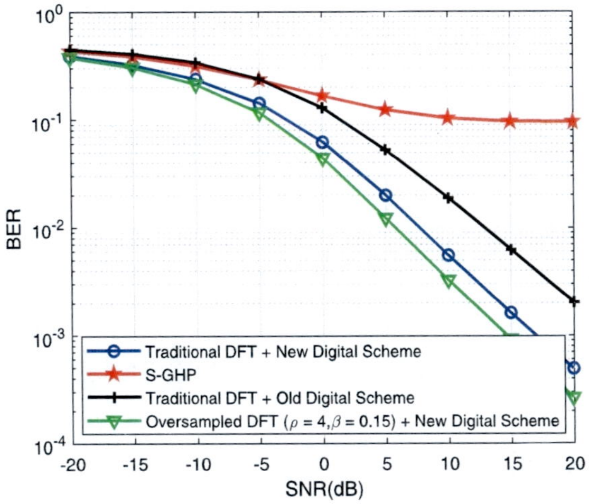

Fig. 4.10 BER performance verse SNR. $U = 2$, $N_s = 2$, and $\rho = 4$, $\beta = 0.15$ for the proposed scheme

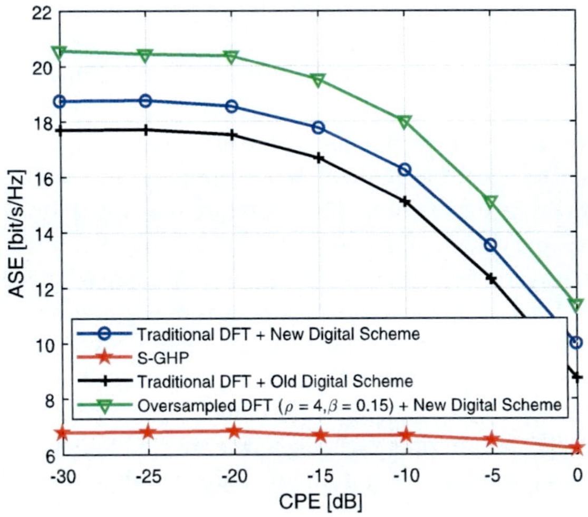

Fig. 4.11 ASE performance verse CPE. SNR = 10 dB, $U = 4$, and $N_s = 1$

4.4 Hybrid Beamforming Design with Partially-Connected Architecture

In the previous text, we have discussed narrowband or wideband beamforming design under the fully-connected hybrid architecture, where each RF chain is connected to all antennas and each connected link requires a dedicated phase shifter. The fully-connected structure requires a large number of phase shifters, which brings unaffordable hardware cost and power consumption. As a remedy, the partially-connected structure is deemed as a more feasible solution for mmWave massive MIMO. In the partially-connected structure, all the antennas are divided into several non-overlapped

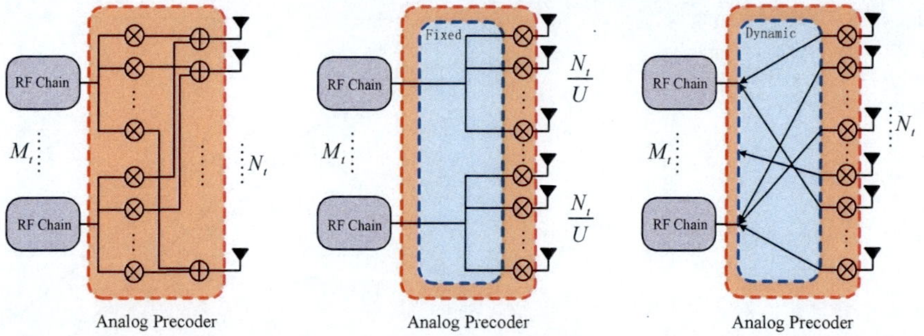

(a) Fully-connected (b) Fixed Partially-connected (c) Dynamic Partially-connected

Fig. 4.12 Three different hybrid beamforming architectures for mmWave massive MIMO systems

groups, and each antenna group is only connected to a dedicated RF chain. In this way, the number of phase shifters required in the hybrid beamforming architecture can be greatly reduced.

The partially-connected structure can be further divided into two types: (1) fixed partially-connected structure and (2) dynamic partially-connected structure. Unlike the former where the connection scheme between the antennas and RF chains is fixed and unadjustable, the latter is capable of dynamically designing the antenna grouping based on the real-time CSI, therefore providing extra DoF for hybrid beamforming design without increasing the number of phase shifters. Figure 4.12 lists three different hybrid beamforming architectures for mmWave massive MIMO systems.

This section considers the hybrid beamforming design under the partially-connected structure. The main difference between the partially-connected structure and the fully-connected structure lies in the analog beamforming design, so we will not repeat the digital beamforming scheme and it can be directly applied after the analog beamforming scheme is determined. Without loss of generality, we focus only on the wideband beamforming design in this section.

4.4.1 System Model

In general, we follow the symbol definition in the previous section. For simplicity, we assume that each user adopts a single RF chain connecting to all antennas via phase shifters (i.e., $N_s = M_r = 1$). At the BS, the partially-connected structure is considered and $M_t = U$. The received signal of the uth user at the kth sub-carrier can be expressed as

$$y_u[k] = \boldsymbol{m}_u^{\mathrm{H}} \left(\boldsymbol{H}_u[k] \boldsymbol{F} \boldsymbol{W}[k] \boldsymbol{x}[k] + \boldsymbol{n}_u[k] \right) \tag{4.30}$$

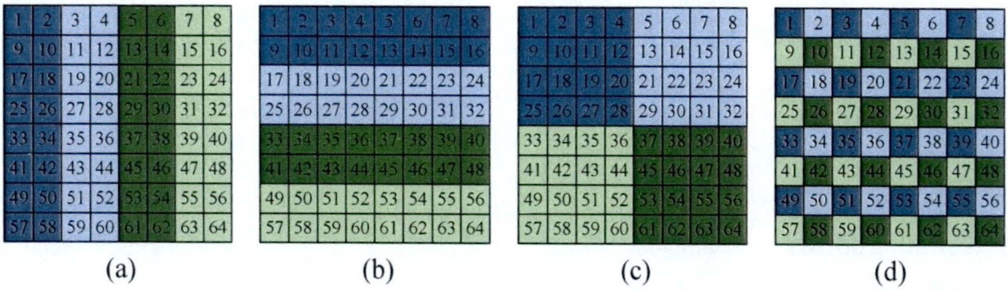

Fig. 4.13 Four fixed antenna grouping schemes. **a** Vertical type. **b** Horizontal type. **c** Squared type. **d** Interlaced type

where $\boldsymbol{m}_u \in \mathbb{C}^{N_r}$ is the analog combiner at the uth user, and $\boldsymbol{F} = \left[\boldsymbol{f}_1, \boldsymbol{f}_2, ..., \boldsymbol{f}_U\right]$ is the analog precoder at the BS. Note that each user has only one RF chain, so some of the precoders/combiners have been represented by vectors instead of matrices. Given the partially-connected structure, there may exist $\left|[\boldsymbol{F}]_{i,j}\right| = 0$, which indicates that there is no connection between the ith antenna and the jth RF chain. For convenience, we denote the set of all BS antennas as $\mathcal{S} = \{1, 2, ..., N_t\}$, while $\mathcal{S}_u \subseteq \mathcal{S}$, $u = 1, 2, ..., U$, stands for the set of antennas dedicated for the uth user (i.e., the antennas connected to the uth RF chain).

4.4.2 *Analog Beamforming Deign Under Fixed Partially-Connected Structure*

For fixed partially-connected structure, the antenna grouping $\{\mathcal{S}_u\}_{u=1}^{U}$ is fixed regardless of the CSI. We present four fixed antenna grouping schemes as shown in Fig. 4.13, where the 8×8 UPA and 4 users are considered.

Once the grouping scheme is chosen, the sets $\{\mathcal{S}_u\}_{u=1}^{U}$ can be determined accordingly. For a certain antenna grouping scheme, we define a diagonal matrix $\boldsymbol{D}_u \in \mathbb{C}^{N_t \times N_t}$ as the antenna grouping matrix

$$[\boldsymbol{D}_u]_{i,i} = \begin{cases} 1, \ i \in \mathcal{S}_u \\ 0, \ i \notin \mathcal{S}_u \end{cases} \tag{4.31}$$

Based on matrices $\{\boldsymbol{D}_u\}_{u=1}^{U}$, we can extend Algorithm 4.3 to the scenario with fixed partially-connected structure, that is Algorithm 4.4.

Algorithm 4.4: Wideband Analog Beamforming Design with Partially-connected Structure

Input: Correlation factor β, BS codebook $\mathcal{D}_t$, user codebooks $\mathcal{D}_r^u$, $\forall u$, antenna grouping matrices $\boldsymbol{D}_u$, $\forall u$, and the channels $\boldsymbol{H}_u[k]$, $\forall u, k$

Output: Analog precoder $\boldsymbol{F} = \left[\boldsymbol{f}_1, \boldsymbol{f}_2, \cdots, \boldsymbol{f}_U\right]$ and combiners $\boldsymbol{M}_u$, $\forall u$

Initialization: $\mathcal{U} = \{1, 2, ..., U\}$, and let $\boldsymbol{F}$ and $\boldsymbol{M}_u$, $\forall u$, be empty matrices

while *the number of columns in* $\boldsymbol{F}$ *is less than* UN_s **do**

$\quad \{u^{\mathrm{opt}}, \boldsymbol{d}_r^{\mathrm{opt}}, \boldsymbol{d}_t^{\mathrm{opt}}\} = \underset{u \in \mathcal{U}, \boldsymbol{d}_r \in \mathcal{D}_r^u, \boldsymbol{d}_t \in \mathcal{D}_t}{\arg\max} \sum_{k=1}^{K} \left|\boldsymbol{d}_r^{\mathrm{H}} \boldsymbol{H}_u[k] \boldsymbol{D}_u \boldsymbol{d}_t\right|^2$;

$\quad \boldsymbol{M}_{u^{\mathrm{opt}}} = \left[\boldsymbol{M}_{u^{\mathrm{opt}}} | \boldsymbol{d}_r^{\mathrm{opt}}\right]$, $\boldsymbol{f}_{u^{\mathrm{opt}}} = \boldsymbol{D}_u \boldsymbol{d}_t^{\mathrm{opt}}$;

$\quad \mathcal{D}_t = \mathcal{D}_t - \left\{\boldsymbol{d}_t | \boldsymbol{d}_t \in \mathcal{D}_t, \left|\boldsymbol{d}_t^{\mathrm{H}} \boldsymbol{d}_t^{\mathrm{opt}}\right| \geq \beta\right\}$;

$\quad \mathcal{U} = \mathcal{U} - u^{\mathrm{opt}}$;

end

Return: $\boldsymbol{F} = \left[\boldsymbol{f}_1, \boldsymbol{f}_2, \cdots, \boldsymbol{f}_U\right]$ and $\boldsymbol{M}_u$, $\forall u$

4.4.3 Analog Beamforming Deign Under Dynamic Partially-Connected Structure

To tackle the problem that the fixed partially-connected structure greatly reduces the DoF for beamforming design, we introduce a antenna selecting network between the antennas and RF chains, which is called dynamic partially-connected structure as shown in Fig. 4.12c. The connection scheme of the antenna selecting network can be dynamically adjusted based on the CSI, therefore providing an extra DoF for beamforming without increasing the number of phase shifters. To be specific, we assume that $N_{\mathrm{sub}} = N_t/U$ is an integral and each RF chain can connect to only N_{sub} antennas that do not overlap with other antenna, which renders $|\mathcal{S}_u|_c = N_{\mathrm{sub}}$, $\mathcal{S}_u \cap \mathcal{S}_{u'} = \emptyset, \forall u \neq u'$. Obviously, if the connection scheme of antenna selecting network is determined, the analog beamforming deign under dynamic partially-connected structure will be exactly the same as that under fixed partially-connected structure as discussed in the last section. Therefore, in the rest of this section, we focus our attention on how to design the connection scheme according to the CSI.

As we argued in Sect. 4.2.3, we can maximize the total energy of the diagonal elements in the effective baseband channel (which is a scaler in this section since we assume $N_s = M_r = 1$) $\left|\boldsymbol{m}_u^{\mathrm{H}} \boldsymbol{H}_u [k] \boldsymbol{D}_u \boldsymbol{f}_u\right|^2$ to design the analog beamforming that minimizes SMSE. We can write

$$\left|\boldsymbol{m}_u^{\mathrm{H}} \boldsymbol{H}_u [k] \boldsymbol{D}_u \boldsymbol{f}_u\right|^2 \leq \|\boldsymbol{m}_u\|_2^2 \|\boldsymbol{H}_u [k] \boldsymbol{D}_u\|_2^2 \|\boldsymbol{f}_u\|_2^2 = \|\boldsymbol{H}_u [k] \boldsymbol{D}_u\|_2^2 = \sigma_{\max}^2 (\boldsymbol{H}_u [k] \boldsymbol{D}_u) \tag{4.32}$$

where $\|\boldsymbol{H}_u[k]\boldsymbol{D}_u\|_2$ is the L_2 norm (i.e., spectral norm) of the matrix $\boldsymbol{H}_u[k]\boldsymbol{D}_u$, which equals to its maximum singular value $\sigma_{\max}(\boldsymbol{H}_u[k]\boldsymbol{D}_u)$. In the derivation of (4.32) we leverage the compatibility between the matrix L_2 norm and vector l_2 norm [123], $\|\boldsymbol{m}_u\|_2^2 = 1$, and $\|\boldsymbol{f}_u\|_2^2 = 1$. (4.32) inspires us to design $\boldsymbol{D}_u$ so as to maximize $\sigma_{\max}(\boldsymbol{H}_u[k]\boldsymbol{D}_u)$, and further, maximize $|\boldsymbol{m}_u^{\mathrm{H}}\boldsymbol{H}_u[k]\boldsymbol{D}_u\boldsymbol{f}_u|^2$. To gain more insights of $\sigma_{\max}(\boldsymbol{H}_u[k]\boldsymbol{D}_u)$, we introduce the notation $\boldsymbol{H}_u^{\langle j\rangle}[k]$ representing the matrix obtained by nulling all the elements of $\boldsymbol{H}_u[k]$ except for its jth column, and the notation $\boldsymbol{h}_u^{(j)}[k]$ representing the jth column vector of $\boldsymbol{H}_u[k]$. Then we have [123]

$$\sigma_{\max}\left(\boldsymbol{H}_u^{\langle j\rangle}[k]\right) = \left\|\boldsymbol{h}_u^{(j)}[k]\right\|_2 \tag{4.33}$$

and

$$\sigma_{\max}(\boldsymbol{H}_u[k]) \le \sigma_{\max}\left(\boldsymbol{H}_u^{\langle j\rangle}[k]\right) + \sigma_{\max}\left(\boldsymbol{H}_u[k] - \boldsymbol{H}_u^{\langle j\rangle}[k]\right) \tag{4.34}$$

Equations (4.33) and (4.34) further give

$$\begin{aligned}\sigma_{\max}(\boldsymbol{H}_u[k]\boldsymbol{D}_u) &= \sigma_{\max}\left(\boldsymbol{H}_u[k] - \sum_{j\notin S_u}\boldsymbol{H}_u^{\langle j\rangle}[k]\right)\\ &\ge \sigma_{\max}(\boldsymbol{H}_u[k]) - \sum_{j\notin S_u}\sigma_{\max}\left(\boldsymbol{H}_u^{\langle j\rangle}[k]\right)\\ &= \sigma_{\max}(\boldsymbol{H}_u[k]) - \sum_{j\notin S_u}\left\|\boldsymbol{h}_u^{(j)}[k]\right\|_2 \end{aligned}\tag{4.35}$$

Clearly, instead of maximizing $\sigma_{\max}(\boldsymbol{H}_u[k]\boldsymbol{D}_u)$, we can maximize its lower bound $\sigma_{\max}(\boldsymbol{H}_u[k]) - \sum_{j\notin S_u}\left\|\boldsymbol{h}_u^{(j)}[k]\right\|_2$, which can be easily realized by nulling $N_t - N_{\text{sub}}$ columns of $\boldsymbol{H}_u[k]$ with the smallest l_2 norms. In other word, we should choose N_{sub} columns of $\boldsymbol{H}_u[k]$ with the largest l_2 norms, and allocate the corresponding antennas to the uth user. This is exactly how we perform antenna grouping under dynamic partially-connected structure, as shown in Algorithm 4.5. Note that since we consider wideband scenario, the channels at all sub-carriers should be taken into account, as detailed in *line* 3 of Algorithm 4.5. After the antenna grouping scheme $S_u, \forall u$ is determined by Algorithm 4.5, we can substitute the corresponding antenna grouping matrices $\boldsymbol{D}_u$, $\forall u$, into Algorithm 4.4 to obtain the solutions for analog precoder/combiner. This completes the analog beamforming design under dynamic partially-connected structure.

4.4.4 Simulation Results

During simulations, for the wideband system, we set $K = 32$, $\tau_{il}^u \sim \mathcal{U}(0, \tau_{\max})$, and $\tau_{\max} = 80\,\text{ns}$. The setting of other parameters is the same as that in Sect. 4.3.4.

Algorithm 4.5: Multi-user Antenna Grouping Design

Input: The channels $\boldsymbol{H}_u[k]$, $\forall u, k$

Output: Antenna grouping $\mathcal{S}_u$, $\forall u$

Initialization: $\mathcal{S} = \{1, 2, ..., N_t\}$, $\mathcal{U} = \{1, 2, ..., U\}$, and let $\mathcal{S}_u$, $\forall u$, be empty sets ;

while *$\mathcal{U}$ is not an empty set* **do**

$\{u^{\text{opt}}, i^{\text{opt}}\} = \arg\max_{u \in \mathcal{U}, i \in \mathcal{S}} \sum_{k=1}^{K} \left\| \boldsymbol{h}_u^{(i)}[k] \right\|_2$;

$\mathcal{S}_{u^{\text{opt}}} = \mathcal{S}_{u^{\text{opt}}} \cup \{i^{\text{opt}}\}$;

$\mathcal{S} = \mathcal{S} - i^{\text{opt}}$;

if $\|\mathcal{S}_{u^{\text{opt}}}\|_c = N_{\text{sub}}$ **then**

$\mathcal{U} = \mathcal{U} - u^{\text{opt}}$;

end

end

Return: $\mathcal{S}_u$, $\forall u$

In Figs. 4.14 and 4.15, we provide the ASE performance and the BER performance, respectively, of the proposed narrowband hybrid beamforming (which can be seen as the wideband beamforming with only one sub-carrier) scheme under different structure. Obviously, from Figs. 4.14 and 4.15, we can draw the conclusion that the fully-connected structure using a large number of phase shifters (e.g., 256 in this section) serves as an upper bound for all other structures in terms of both ASE and BER. On the other hand, the (fixed or dynamic) partially-connected structure needs 64 phase shifters, which is only a quarter of that in the fully-connected structure. This sharp reduction of phase shifters makes the performance loss in the partially-connected structure acceptable. Meanwhile, it is observed that the dynamic partially-connected structure significantly outperforms the fixed partially-connected structure, as the former provides more DoF. In Figs. 4.16 and 4.17, the ASE performance and the BER performance are further demonstrated under wideband scenario. It is observed that the superiority of the proposed beamforming scheme can also be guaranteed in wideband mmWave massive MIMO communications.

Moreover, we investigate the ASE performance as a function of the number of users U in Fig. 4.18. For the fixed partially-connected structure, we only reserve the horizontal type and vertical type for simplicity. Similarly to Fig. 4.5, we can see that increasing U (within a certain range) can significantly improve the ASE performance. Also, it is observed the fully-connected structure has the best robustness to the number of users, while the dynamic partially-connected structure outperforms the fixed partially-connected structure as expected.

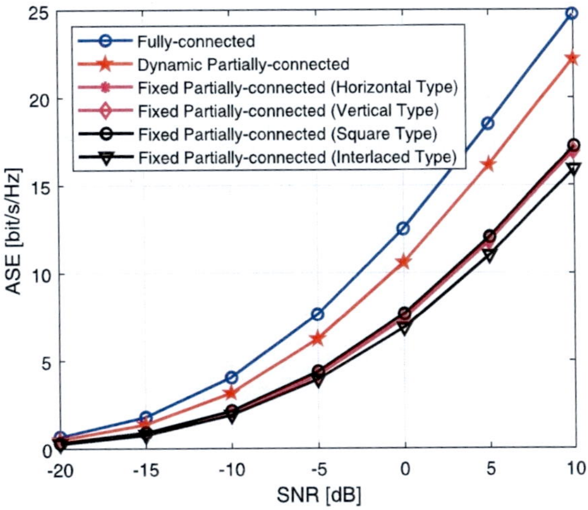

Fig. 4.14 Narrowband ASE performance under different structures. $U = 4$

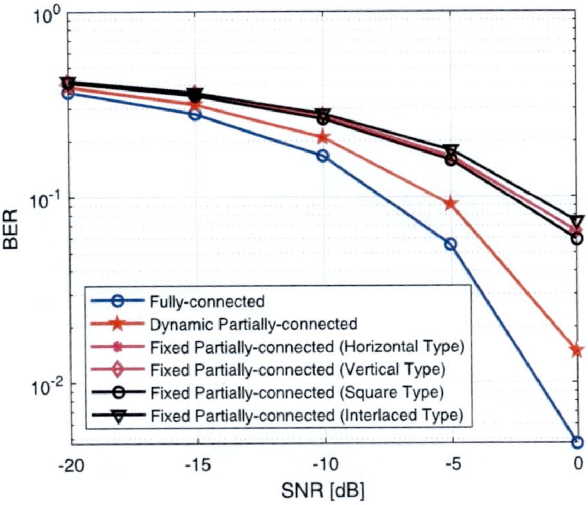

Fig. 4.15 Narrowband BER performance under different structures. $U = 4$

4.5 Summary

In this chapter, we discuss the hybrid beamforming problem in mmWave massive MIMO under narrowband/wideband systems and fully-/partially-connected structures. Unlike the most of previous works that directly optimizing the spectral efficiency of systems, we apply the min-SMSE criterion to beamforming design. Based on the different properties in different scenarios, we design a series of hybrid beamforming schemes, and prove their rationality via mathematical derivation. In addition, we conduct comprehensive simulations and compare the proposed scheme with other existing schemes, which verifies the superiority of the proposed scheme over

Fig. 4.16 Wideband ASE performance under different structures. $U = 4$

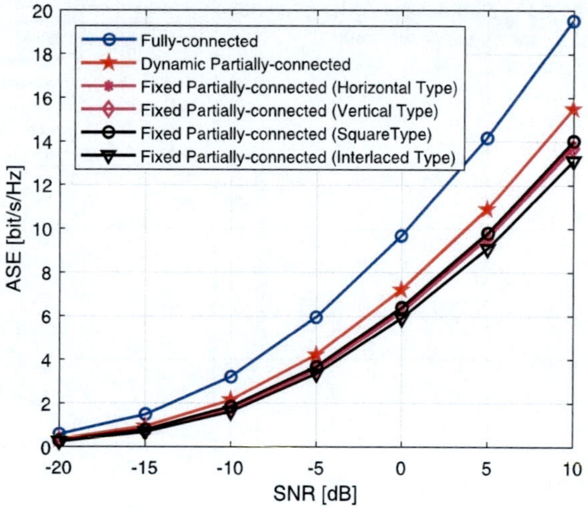

Fig. 4.17 Wideband BER performance under different structures. $U = 4$

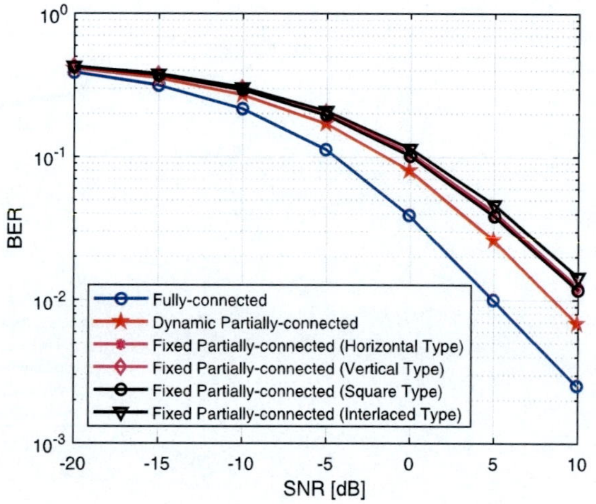

its counterparts. The proposed hybrid beamforming scheme is able to trade off the system performance, hardware complexity and power consumption.

Undoubtedly, there are still many open problems in the area of hybrid beamforming design. For example, the channel we consider in this chapter is static and ignores the mobility of users. How to design the fast and good beamforming scheme under time-varying channels is a very challenging problem. On the other hand, as the number of antennas in massive MIMO increases, the dimension of CSI becomes larger, and the CSI feedback with affordable overhead is a prerequisite for good beamforming design. A powerful tool that can be applied to beamforming design is *deep learning*, which facilitates beamforming design by leveraging the pre-collected data. The *beam squint* effect in the massive MIMO with extremely large aperture

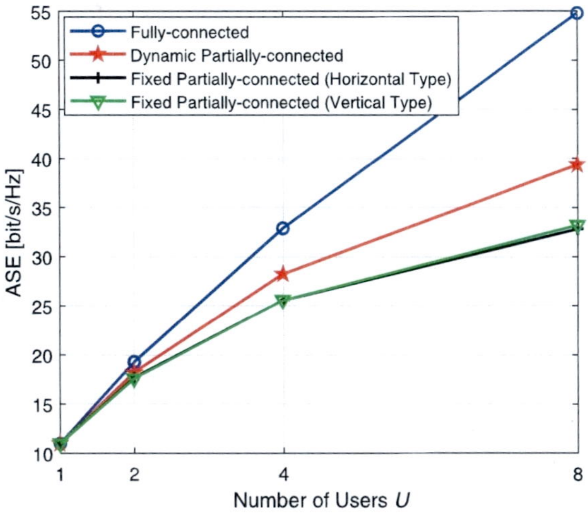

Fig. 4.18 Wideband ASE performance as a function of the number of users U under different structures

is also a non-negligible factor when designing beamforming scheme. In a nutshell, there is still a long way to go for the research of hybrid beamforming in massive MIMO systems.

Chapter 5
Compressive Sensing Based Joint Active User Detection and Channel Estimation in MmWave XL-MIMO Systems

Abstract This chapter proposes a compressive sensing (CS)-based joint active user detection (AUD) and channel estimation (CE) scheme for the indoor near-field massive IoT access scenarios constructed by millimeter-wave (mmWave) extra-large scale MIMO (XL-MIMO) arrays. Specifically, we model an mmWave XL-MIMO channel involving near-field spherical waves based on the ultra-large aperture structure of mmWave XL-MIMO arrays. This mmWave XL-MIMO channel exhibits significant spatial non-stationarity and has the coexistence of far-field and near-field conditions. By exploiting the frequency-domain common support set of the mmWave XL-MIMO channels, the joint AUD and CE problems can be then formulated as multiple measurement vector (MMV)-CS and generalized MMV (GMMV)-CS problems, where the sensing matrices associated with the MMV-CS and GMMV-CS can be designed respectively. Furthermore, based on the common block sparsity of mmWave XL-MIMO channels in the spatial and angular domains, we propose two CS-based algorithms, where a preset iteration termination threshold is considered to adaptively obtain the active user support set for the unknown number of active user equipments. Finally, our simulation results verify that the proposed solutions exhibit better AUD and CE performance for near-field massive IoT access compared with the state-of-the-art greedy CS-based schemes.

5.1 Introduction

Massive access and its typical application scenarios Internet of Things (IoT) aim to realize the future massive machine-type communication (mMTC) in the beyond fifth-generation (B5G) and sixth-generation (6G) wireless communication networks by connecting millions of devices per square kilometer [8, 9]. Based on the capability of broadband multiple access, massive IoT access technologies will be widely used in all aspects of future smart cities, including smart factories, smart homes, modern agriculture, smart transportation, etc. [124, 125]. With the explosive growth of the number of indoor and outdoor IoT devices in recent years, massive IoT access is also playing an increasingly important role in the current 5G mobile communication systems [126]. To ensure the reliable access and service quality of tremendous user

Z. Gao et al., *Millimeter-Wave/Sub-Terahertz Ultra-Massive MIMO Transmission Technology*, https://doi.org/10.1007/978-981-99-2388-5_5

equipments (UEs) in the IoT scenarios, the massive IoT access systems need to realize the accurate ultra reliable low-latency signal detection and channel estimation (CE) at the base station (BS) [127, 128]. There are two random access methods in the current IoT scenarios, namely traditional grant-based random access protocols and emerging grant-free random access protocols [124, 127, 128]. Specifically, in grant-based random access protocols, UEs need to perform multiple control signaling information interactions before accessing the network to achieve the scheduling and authorization of physical resources. However, in grant-free random access protocols, UEs directly send pilot and data signals to the BS without access grants in advance, thereby avoiding extra signaling overhead [127]. Although grant-free-based massive IoT access systems support flexible access for massive UEs, it will still bring great pressure on the systems if a large number of UEs are connected at the same time. Fortunately, most UEs in IoT present sporadic transmission characteristics, which means that only several potential UEs are activated within any given coherent time [69].

The joint active user detection (AUD) and CE solutions in the current large-scale IoT access scenarios consider the low-frequency outdoor far-field conditions, where the fully-digital transceiver with multiple-input multiple-output (MIMO) or massive MIMO architecture are utilized. These schemes are matched with the specific outdoor uplink scenarios, which is, however, unavailable in indoor scenarios. Besides, the spectral resource available in sub-6GHz becomes more and more scarce for wireless communication systems. Therefore, millimeter-wave (mmWave) and even terahertz (THz) technologies which provide rich spectral resources should be introduced in massive IoT access scenarios to meet the ultra-high data rate requirements for indoor intelligent IoT devices, such as augmented reality, mixed reality, digital twin, and holographic imaging scenes.

However, when the electromagnetic (EM) signal is transmitted in the free space, there exist many problems such as higher path loss, more vulnerability to obstacles, and larger reflection attenuation [3]. Short-range indoor transmission can avoid the problem of excessive path loss compared to the long-distance outdoor propagation of high-frequency signals. Note that the Rayleigh distance is proportional to the array aperture and inversely proportional to the carrier wavelength. Therefore, the indoor short-distance uplink access may face obvious near-field effects, which means that the planar wave assumption of the far-field EM signal propagation is no longer valid [129]. In addition, recent researches on extra-large (XL)-MIMO under near-field spatial non-stationary (SNS) conditions [31, 32, 80–86, 130] did not address the issue of massive IoT access. [131] considers the joint AUD and CE problems of grant-free massive access in a spatially non-stationary XL-MIMO system, but the channel model is set as the Bernoulli-Gaussian distribution under the assumption of planar waves, rather than spherical wave. Currently, few studies have considered indoor massive IoT access under near-field SNS conditions in mmWave XL-MIMO systems and corresponding joint AUD and CE schemes.

Based on the analysis above, this chapter aims at the uplink massive IoT access scenarios in the typical indoor smart factory environment in the future 6G industrial

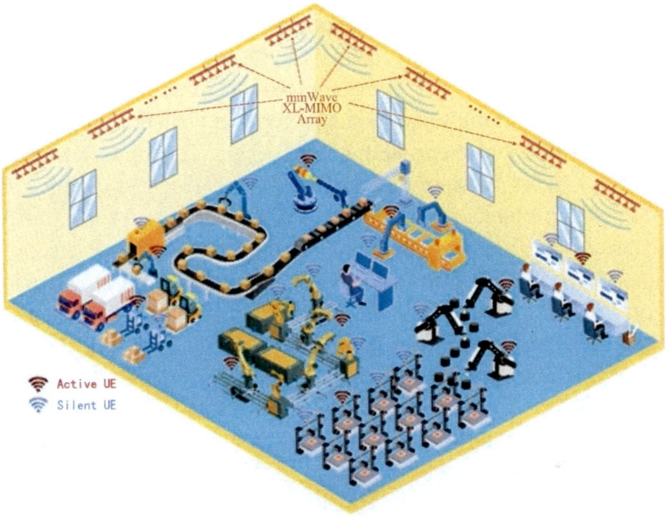

Fig. 5.1 The uplink massive IoT access scenario in a typical indoor smart factory, in which the mmWave XL-MIMO array deployed on the indoor wall can be realized by a simple and flexible wireless strip antenna [132, 133], and the BS equipped with the mmWave XL-MIMO array can provide access services for massive UEs (including active and silent UEs)

IoT that requires ultra-high system throughput. As shown in Fig. 5.1, the broadband mmWave communication technology and XL-MIMO are used to enhance the data transmission rate in massive access. Due to the vulnerability of indoor mmWave signals to obstacles and high path loss, the mmWave XL-MIMO array at the BS will be designed with a distributed array structure, where the whole array is split into multiple subarrays separated by ultra-wide spacing. It is worth noting that in such deployment, the line of sight (LoS) link between the UE and BS can be ensured.

Obviously, for the entire mmWave XL-MIMO array, there are a large number of indoor IoT devices in the near-field region, but only part of elements in XL-MIMO can receive signals due to factors such as the large array aperture and indoor obstacles, which results in SNS characteristics in mmWave XL-MIMO channels. In addition, the BS employs a hybrid analog-digital MIMO with partially-connected structure to reduce the hardware cost and power consumption [31]. It is worth mentioning that a simple and feasible implementation of the distributed mmWave XL-MIMO array designed to use here is a flexible and easy-to-deploy wireless strip antenna, that is, a strip-shaped active ultra-large-scale array [132, 133].

Specifically, this chapter designs joint AUD and CE schemes based on compressive sensing (CS) and distributed compressive sensing (DCS) for uplink massive IoT access scenarios in indoor near-field XL-MIMO systems, in which the near-field SNS of mmWave XL-MIMO channels are considered. Firstly, a signal transmission model for massive access of uplink UEs with the hybrid analog-digital XL-MIMO is established. Secondly, according to the ultra-large array aperture of the mmWave XL-MIMO array, the corresponding channel model considering near-field spherical waves is given, where the XL-MIMO channel exhibits obvious SNS. Besides, there exist cases where far-field multipath components and near-field multipath components coexist due to the distributed antenna array design. The joint AUD and CE problems can be formulated as multiple-measurement-vector (MMV)-CS and generalized MMV (GMMV)-CS problems according to the common support set of mmWave XL-MIMO channels in the frequency domain, and the corresponding sensing matrices can be designed for MMV-CS and GMMV-CS correspondingly. Then, two joint AUD and CE algorithms are proposed for MMV-CS and GMMV-CS problems, namely, MMV-blocked simultaneous orthogonal matching pursuit (BSOMP) algorithm and the GMMV-BSOMP algorithm. By utilizing a preset iteration threshold, the proposed algorithms are capable of adaptively retrieving the active user support set when the number of active UEs is unknown. Finally, the simulation results demonstrate that the AUD and CE performance of the proposed MMV-BSOMP and GMMV-BSOMP algorithms are better than those of the existing algorithms.

Compared with the existing joint AUD and CE schemes based on greedy algorithms [71–74], the main contributions of the scheme designed in this chapter are summarized as the following three points:

- **An indoor near-field massive IoT access scenario with the proposed mmWave XL-MIMO array is established**. Since most of the existing joint AUD and CE solutions [63–79] consider massive IoT access under outdoor far-field conditions, the proposed joint AUD and CE solution combines indoor mmWave communication with XL-MIMO to construct indoor near-field massive IoT access scenarios. At the same time, most of the existing solutions employ single-antenna or fully-digital MIMO transceivers at the BS, which leads to the unaffordable hardware cost and power consumption overhead in mmWave band. Thus, the proposed mmWave XL-MIMO BS transceiver adopts a partially-connected hybrid analog-digital architecture that is rarely involved in the current research on massive IoT access. The entire mmWave XL-MIMO array is split into multiple subarrays with an ultra-wide spacing to ensure LoS transmission.
- **MmWave XL-MIMO channel modeling for a near-field massive IoT access system with large array apertures**. Since the mmWave XL-MIMO array is designed as a distributed array, the mmWave XL-MIMO channel model exhibits obvious SNS and the coexistence of far-field propagation and near-field propagation. Note that the far-field beam steering vectors are generated within each subarray and the near-field steering vectors corresponds to spherical-wave phase difference from the distance among different subarrays.
- **The designed joint AUD and CE algorithms take full advantage of the common structured block sparsity of mmWave XL-MIMO channels in both spatial and angular domains**. The joint AUD and CE problem is firstly described as an MMV-CS and GMMV-CS problem according to the common support set on multiple pilot subcarriers, and the corresponding sensing matrix is designed. Then, the proposed

MMV-BSOMP algorithm and GMMV-BSOMP algorithm by utilizing the common structured block sparsity exhibited by the mmWave XL-MIMO channel in the spatial and angular domains have been proved to present more robustness for AUD and CE performance than existing CS algorithms. In addition, when the number of active UEs is unavailable, the proposed algorithm designs an optimal iteration termination threshold after theoretical analysis based on maximum likelihood (ML) criterion, so as to adaptively obtain the active user support set. The simulation results also verify the effectiveness of the preset threshold.

5.2 System Model

In order to combat the frequency selective fading caused by the multipath propagation in mmWave systems, the orthogonal frequency division multiplexing (OFDM) technology is utilized. This section firstly models the signal transmission of uplink massive IoT access in the indoor mmWave XL-MIMO-OFDM system and the mmWave XL-MIMO channel. Figure 5.2 depicts a mmWave XL-MIMO-OFDM-based transceiver structure, where a mmWave BS utilizes an XL-MIMO transceiver to provide access services for a large number of potential UEs. Considering the severe path loss and vulnerability to obscales characteristics of high-frequency mmWave signals, the mmWave XL-MIMO BS can be divided into multiple subarrays with ultra-wide spacing to ensure that the BS can receive the LoS signal from each UE.

The mmWave XL-MIMO array at the BS adopts a uniform linear array (ULA) structure with N_{BS} antennas. The N_{BS} antennas can be divided into M subarrays equally, and the number of antennas in each subarray is N_{s}, that is, the total number of antennas is $N_{\mathrm{BS}} = MN_{\mathrm{s}}$. In addition, the BS is equipped with M radio frequency

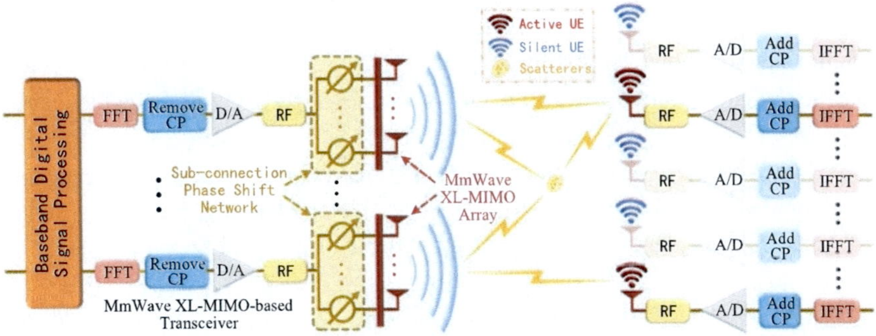

Fig. 5.2 The uplink massive IoT access mmWave XL-MIMO-OFDM transceiver structure, in which the XL-MIMO array is divided into multiple subarrays, and there is a sub-connection phase shift network (PSN) between each subarray and the corresponding RF chain. The terms FFT, IFFT, CP, D/A and A/D denote fast fourier transformation, inverse FFT, cyclic prefix, digital-to-analog and analog-to-digital, respectively

(RF) chains (corresponding to M subarrays), where each RF chain is connected to the corresponding subarray by a partially-connected PSN. The adjacent antenna spacing d in the subarray is set to half of the wavelength $d=\lambda/2$, where λ denotes the wavelength. The ultra-large spacing between adjacent sub-arrays is defined as Δ ($\Delta \gg d$).

5.2.1 *Signal Transmission Model*

According to the key characteristics of the sporadic payload of UEs in massive IoT access, in a certain coherence time, only a small part of K potential UEs are active and expect to access the communication network, that is, there are only K_{a} ($K_{\mathrm{a}} \ll K$) active UEs [69]. Considering that there are G consecutive OFDM symbols used in the joint AUD and CE stage, the received baseband signal vector $\boldsymbol{y}_p^{(g)} \in \mathbb{C}^M$ on the pth pilot subcarrier of the gth ($1 \leq g \leq G$) OFDM symbol from K_{a} active UEs can be written as

$$\begin{aligned} \boldsymbol{y}_p^{(g)} &= \left(\boldsymbol{W}_{\mathrm{RF}}^{(g)} \boldsymbol{W}_{\mathrm{BB},p}^{(g)}\right)^{\mathrm{H}} \sum_{k=1}^{K} \zeta_k \boldsymbol{h}_{k,p} s_{k,p}^{(g)} + \boldsymbol{n}_p^{(g)} \\ &= \left(\boldsymbol{W}_p^{(g)}\right)^{\mathrm{H}} \boldsymbol{H}_p \boldsymbol{s}_p^{(g)} + \boldsymbol{n}_p^{(g)} \end{aligned} \tag{5.1}$$

where $\boldsymbol{W}_p^{(g)} = \boldsymbol{W}_{\mathrm{RF}}^{(g)} \boldsymbol{W}_{\mathrm{BB},p}^{(g)} \in \mathbb{C}^{N_{\mathrm{BS}} \times M}$ denotes the cascaded hybrid combiner matrix composed of the analog combiner matrix $\boldsymbol{W}_{\mathrm{RF}}^{(g)} \in \mathbb{C}^{N_{\mathrm{BS}} \times M}$ and baseband combiner matrix $\boldsymbol{W}_{\mathrm{BB},p}^{(g)} \in \mathbb{C}^{M \times M}$. It is worth mentioning that ζ_k is a binary activity indicator factor where $\zeta_k = 1$ when the kth UE is activated and $\zeta_k = 0$ otherwise. In Eq. (5.1), $\boldsymbol{h}_{k,p} \in \mathbb{C}^{N_{\mathrm{BS}}}$ and $\boldsymbol{s}_p^{(g)} = \left[s_{1,p}^{(g)}, s_{2,p}^{(g)}, \cdots, s_{K,p}^{(g)}\right]^{\mathrm{T}} \in \mathbb{C}^K$ denote the XL-MIMO channel model and pilot signal vector, respectively. $\boldsymbol{n}_p^{(g)} = (\boldsymbol{W}_p^{(g)})^{\mathrm{H}} \bar{\boldsymbol{n}}_p^{(g)}$ is the equivalent noise vector where $\bar{\boldsymbol{n}}_p^{(g)}$ denotes the additive white Gaussian noise (AWGN) following the distribution of $\bar{\boldsymbol{n}}_p^{(g)} \sim \mathcal{CN}(\boldsymbol{0}_{N_{\mathrm{BS}}}, \sigma_{\mathrm{n}}^2 \boldsymbol{I}_{N_{\mathrm{BS}}})$. In addition, $\boldsymbol{H}_p = \left[\zeta_1 \boldsymbol{h}_{1,p}, \zeta_2 \boldsymbol{h}_{2,p}, \cdots, \zeta_K \boldsymbol{h}_{K,p}\right] \in \mathbb{C}^{N_{\mathrm{BS}} \times K}$ in Eq. (5.1) denotes the stacked channel matrix after combiner. Defining the activity indicator vector set as $\boldsymbol{\zeta} = [\zeta_1, \zeta_2, \cdots, \zeta_K]^{\mathrm{T}} \in \mathbb{R}^K$ and $|\boldsymbol{\zeta}|_0 = K_{\mathrm{a}}$, $\boldsymbol{H}_p$ can be written as $\boldsymbol{H}_p = \bar{\boldsymbol{H}}_p \boldsymbol{D}_\zeta$, where $\bar{\boldsymbol{H}}_p = \left[\boldsymbol{h}_{1,p}, \boldsymbol{h}_{2,p}, \cdots, \boldsymbol{h}_{K,p}\right]$ and $\boldsymbol{D}_\zeta = \mathrm{diag}\left(\boldsymbol{\zeta}\right)$.

5.2.2 *MmWave XL-MIMO Channel Model*

Due to the large system bandwidth in the mmWave band and the different transmission distances from UEs to each subarray in the near-field XL-MIMO communication scenario, OFDM can be used to combat the frequency-selective fading in the uplink massive IoT access. In order to improve the efficiency of data transmission, P pilot subcarriers from the total N_c subcarriers are uniformly selected to transmit the required pilot signals in the joint AUD and CE stage. Given the system bandwidth f_s and the maximum delay spread $\tau_{\max}$, the adjacent subcarrier spacing and coherence bandwidth can be calculated as $\Delta f = f_s/N_c$ and $\Delta B = 1/\tau_{\max}$. Therefore, the intervals between two adjacent pilot subcarriers are $\bar{P} = \lceil \Delta B/\Delta f \rceil$, and there are $P = \lceil N_c/\bar{P} \rceil$ pilot subcarriers in total. In the following text, the specific modeling of the mmWave XL-MIMO channel will be introduced.

For mmWave XL-MIMO arrays with large array apertures, all the elements in the array may not receive the same signal. Therefore, mmWave XL-MIMO channels exhibit inherent near-field SNS [31, 82, 85, 130]. Considering that the channel between the kth $(1 \le k \le K)$ UE and the BS consists of L_k multi-path components (MPCs), then the uplink near-field SNS channel vector on the pth pilot subcarrier $\boldsymbol{h}_{k,p} \in \mathbb{C}^{N_{\mathrm{BS}}}$ can be expressed as

$$\boldsymbol{h}_{k,p} = \sqrt{\frac{\gamma_k}{\gamma_k + 1}} \alpha_{k,1} \boldsymbol{a}_{k,p,1} + \sqrt{\frac{1}{\gamma_k + 1}} \sum_{l=2}^{L_k} \alpha_{k,l} \boldsymbol{a}_{k,p,l}, \tag{5.2}$$

where γ_k is the Rician factor for the power distribution between the LoS path and multiple NLoS MPCs. $\alpha_{k,l} \sim \mathcal{CN}(0, \sigma_\alpha^2)$ and $\sigma_\alpha^2 = 1$ represent the small-scale fading coefficient of the lth $(1 \le l \le L_k)$ MPC for the kth UE, and $\boldsymbol{a}_{k,p,l} \in \mathbb{C}^{N_{\mathrm{BS}}}$ is the mmWave XL-MIMO SNS array steering vector on the pth subcarrier corresponding to the lth $(1 \le l \le L_k)$ MPC.

As shown in Fig. 5.3a, affected by ubiquitous obstacles in indoor massive IoT access scenarios, mmWave XL-MIMO channels exhibit inherent near-field SNS characteristics, that is, the signal propagated by the LoS path and/or NLoS MPCs from

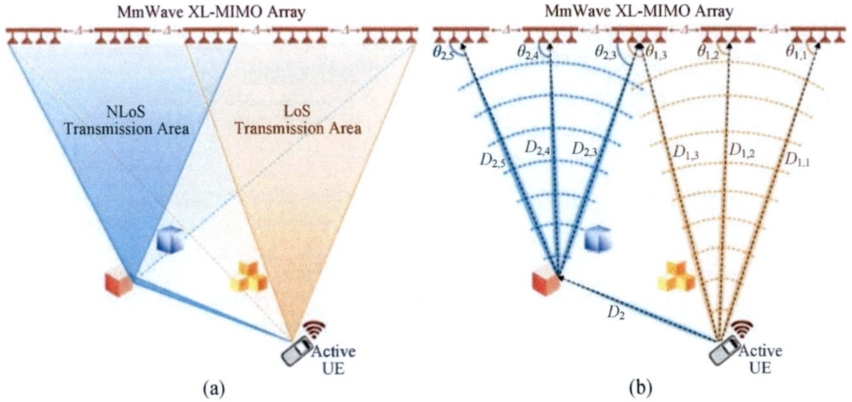

Fig. 5.3 Schematic diagram of inherent characteristics in mmWave XL-MIMO channels consisting of the LoS path and NLoS MPCs: **a** the near-field SNS; **b** the coexistence of near-field and far-field components

the same active UE may not be received by the entire mmWave XL-MIMO array consisting of multiple subarrays. Therefore, the SNS array steering vector $\boldsymbol{a}_{k,p,l}$ corresponding to the lth $(1 \le l \le L_k)$ MPC in Eq. (5.2) can be further written as

$$\boldsymbol{a}_{k,p,l} = \left(\boldsymbol{b}_{k,l} \otimes \mathbf{1}_{N_s}\right) \circ \left[\boldsymbol{c}_{k,p,l,1}^{\mathrm{T}}, \boldsymbol{c}_{k,p,l,2}^{\mathrm{T}}, \cdots, \boldsymbol{c}_{k,p,l,M}^{\mathrm{T}}\right]^{\mathrm{T}}, \tag{5.3}$$

where $\boldsymbol{c}_{k,p,l,m} \in \mathbb{C}^{N_s}$ denotes the steering vector related to the mth $(1 \le m \le M)$ subarray. Furthermore, considering the impact of Rayleigh distance (also known as far-field distance), both far-field and near-field MPCs coexist in the indoor near-field massive IoT access scenario in Fig. 5.3b [85, 86]. In other words, in an indoor area with the length and width of tens of meters, for a single subarray, the UEs and the scatterers can be regarded as far-field MPCs. However, for an XL-MIMO array containing multiple subarrays, the UEs and scatterers are located in the near-field region.[1]

Therefore, the array steering subvector $\boldsymbol{c}_{k,p,l,m}$ in Eq. (5.3) includes not only the far-field steering vector corresponding to each subarray, but also the phase difference of the spherical-wave propagation in the near-field region for the entire mmWave XL-MIMO array. Specifically, there are different expressions for $\boldsymbol{c}_{k,p,l,m}$ for the LoS path and NLoS MPCs

$$\boldsymbol{c}_{k,p,1,m} = \sqrt{\beta_{k,1,m}} \underbrace{e^{\mathrm{j}\frac{2\pi D_{k,1,m}}{\lambda}} \qquad e^{-\mathrm{j}2\pi \tau_{k,1,m}\left(-\frac{f_s}{2} + \frac{[(p-1)\bar{P}+1] f_s}{N_c}\right)}}_{\textbf{near-field components}} \underbrace{\boldsymbol{e}\left(\theta_{k,1,m}\right)}_{\textbf{far-field array steering vector}}, \tag{5.4}$$

where $D_{k,1,m}$ and $\theta_{k,1,m}$ denote the transmission distance and the corresponding angle of arrival (AoA) between the mth subarray and the kth UE, as shown in Fig. 5.3b. $\beta_{k,1,m} = \frac{G_s \lambda^2}{\left(4\pi D_{k,1,m}\right)^2}$ and $\tau_{k,1,m} = \frac{D_{k,1,m}}{v_c}$ denote the large-scale fading coefficient and path delay, respectively. G_s and v_c are the antenna gain and the speed of light, respectively. In Eq. (5.4), $e^{\mathrm{j}\frac{2\pi D_{k,1,m}}{\lambda}}$ corresponds to the effect of the near-field spherical-wave propagation, while $\boldsymbol{e}\left(\theta_{k,1,m}\right) \in \mathbb{C}^{N_s}$ is the far-field array steering vector formulated by the AoA, which can be expressed as [69]

$$\boldsymbol{e}(\theta_{k,1,m}) = \left[1, e^{\mathrm{j}\frac{2\pi d}{\lambda}\sin(\theta_{k,1,m})}, \cdots, e^{\mathrm{j}(N_s-1)\frac{2\pi d}{\lambda}\sin(\theta_{k,1,m})}\right]^{\mathrm{T}}, \tag{5.5}$$

For NLoS MPCs $(2 \le l \le L_k)$, $\boldsymbol{c}_{k,p,l,m}$ can be further expressed as

---

[1] Let us take an example to explain the case with both far-field and near-field conditions. According to [129], the Rayleigh distance L_{Rd} can be calculated as $L_{\mathrm{Rd}} = 2A^2/\lambda$, where A represents the array aperture. Consider the following system parameters, the carrier frequency is $f_c = 28$ GHz, the number of antennas in each subarray is $N_s = 8$, the antenna spacing is half of the wavelength, the spacing between two adjacent subarrays is $\Delta = 5$ m. Then, the apertures of the XL-MIMO array composed by one array and two array are $A_1 = N_s\lambda/2 \approx 0.04286$ m and $A_2 = \Delta + 2A_1 \approx 5.0857$ m, respectively. Therefore, the corresponding Rayleigh distances can be calculated as $L_{\mathrm{Rd},1} = 2A_1^2/\lambda \approx 0.34286$ m, and $L_{\mathrm{Rd},2} = 2A_1^2/\lambda \approx 4828$ m. Obviously, for a UE in an indoor space with the length and width of tens of meters, the distances from the UE and the scatterers to a single subarray will be greater than $L_{\mathrm{Rd},1}$ and much smaller than $L_{\mathrm{Rd},2}$.

$$\boldsymbol{c}_{k,p,l,m} = \underbrace{\sqrt{\beta_{k,l}^{(1)}} e^{\mathrm{j}\frac{2\pi D_{k,l}^{(1)}}{\lambda}}}_{\textbf{UE-scatterer}} \cdot \underbrace{\sqrt{\beta_{k,l,m}^{(2)}} e^{\mathrm{j}\frac{2\pi D_{k,l,m}^{(2)}}{\lambda}}}_{\textbf{scatterer-subarray}} e^{-\mathrm{j}2\pi \tau_{k,l,m}\left(-\frac{f_s}{2}+\frac{[(p-1)\bar{P}+1]f_s}{N_c}\right)} \boldsymbol{e}\left(\theta_{k,l,m}\right), \quad (5.6)$$

where $D_{k,l}^{(1)}$ and $D_{k,l,m}^{(2)}$ are the distance between the kth UE and the $(l-1)$th scatterer and the distance between this scatterer and the mth subarray, $\beta_{k,l}^{(1)} = \lambda^2 \setminus \left(4\pi D_{k,l}^{(1)}\right)^2$ and $\beta_{k,l,m}^{(2)} = (G_s\lambda^2) \setminus \left(4\pi D_{k,l,m}^{(2)}\right)^2$ represent the corresponding large-scale fading coefficients, and $\tau_{k,l,m} = \frac{D_{k,l}^{(1)} + D_{k,l,m}^{(2)}}{v_c}$ is the total path delay of the lth MPCs between the kth UE and the mth subarray. In Eq. (5.6), $\sqrt{\beta_{k,l}^{(1)}} e^{\mathrm{j}\frac{2\pi D_{k,l}^{(1)}}{\lambda}}$ corresponds to the distance from the kth UE to the $(l-1)$th scatterer and $\sqrt{\beta_{k,l,m}^{(2)}} e^{\mathrm{j}\frac{2\pi D_{k,l,m}^{(2)}}{\lambda}}$ indicates the distance between the $(l-1)$th scatterer and the mth subarray. Note that the far-field steering vector $\boldsymbol{e}\left(\theta_{k,l,m}\right)$ and the angle $\theta_{k,l,m}$ can be formulated similarly with the corresponding variables which is defined in the previous expressions, which will not be repeated here.

5.3 Joint AUD and CE Scheme Design Based on Compressed Sensing

5.3.1 *Compressed Sensing Problem Description*

First, we vectorize the received signal vector $\boldsymbol{y}_p^{(g)}$ of gth $(1 \le g \le G)$ OFDM symbols to the pth $(1 \le p \le P)$ pilot subcarrier in Eq. (5.1) as

$$\boldsymbol{y}_p^{(g)} = \boldsymbol{F}_p^{(g)} \boldsymbol{h}_p + \boldsymbol{n}_p^{(g)}, \quad (5.7)$$

where $\boldsymbol{F}_p^{(g)} = (\boldsymbol{s}_p^{(g)})^{\mathrm{T}} \otimes (\boldsymbol{W}_p^{(g)})^{\mathrm{H}} \in \mathbb{C}^{M \times K N_{\mathrm{BS}}}$, $\boldsymbol{h}_p = \mathrm{vec}\left(\boldsymbol{H}_p\right) = \left(\boldsymbol{D}_\zeta \otimes \boldsymbol{I}_{N_{\mathrm{BS}}}\right) \bar{\boldsymbol{h}}_p \in \mathbb{C}^{K N_{\mathrm{BS}}}$ and $\bar{\boldsymbol{h}}_p = \mathrm{vec}\left(\bar{\boldsymbol{H}}_p\right) = \left[\boldsymbol{h}_{1,p}^{\mathrm{T}}, \boldsymbol{h}_{2,p}^{\mathrm{T}}, \cdots, \boldsymbol{h}_{K,p}^{\mathrm{T}}\right]^{\mathrm{T}} \in \mathbb{C}^{K N_{\mathrm{BS}}}$.

From Eq. (5.1), in order to detect the activity indicator vector $\boldsymbol{\zeta}$ from the received signal and estimate the channel matrix $\boldsymbol{H}_p$ corresponding to active UEs, Eq. (5.7) should be further described as a joint AUD and CE problem. According to the sporadic activity of UEs in massive IoT access [66, 69], the channel matrix $\boldsymbol{H}_p$ exhibits the structural block sparsity in the spatial domain, and the described joint AUD and CE can also be considered as a CS-based sparse signal recovery problem. Therefore, this subsection will give a detailed formulation of the CS problem of the joint AUD and CE. In the frequency domain, for multiple pilot subcarriers there is a common sparse support set in the UE activity sparsity of XL-MIMO systems, which will be described in detail in Sect. 5.3.2. Therefore, by jointly utilizing the observations on all pilot subcarriers, the CS problem can be further described as an MMV-CS problem or a GMMV-CS problem, which can provide more robustness than the single measurement vector (SMV)-CS problem in the AUD and CE performance [134]. In the following text, according to the different design methods of the sensing matrix, the two problem construction methods of MMV-CS and GMMV-CS will be described in detail, respectively.

5.3.1.1 Problem Formulation Based on MMV-CS

Firstly, let us consider the case utilizing the same signal vector and hybrid combining matrix on all pilot subcarriers[2], that is, $\forall p$, $\boldsymbol{s}^{(g)} = \boldsymbol{s}_p^{(g)}$ and $\boldsymbol{W}^{(g)} = \boldsymbol{W}_p^{(g)}$ such that $\boldsymbol{F}^{(g)} = \boldsymbol{F}_p^{(g)}$. Collect the signal vectors on the P pilot subcarriers in the gth OFDM symbol, the received signal matrix $\boldsymbol{Y}^{(g)} \in \mathbb{C}^{M \times P}$ can be expressed as

$$\begin{aligned} \boldsymbol{Y}^{(g)} &= \left[\boldsymbol{y}_1^{(g)}, \boldsymbol{y}_2^{(g)}, \cdots, \boldsymbol{y}_P^{(g)}\right] \\ &= \boldsymbol{F}^{(g)} \boldsymbol{H} + \boldsymbol{N}^{(g)}, \end{aligned} \tag{5.8}$$

where $\boldsymbol{H} = [\mathbf{h}_1, \mathbf{h}_2, \cdots, \mathbf{h}_P] = \left(\boldsymbol{D}_\zeta \otimes \boldsymbol{I}_{N_{\text{BS}}}\right) \bar{\boldsymbol{H}} \in \mathbb{C}^{K N_{\text{BS}} \times P}$ represents the channel matrix and $\bar{\boldsymbol{H}} = \left[\bar{\boldsymbol{h}}_1, \bar{\boldsymbol{h}}_2, \cdots, \bar{\boldsymbol{h}}_P\right] \in \mathbb{C}^{K N_{\text{BS}} \times P}$, and $\boldsymbol{N}^{(g)}$ is the noise matrix after collection. Considering that the BS adopts a hybrid analog-digital architecture with the partially-connected structure, the receiver can only obtain M ($M \ll N_{\text{BS}}$) effective observations in one OFDM symbol. Therefore, it is necessary to jointly utilize the signals collected by G OFDM symbols to improve the performance of joint AUD and CE. By jointly utilizing the received signal matrices of all G OFDM symbols $\{\boldsymbol{Y}^{(g)}\}_{g=1}^{G}$, the stacked signal matrix $\boldsymbol{Y} \in \mathbb{C}^{GM \times P}$ can be obtained as

$$\begin{aligned} \boldsymbol{Y} &= \left[(\boldsymbol{Y}^{(1)})^{\mathrm{T}}, (\boldsymbol{Y}^{(2)})^{\mathrm{T}}, \cdots, (\boldsymbol{Y}^{(G)})^{\mathrm{T}}\right]^{\mathrm{T}} \\ &= \boldsymbol{F}\boldsymbol{H} + \boldsymbol{N}. \end{aligned} \tag{5.9}$$

where $\boldsymbol{F} = \left[(\boldsymbol{F}^{(1)})^{\mathrm{T}}, (\boldsymbol{F}^{(2)})^{\mathrm{T}}, \cdots, (\boldsymbol{F}^{(G)})^{\mathrm{T}}\right]^{\mathrm{T}} \in \mathbb{C}^{GM \times K N_{\text{BS}}}$ is the sensing matrix to be designed and its dimension satisfies $GM \ll K N_{\text{BS}}$, and $\boldsymbol{N}$ is the corresponding noise matrix. Therefore, Eq. (5.9) is the formulaic representation of the joint AUD and CE problem based on MMV-CS.

[2] Since the same signal is used on all pilot subcarriers in the OFDM system, resulting in an excessively high peak-to-average power ratio (PAPR), the method in the literature [101] can be used. Specifically, a pseudo-random scrambling sequence is designed to preprocess these subcarrier signals in order to effectively reduce the high PAPR.

5.3.1.2 Problem Formulation Based on GMMV-CS

Before we describe the joint AUD and CE problem based on GMMV-CS, we firstly discuss the characteristics of the mmWave XL-MIMO channel in the virtual angular domain. For a mmWave XL-MIMO array divided into M subarrays, a block discrete fourier transform (DFT) matrix is defined here as $\boldsymbol{D}_{\mathrm{F}} = \boldsymbol{I}_M \otimes \boldsymbol{F}_{N_s} \in \mathbb{C}^{N_{\mathrm{BS}} \times N_{\mathrm{BS}}}$, where $\boldsymbol{F}_{N_s} \in \mathbb{C}^{N_s \times N_s}$ is N_s-dimensional normalized DFT matrix. The vector $\boldsymbol{h}_{k,p}$ can be further transformed into a virtual angular domain channel vector $\widetilde{\boldsymbol{h}}_{p,k} \in \mathbb{C}^{N_{\mathrm{BS}}}$

$$\widetilde{\boldsymbol{h}}_{p,k} = \boldsymbol{D}_{\mathrm{F}}^{\mathrm{H}} \boldsymbol{h}_{k,p}. \tag{5.10}$$

Here, Eq. (5.10) is also equivalent to $\boldsymbol{h}_{k,p} = \boldsymbol{D}_{\mathrm{F}} \widetilde{\boldsymbol{h}}_{p,k}$. Considering the above transformation, the received signal vector $\boldsymbol{y}_p^{(g)}$ in Eq. (5.1) can be further written as

$$\boldsymbol{y}_p^{(g)} = \left(\boldsymbol{W}_p^{(g)}\right)^{\mathrm{H}} \boldsymbol{D}_{\mathrm{F}} \widetilde{\boldsymbol{H}}_p \boldsymbol{s}_p^{(g)} + \boldsymbol{n}_p^{(g)}, \tag{5.11}$$

where $\widetilde{\boldsymbol{H}}_p = \boldsymbol{D}_{\mathrm{F}}^{\mathrm{H}} \boldsymbol{H}_p = [\zeta_1 \widetilde{\boldsymbol{h}}_{1,p}, \zeta_2 \widetilde{\boldsymbol{h}}_{2,p}, \cdots, \zeta_K \widetilde{\boldsymbol{h}}_{K,p}]$ denotes the transformed matrix of the channel matrix $\boldsymbol{H}_p$ in the virtual angular domain. After vectorizing $\boldsymbol{y}_p^{(g)}$ in Eq. (5.11), we have

$$\boldsymbol{y}_p^{(g)} = \boldsymbol{Z}_p^{(g)} \widetilde{\boldsymbol{h}}_p + \boldsymbol{n}_p^{(g)}, \tag{5.12}$$

where $\boldsymbol{Z}_p^{(g)} = (\boldsymbol{s}_p^{(g)})^{\mathrm{T}} \otimes \left((\boldsymbol{W}_p^{(g)})^{\mathrm{H}} \boldsymbol{D}_{\mathrm{F}}\right) \in \mathbb{C}^{M \times K N_{\mathrm{BS}}}$ and $\widetilde{\boldsymbol{h}}_p = \mathrm{vec}(\widetilde{\boldsymbol{H}}_p) = \left(\boldsymbol{D}_{\zeta} \otimes \boldsymbol{D}_{\mathrm{F}}^{\mathrm{H}}\right) \bar{\boldsymbol{h}}_p \in \mathbb{C}^{K N_{\mathrm{BS}}}$. By collecting the received signal vectors of all G OFDM symbols $\{\boldsymbol{y}_p^{(g)}\}_{g=1}^G$, the stacked signal vector $\boldsymbol{y}_p \in \mathbb{C}^{GM}$ can be obtained as

$$\begin{aligned} \boldsymbol{y}_p &= \left[(\boldsymbol{y}_p^{(1)})^{\mathrm{T}}, (\boldsymbol{y}_p^{(2)})^{\mathrm{T}}, \cdots, (\boldsymbol{y}_p^{(G)})^{\mathrm{T}}\right]^{\mathrm{T}} \\ &= \boldsymbol{Z}_p \widetilde{\boldsymbol{h}}_p + \boldsymbol{n}_p, \end{aligned} \tag{5.13}$$

where $\boldsymbol{Z}_p = \left[(\boldsymbol{Z}_p^{(1)})^{\mathrm{T}}, (\boldsymbol{Z}_p^{(2)})^{\mathrm{T}}, \cdots, (\boldsymbol{Z}_p^{(G)})^{\mathrm{T}}\right]^{\mathrm{T}} \in \mathbb{C}^{GM \times K N_{\mathrm{BS}}}$ is the sensing matrix corresponding to the pth pilot subcarrier to be designed and it should be satisfied that $GM \ll K N_{\mathrm{BS}}$, and $\boldsymbol{n}_p = \left[(\boldsymbol{n}_p^{(1)})^{\mathrm{T}}, (\boldsymbol{n}_p^{(2)})^{\mathrm{T}}, \cdots, (\boldsymbol{n}_p^{(G)})^{\mathrm{T}}\right]^{\mathrm{T}} \in \mathbb{C}^{GM}$ is the stacked noise vector. Considering that P pilot subcarriers can be used to realize joint AUD and CE, then, P signal vectors $\{\boldsymbol{y}_p\}_{p=1}^P$ and corresponding sensing matrices $\{\boldsymbol{Z}_p\}_{p=1}^P$ can be obtained.

5.3.2 Common Structured Block Sparsity for XL-MIMO Channels

Since the number of active UEs is only a small fraction of the total number of UEs (i.e., $|\zeta|_0 = K_{\mathrm{a}} \ll K$), some columns contain non-zero elements of the collected channel matrix $\boldsymbol{H}_p$ in Eq. (5.1) which corresponds to the active UE support set supp$\{\boldsymbol{\zeta}\}$. Therefore, for a massive IoT access system based on mmWave XL-MIMO, the XL-MIMO channel vector $\boldsymbol{h}_p = (\boldsymbol{D}_\zeta \otimes \boldsymbol{I}_{N_{\text{BS}}})\,\bar{\boldsymbol{h}}_p$ in the spatial domain and the virtual angular domain channel vector $\widetilde{\boldsymbol{h}}_p = (\boldsymbol{D}_\zeta \otimes \boldsymbol{D}_{\text{F}}^{\text{H}})\,\bar{\boldsymbol{h}}_p$ in Eq. (5.12) exhibit unique structured block sparsity properties. In addition, by combining the virtual angular domain channel vectors $\{\widetilde{\boldsymbol{h}}_p\}_{p=1}^{P}$ on P subcarriers, the corresponding angular domain channel matrix can be expressed as $\widetilde{\boldsymbol{H}} = [\widetilde{\mathbf{h}}_1, \widetilde{\mathbf{h}}_2, \cdots, \widetilde{\mathbf{h}}_P] = (\boldsymbol{D}_\zeta \otimes \boldsymbol{D}_{\text{F}}^{\text{H}})\,\bar{\boldsymbol{H}} \in \mathbb{C}^{KN_{\text{BS}} \times P}$. Since the support set supp$\{\boldsymbol{\zeta}\}$ is the same for all P subcarriers, the columns of the angular-domain channel matrix $\boldsymbol{H} = (\boldsymbol{D}_\zeta \otimes \boldsymbol{I}_{N_{\text{BS}}})\,\bar{\boldsymbol{H}}$ and spatial-domain channel matrix $\widetilde{\boldsymbol{H}}$ in Eq. (5.8) exhibit the common sparse support set in the frequency domain, which is

$$\text{supp}\{\boldsymbol{h}_1\} = \text{supp}\{\boldsymbol{h}_2\} = \cdots = \text{supp}\{\boldsymbol{h}_P\}, \tag{5.14}$$

$$\text{supp}\{\widetilde{\boldsymbol{h}}_1\} = \text{supp}\{\widetilde{\boldsymbol{h}}_2\} = \cdots = \text{supp}\{\widetilde{\boldsymbol{h}}_P\}. \tag{5.15}$$

It should be pointed out that in the indoor massive IoT access scenario, influenced by the near-field SNS of the mmWave XL-MIMO channel, the support sets of $\{\boldsymbol{h}_p\}_{p=1}^{P}$ and $\{\widetilde{\boldsymbol{h}}_p\}_{p=1}^{P}$ satisfy the following relation:

$$\left|\text{supp}\{\widetilde{\boldsymbol{h}}_p\}\right|_c \le \left|\text{supp}\{\boldsymbol{h}_p\}\right|_c \le K_{\text{a}} N_{\text{BS}} \ll K N_{\text{BS}}, \forall p. \tag{5.16}$$

Figure 5.4 depicts the common structured block sparsity of the spatial and angular domain channel matrices, where the number of subarrays is $M = 10$, the number of subarray antennas $N_{\text{s}} = 8$, the total number of UEs is $K = 20$, the number of active UEs is $K_{\text{a}} = 4$, and the number of pilot subcarriers is $P = 67$. Obviously, from Fig. 5.4, it can be observed that the channels on all pilot subcarriers share a common sparse support set in the spatial and angular domain. In addition, as shown in Fig. 5.4b, the spatial domain XL-MIMO channel of a single UE exhibits the near-field SNS characteristics. In other words, for the channel corresponding to the mmWave XL-MIMO array where $N_{\text{BS}} = 80$, the channel gains of the elements from different subarrays are different, and the channel gains of elements within each subarray are almost the same, which exhibits stepped SNS. On the other hand, the angular-domain XL-MIMO channel of a single UE in Fig. 5.4d presents better sparsity than the spatial-domain channel. Therefore, according to the aforementioned common structured block sparsity of mmWave XL-MIMO channels, the joint AUD and CE problems in Eqs. (5.9) and (5.13) are MMV-CS and GMMV-CS problems under the CS and DCS theoretical frameworks, respectively [134, 135].

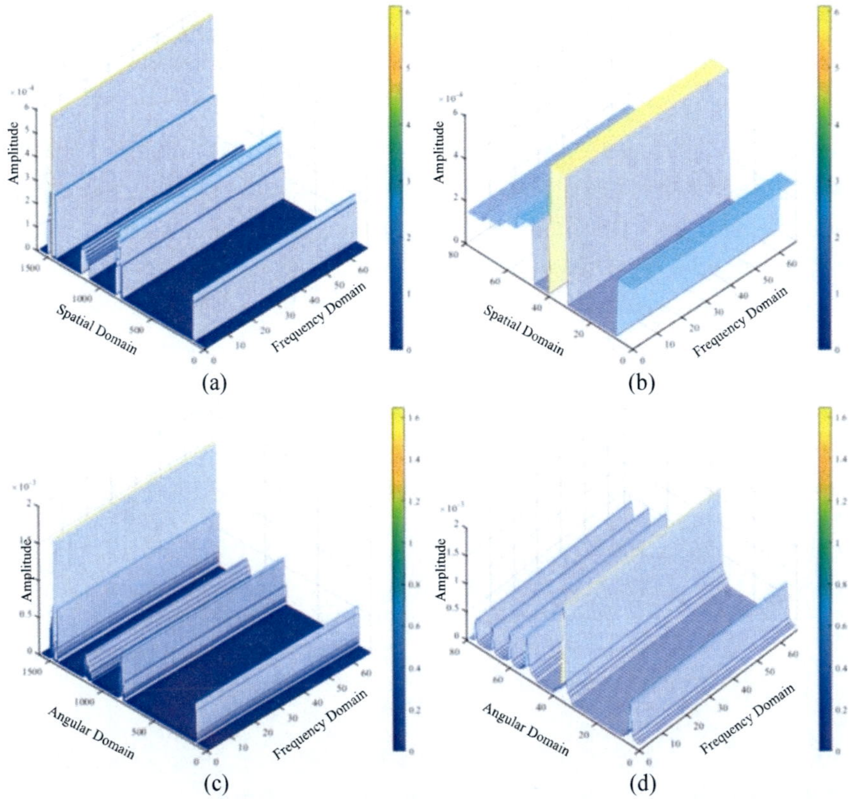

Fig. 5.4 Common structured block sparsity of mmWave XL-MIMO channel in the spatial and angular domains: **a** the spatial-domain XL-MIMO channel matrix $\boldsymbol{H}$; **b** the spatial-domain channel matrix of corresponding to the 4th active UE; **c** the angular-domain XL-MIMO channel matrix $\widetilde{\boldsymbol{H}}$; and **d** the angular-domain channel matrix corresponding to the 4th active UE

5.3.3 Sensing Matrices Design

In order to obtain good compression and reconstruction performance for sparse signals, it is necessary to carefully design the sensing matrices $\boldsymbol{F}$ and $\{\boldsymbol{Z}_p\}_{p=1}^{P}$ in Eqs. (5.9) and (5.13) to ensure the restricted isometry property (RIP) [135]. Specifically, for the gth ($1 \le g \le G$) OFDM symbol on the pth ($1 \le p \le P$) pilot subcarrier, the pilot signal vector $\boldsymbol{s}_p^{(g)}$ transmitted by all UEs and the hybrid combining matrix $\boldsymbol{W}_p^{(g)}$ at the BS will jointly form the sensing matrix $\boldsymbol{F}_p^{(g)}$ and $\boldsymbol{Z}_p^{(g)}$. Their designs are detailed as follow, first we randomly set k items ($1 \le k \le K$) of $\boldsymbol{s}_p^{(g)}$, namely $s_{k,p}^{(g)}$ as $s_{k,p}^{(g)} = \sqrt{P_{\text{tx}}} e^{\mathrm{j}\phi_{k,p}^{(g)}}$, where P_{tx} is the transmit power of each UE and the phase is $\phi_{k,p}^{(g)} \sim \mathcal{U}(0, 2\pi)$. Then, $\boldsymbol{W}_p^{(g)} = \boldsymbol{W}_{\text{RF}}^{(g)} \boldsymbol{W}_{\text{BB},p}^{(g)}$ is composed by analog and digital matrices $\boldsymbol{W}_{\text{RF}}^{(g)}$ and $\boldsymbol{W}_{\text{BB},p}^{(g)}$.

On the one hand, since the transceiver of the mmWave BS adopts a hybrid analog-digital architecture composed of a partially-connected PSN, the analog combiner matrix $\boldsymbol{W}_{\mathrm{RF}}^{(g)}$ is block diagonal, that is, $\boldsymbol{W}_{\mathrm{RF}}^{(g)} = \mathrm{Bdiag}\left([\boldsymbol{w}_{\mathrm{RF},1}^{(g)}, \boldsymbol{w}_{\mathrm{RF},2}^{(g)}, \cdots, \boldsymbol{w}_{\mathrm{RF},M}^{(g)}]\right)$, where the nth $(1 \leq n \leq N_{\mathrm{s}})$ term of the analog combining vector $\boldsymbol{w}_{\mathrm{RF},m}^{(g)} \in \mathbb{C}^{N_{\mathrm{s}}}$ corresponding to the mth $(1 \leq m \leq M)$ subarray can be designed as $\{\boldsymbol{w}_{\mathrm{RF},m}^{(g)}\}_n = \frac{1}{\sqrt{N_{\mathrm{s}}}} e^{\mathrm{j}\varphi_{m,n}^{(g)}}$ and $\varphi_{m,n}^{(g)} \sim \mathcal{U}(0, 2\pi)$. Obviously, $\boldsymbol{W}_{\mathrm{RF}}^{(g)}$ is a semi-unitary matrix. In other words, $(\boldsymbol{W}_{\mathrm{RF}}^{(g)})^{\mathrm{H}} \boldsymbol{W}_{\mathrm{RF}}^{(g)} = \boldsymbol{I}_M, \forall g$. On the other hand, the baseband combining matrix $\boldsymbol{W}_{\mathrm{BB},p}^{(g)}$ can be designed as an identity matrix, i.e., $\boldsymbol{W}_{\mathrm{BB},p}^{(g)} = \boldsymbol{I}_M, \forall p, g$. Therefore, the hybrid combining matrix designed above satisfies $\boldsymbol{W}^{(g)} = \boldsymbol{W}_p^{(g)} = \boldsymbol{W}_{\mathrm{RF}}^{(g)}, \forall p$, and $\boldsymbol{W}^{(g)}$ is also a semi-unitary matrix.

It should be pointed out that in the MMV-CS problem, all pilot subcarriers adopt the same sensing matrix, i.e., $\boldsymbol{F}^{(g)} = \boldsymbol{F}_p^{(g)} = (\boldsymbol{s}_p^{(g)})^{\mathrm{T}} \otimes (\boldsymbol{W}^{(g)})^{\mathrm{H}}, \forall p$. In the GMMV-CS problem, the sensing matrices $\boldsymbol{Z}_p^{(g)} = (\boldsymbol{s}_p^{(g)})^{\mathrm{T}} \otimes \left((\boldsymbol{W}^{(g)})^{\mathrm{H}} \boldsymbol{D}_{\mathrm{F}}\right), \forall p$ are different for all pilot subcarriers, and the sensing matrices $\boldsymbol{F}$ and $\boldsymbol{Z}_p$ in Eqs. (5.9) and (5.13) are formed by $\{\boldsymbol{F}^{(g)}\}_{p=1}^{P}$ and $\{\boldsymbol{Z}_p^{(g)}\}_{p=1}^{P}$. Therefore, according to the CS theory in [134], the designed $\boldsymbol{s}_p^{(g)}$ and $\boldsymbol{W}_p^{(g)}$ can make the sensing matrix elements in $\boldsymbol{F}$ and $\boldsymbol{Z}_p$ follow independent and identically distributed (i.i.d.) complex Gaussian distribution with zero mean and unit variance [135], that is, i.i.d. $\mathcal{CN}(0, 1)$, which can prompt more accurate support set detection and signal reconstruction results.

5.3.4 The Proposed Joint AUD and CE Algorithm

According to the CS problem modeling (including MMV-CS problem and GMMV-CS problem) in Sect. 5.3.1 and the sensing matrix designed in Sect. 5.3.3, the characteristics of common structured block sparsity of the mmWave XL-MIMO channel in the spatial and angular domains as described in Sect. 5.3.2 can be fully utilized here. Based on the traditional simultaneous orthogonal matching pursuit (SOMP) algorithm [134], two joint AUD and CE algorithms based on the mmWave XL-MIMO channel are further proposed, namely the MMV-BSOMP algorithm and GMMV-BSOMP algorithm.

5.3.4.1 Joint AUD and CE Algorithm Based on MMV-BSOMP

According to the aforementioned analysis, the MMV-BSOMP algorithm proposed in Algorithm 5.1 can be used to achieve joint AUD and CE for the MMV-CS problem in Eq. (5.9), when the sparsity, i.e., the number of active UEs K_{a}, is assumed to be known. Specifically, the proposed MMV-BSOMP algorithm computes the structured block correlation between the sensing block matrix $\boldsymbol{F}_k$ and the residual matrix $\boldsymbol{R}^{i-1}$ in *line* 6 by exploiting the spatially structured block sparsity and common

Algorithm 5.1: Proposed MMV-BSOMP Algorithm

Input: Received signal matrix $\boldsymbol{Y}$, sensing matrix $\boldsymbol{F}$, and number of active UEs K_{a}

Output: Estimated activity indicator vector $\widehat{\boldsymbol{\zeta}}$ and channel matrix $\widehat{\boldsymbol{H}}$

Initialization: Set iterative index $i=1$, active UE support set $\mathcal{S}^0=\emptyset$, selected index set $\mathcal{I}^0=\emptyset$, residual matrix $\boldsymbol{R}^0=\boldsymbol{Y}$;
% Split $\boldsymbol{F}$ into K block matrices of the same dimension %
$\boldsymbol{F}_k \in \mathbb{C}^{GM\times N_{\mathrm{BS}}}$, $1\leq k\leq K$, i.e., $\boldsymbol{F}=[\boldsymbol{F}_1, \boldsymbol{F}_2,\cdots, \boldsymbol{F}_K]$;
while $i\leq K_{\mathrm{a}}$ **do**
  % Calculate the structured block correlation %
  for $k=1,2,\cdots,K$ **do**
    $d_k=\sum_j \sum_n \left|\left\{\boldsymbol{F}_k^{\mathrm{H}}\boldsymbol{R}^{i-1}\right\}_{j,n}\right|$;
  end
  % Find the block index with the largest projection %
  $k^\star=\max\left\{ [d_1, d_2,\cdots, d_K]^{\mathrm{T}} \right\}$;
  % Set block index collection %
  $\widetilde{\mathcal{I}}=\left\{(k^\star-1)N_{\mathrm{BS}}+1, (k^\star-1)N_{\mathrm{BS}}+2,\cdots, k^\star N_{\mathrm{BS}}\right\}$;
  % Update the active UE support set and the selected index set %
  $\mathcal{S}^i=\mathcal{S}^{i-1}\cup k^\star$ and $\mathcal{I}^i=\mathcal{I}^{i-1}\cup\widetilde{\mathcal{I}}$;
  % Estimate the block channel matrix with LS %
  $\boldsymbol{H}_{\mathrm{b}}=\mathbf{F}^{\dagger}_{\{:,\mathcal{I}^i\}}\boldsymbol{Y}$;
  % Update residual matrix %
  $\boldsymbol{R}^i=\boldsymbol{Y}-\mathbf{F}_{\{:,\mathcal{I}^i\}}\boldsymbol{H}_{\mathrm{b}}$;
  $i=i+1$;
end
% Obtain the estimated activity indicator vector %
Initialize $\widehat{\boldsymbol{\zeta}}=\mathbf{0}_K$, then let $\widehat{\boldsymbol{\zeta}}_{\{\mathcal{S}^{K_{\mathrm{a}}}\}}=\mathbf{1}_{K_{\mathrm{a}}}$;
% Obtain the estimated channel matrix %
Split $\boldsymbol{H}_{\mathrm{b}}$ into stacked K_a block matrices of the same dimension $\mathbf{H}_{\mathrm{b,k}}\in\mathbb{C}^{N_{\mathrm{BS}}\times P}$, $1\leq k\leq K_{\mathrm{a}}$, that is, $\boldsymbol{H}_{\mathrm{b}}=\left[\boldsymbol{H}^{\mathrm{T}}_{\mathrm{b},1}, \boldsymbol{H}^{\mathrm{T}}_{\mathrm{b},2},\cdots, \boldsymbol{H}^{\mathrm{T}}_{\mathrm{b},K_{\mathrm{a}}}\right]^{\mathrm{T}}$;
Initialize $\widehat{\mathbf{H}}=\boldsymbol{O}_{KN_{\mathrm{BS}}\times P}$;
for $k=1,2,\cdots K_{\mathrm{a}}$ **do**
  Define $\widetilde{k}=\{\mathcal{S}^{K_{\mathrm{a}}}\}_k$, and $\mathcal{I}_k=\left\{(\widetilde{k}-1)N_{\mathrm{BS}}+1, (\widetilde{k}-1)N_{\mathrm{BS}}+2,\cdots, \widetilde{k}N_{\mathrm{BS}}\right\}$;
  $\widehat{\boldsymbol{H}}_{\{\mathcal{I}_k,:\}}=\boldsymbol{H}_{\mathrm{b},k}$;
end
Return: $\widehat{\boldsymbol{\zeta}}$ and $\widehat{\boldsymbol{H}}$

frequency-domain sparse support set of mmWave XL-MIMO channels, which can significantly improve the detection accuracy of active UEs. According to the obtained block index $k^\star$ and the corresponding block index set $\mathcal{I}^i$, the active UE support set $\mathcal{S}^i$ and the selected index set $\mathcal{I}^i$ are updated, respectively. Then, in *line* 15, the block channel matrix $\boldsymbol{H}_\text{b}$ is computed by the least square (LS) estimator. Obviously, in order to obtain good channel estimation performance, the LS estimation of the last iteration (i.e., $i = K_\text{a}$) in *line* 15 requires the extracted sensing block matrix $\mathbf{F}_{\{:,\mathcal{I}^{K_\text{a}}\}} \in \mathbb{C}^{GM \times K_\text{a} N_\text{BS}}$ from $\mathcal{I}^{K_\text{a}}$ to be a square or "fat" matrix, that is, $G \geq K_\text{a} N_\text{s}$, to avoid the situation of lack of rank in the matrix inversion of the LS, which will seriously affect the channel estimation performance.

Finally, after assigning values, the estimated block index set $\mathcal{S}^{K_\text{a}}$ and block matrix $\boldsymbol{H}_\text{b}$ can be obtained. The output of Algorithm 5.1 are the estimated value $\widehat{\zeta}$ of the activity indicator vector ζ and the estimated value $\widehat{\mathbf{H}}$ of the channel matrix $\mathbf{H}$. Note that the support set of the estimated activity indicator vector is written as supp$\{\zeta\}$.

5.3.4.2 Joint AUD and CE Algorithm Based on GMMV-BSOMP

Algorithm 5.2 exhibits the proposed GMMV-BSOMP algorithm, which is utilized in the GMMV-CS problem in Eq. (5.13). Compared with the proposed MMV-BSOMP algorithm in Algorithm 5.1, the proposed GMMV-BSOMP algorithm utilizes different sensing matrices on different subcarriers to more reliable recovery performance in GMMV-CS problem. Specifically, the proposed GMMV-BSOMP algorithm leverages the angular-domain structured block sparsity of mmWave XL-MIMO channels and the common support set on different subcarriers. The distributed structured block correlation is calculated in *line* 4. Then, in *line* 6, the block index corresponding to the maximum projection, i.e., the active user index $k^\star$, is obtained. After updating the support set and index set, LS is utilized individually on different subcarriers to estimate the block channel vector and update the residual vector in *line* 13 and *line* 15, respectively. Different from Algorithm 5.1, the proposed GMMV-BSOMP algorithm implements the residual-based adaptive termination criterion by calculating the residual mean ϵ in *line* 18, rather than leveraging a known number of active UEs. After enough iterations, the combined angular-domain block channel matrix $\widetilde{\boldsymbol{H}}_\text{b}$ in *line* 22 can be obtained. *line* 24 determines the estimated number of active UEs, which is denoted as $\widehat{K}_\text{a}$. Finally, for the obtained block index set $\mathcal{S}^{\widehat{K}_\text{a}}$ and block channel matrix $\widetilde{\boldsymbol{H}}_\text{b}$, we convert the angular-domain channel to the spatial-domain channel for the estimated activity indicator vector $\widehat{\zeta}$ and the channel matrix $\widehat{\mathbf{H}}$.

According to the DCS theory, the aforementioned distributed processing in the proposed GMMV-BSOMP algorithm can utilize different sensing matrices for sufficient diversity gain. Compared with the MMV-BSOMP algorithm, the GMMV-BSOMP algorithm can achieve better support set detection results and LS estimation performance under the same number of observations and sparsity conditions [135]. In other words, the joint AUD and CE performs better when GMMV-BSOMP is adopted, which can also be verified in following simulation results. In addition, the angular-domain XL-MIMO channel exhibits a sparser common support set than the

Algorithm 5.2: Proposed GMMV-BSOMP Algorithm

Input: Received signal matrix $\{\boldsymbol{y}_p\}_{p=1}^{P}$, sensing matrix $\{\boldsymbol{Z}_p\}_{p=1}^{P}$, predefined iteration termination threshold ϵ_{th}, and block DFT matrix $\boldsymbol{D}_{\text{F}}$

Output: Estimated activity indicator vector $\widehat{\boldsymbol{\zeta}}$ and the channel matrix $\widehat{\boldsymbol{H}}$

Initialization: Set iteration index $i=1$, residual mean $\epsilon=2\epsilon_{\text{th}}$, active UE support set $\mathcal{S}^0=\emptyset$, selected index set $\mathcal{I}^0=\emptyset$, residual vector $\boldsymbol{r}_p^0=\boldsymbol{y}_p$, $1\le p\le P$;

while $\epsilon>\epsilon_{\text{th}}$ **do**

    % Compute distributed structured block correlations %

    $\boldsymbol{d}=\sum_{p=1}^{P}\left|\boldsymbol{Z}_p^{\text{H}}\boldsymbol{r}_p^{i-1}\right|$ and $\boldsymbol{D}=\text{mat}(\boldsymbol{d};\,N_{\text{BS}},\,K)$;

    % Find the block index of the largest projection %

    $k^\star=\max\left\{\sum_{n=1}^{N_{\text{BS}}}\boldsymbol{D}_{\{n,:\}}\right\}$;

    % Set block index collection %

    $\widetilde{\mathcal{I}}=\left\{(k^\star-1)N_{\text{BS}}+1,\,(k^\star-1)N_{\text{BS}}+2,\cdots,\,k^\star N_{\text{BS}}\right\}$;

    % Update active UE support set and selected index set %

    $\mathcal{S}^i=\mathcal{S}^{i-1}\cup k^\star$ and $\mathcal{I}^i=\mathcal{I}^{i-1}\cup\widetilde{\mathcal{I}}$;

    for $p=1,2,\cdots P$ **do**

        % Estimate block channel vector using LS %

        $\widetilde{\boldsymbol{h}}_{\text{b},p}^{i}=\boldsymbol{Z}_{p_{\{:,\mathcal{I}^i\}}}^{\dagger}\boldsymbol{y}_p$;

        % Update the residual vector %

        $\boldsymbol{r}_p^i=\boldsymbol{y}_p-\boldsymbol{Z}_{p_{\{:,\mathcal{I}^i\}}}\widetilde{\boldsymbol{h}}_{\text{b},p}^{i}$;

    end

    % Calculate the residual mean %

    $\epsilon=\frac{1}{GMP}\sum_{p=1}^{P}\left(\boldsymbol{r}_p^i\right)^{\text{H}}\boldsymbol{r}_p^i$;

    $i=i+1$;

end

% Combine angular-domain block channel matrices %

$\widetilde{\boldsymbol{H}}_{\text{b}}=\left[\widetilde{\boldsymbol{h}}_{\text{b},1}^{i-1},\,\widetilde{\boldsymbol{h}}_{\text{b},2}^{i-1},\cdots,\,\widetilde{\boldsymbol{h}}_{\text{b},P}^{i-1}\right]$;

% Get the estimated activity indicator vector %

Determine the number of active UEs $\widehat{K}_{\text{a}}=\left|\mathcal{S}^{i-1}\right|_c$, initialize $\widehat{\boldsymbol{\zeta}}=\boldsymbol{0}_K$, then set $\widehat{\boldsymbol{\zeta}}_{\{\mathcal{S}^{\widehat{K}_{\text{a}}}\}}=\boldsymbol{1}_{\widehat{K}_{\text{a}}}$;

% Get the estimated channel matrix %

%Split $\widetilde{\boldsymbol{H}}_{\text{b}}$ into stacked $\widehat{K}_{\text{a}}$ block matrices of the same dimension %

$\widetilde{\mathbf{H}}_{\text{b},k}\in\mathbb{C}^{N_{\text{BS}}\times P}$, $1\le k\le\widehat{K}_{\text{a}}$, that is, $\widetilde{\boldsymbol{H}}_{\text{b}}=\left[\widetilde{\boldsymbol{H}}_{\text{b},1}^{\text{T}},\,\widetilde{\boldsymbol{H}}_{\text{b},2}^{\text{T}},\cdots,\,\widetilde{\boldsymbol{H}}_{\text{b},\widehat{K}_{\text{a}}}^{\text{T}}\right]^{\text{T}}$;

Initialize $\widehat{\mathbf{H}}=\boldsymbol{O}_{KN_{\text{BS}}\times P}$;

for $k=1,2,\cdots\widehat{K}_{\text{a}}$ **do**

    Define $\widetilde{k}=\{\mathcal{S}^{\widehat{K}_{\text{a}}}\}_k$, and $\mathcal{I}_k=\left\{(\widetilde{k}-1)N_{\text{BS}}+1,\,(\widetilde{k}-1)N_{\text{BS}}+2,\cdots,\,\widetilde{k}N_{\text{BS}}\right\}$;

    $\widehat{\boldsymbol{H}}_{\{\mathcal{I}_k,:\}}=\boldsymbol{D}_{\text{F}}\widetilde{\boldsymbol{H}}_{\text{b},k}$;

end

Return: $\widehat{\boldsymbol{\zeta}}$ and $\widehat{\boldsymbol{H}}$

spatial-domain channel, which contributes to reducing the computational complexity at the CE stage in Algorithm 5.2. It should be mentioned that since the number of active users K_a is usually unknown at the BS, to adaptively estimate the support set of active UEs, the proposed GMMV-BSOMP algorithm adopts a pre-defined threshold ϵ_{th} which is related to AWGN noise level as the termination condition. How to determine the threshold ϵ_{th} will be explained in detail as below.

Firstly, let us analyze the ML estimate of the noise variance σ_n^2, which is denoted as $\widehat{\sigma}_n^2$. Let $\boldsymbol{x}_p = \boldsymbol{Z}_p \widetilde{\boldsymbol{h}}_p$ be a valid signal vector, then Eq. (5.13) can be written as $\boldsymbol{y}_p = \boldsymbol{x}_p + \boldsymbol{n}_p$. Furthermore, according to the combined noise vector $\boldsymbol{n}_p^{(g)} = (\boldsymbol{W}_p^{(g)})^{\mathrm{H}} \bar{\boldsymbol{n}}_p^{(g)}$ and $\boldsymbol{W}^{(g)} = \boldsymbol{W}_p^{(g)}, \forall p$ in Eq. (5.1), the stacked noise vector $\boldsymbol{n}_p$ in Eq. (5.13) can be further written as $\boldsymbol{n}_p = \bar{\boldsymbol{W}}^{\mathrm{H}} \bar{\boldsymbol{n}}_p$, where $\bar{\boldsymbol{W}} = \mathrm{Bdiag}\left(\left[\boldsymbol{W}^{(1)}, \boldsymbol{W}^{(2)}, \cdots, \boldsymbol{W}^{(G)}\right]\right) \in \mathbb{C}^{GN_{\mathrm{BS}} \times GM}$ and $\bar{\boldsymbol{n}}_p = \left[(\bar{\boldsymbol{n}}_p^{(1)})^{\mathrm{T}}, (\bar{\boldsymbol{n}}_p^{(2)})^{\mathrm{T}}, \cdots, (\bar{\boldsymbol{n}}_p^{(G)})^{\mathrm{T}}\right]^{\mathrm{T}}$ are the block diagonal matrix and the stacked AWGN vector, respectively. Since $\boldsymbol{n}_p (1 \le p \le P)$ is i.i.d., we can further consider stacking all P subcarriers for the measurement vector $\boldsymbol{y} \in \mathbb{C}^{GMP}$. In other words, $\boldsymbol{y} = \left[\boldsymbol{y}_1^{\mathrm{T}}, \boldsymbol{y}_2^{\mathrm{T}}, \cdots, \boldsymbol{y}_P^{\mathrm{T}}\right]^{\mathrm{T}} = \boldsymbol{x} + \boldsymbol{n}$, where $\boldsymbol{x} = \left[\boldsymbol{x}_1^{\mathrm{T}}, \boldsymbol{x}_2^{\mathrm{T}}, \cdots, \boldsymbol{x}_P^{\mathrm{T}}\right]^{\mathrm{T}}$ and $\boldsymbol{n} = \left[\boldsymbol{n}_1^{\mathrm{T}}, \boldsymbol{n}_2^{\mathrm{T}}, \cdots, \boldsymbol{n}_P^{\mathrm{T}}\right]^{\mathrm{T}}$. Note that $\boldsymbol{n}$ can be further expressed as $\boldsymbol{n} = \widetilde{\boldsymbol{W}}^{\mathrm{H}} \bar{\boldsymbol{n}}$, where $\widetilde{\boldsymbol{W}} = \boldsymbol{I}_P \otimes \bar{\boldsymbol{W}} \in \mathbb{C}^{GN_{\mathrm{BS}} P \times GMP}$ and $\bar{\boldsymbol{n}} = \left[\bar{\boldsymbol{n}}_1^{\mathrm{T}}, \bar{\boldsymbol{n}}_2^{\mathrm{T}}, \cdots, \bar{\boldsymbol{n}}_P^{\mathrm{T}}\right]^{\mathrm{T}}$. Obviously, the stacked noise vector $\boldsymbol{n}$ still follows the complex Gaussian distribution with the mean of $\boldsymbol{0}_{GMP}$ and the covariance matrix of $\sigma_n^2 \boldsymbol{C}_n$, i.e., $\boldsymbol{n} \sim \mathcal{CN}(\boldsymbol{0}_{GMP}, \sigma_n^2 \boldsymbol{C}_n)$. Note that the covariance matrix can be decomposed into $\boldsymbol{C}_n = \widetilde{\boldsymbol{W}}^{\mathrm{H}} \widetilde{\boldsymbol{W}} = \boldsymbol{I}_P \otimes \left(\bar{\boldsymbol{W}}^{\mathrm{H}} \bar{\boldsymbol{W}}\right) \in \mathbb{C}^{GMP \times GMP}$. Therefore, taking the noise variance σ_n^2 as an unknown parameter, the likelihood function of the measurement vector $\boldsymbol{y}$, which is denoted as $\mathcal{P}(\boldsymbol{y}; \sigma_n^2)$, can be formulated as

$$
\begin{aligned}
&\mathcal{P}(\boldsymbol{y}; \sigma_n^2) \\
&= \frac{1}{\pi^{GMP} \det\left(\sigma_n^2 \boldsymbol{C}_n\right)} \exp\left(-(\boldsymbol{y} - \boldsymbol{x})^{\mathrm{H}} \left(\sigma_n^2 \boldsymbol{C}_n\right)^{-1} (\boldsymbol{y} - \boldsymbol{x})\right) \\
&\overset{(a)}{=} \frac{1}{\left(\pi \sigma_n^2\right)^{GMP} \left(\det\left(\bar{\boldsymbol{W}}^{\mathrm{H}} \bar{\boldsymbol{W}}\right)\right)^{P}} \exp\left(-\frac{1}{\sigma_n^2} \sum_{p=1}^{P} (\boldsymbol{y}_p - \boldsymbol{x}_p)^{\mathrm{H}} \left(\bar{\boldsymbol{W}}^{\mathrm{H}} \bar{\boldsymbol{W}}\right)^{-1} (\boldsymbol{y}_p - \boldsymbol{x}_p)\right),
\end{aligned}
\tag{5.17}
$$

where the identity transformation $\det(\boldsymbol{A}_{m \times m} \otimes \boldsymbol{B}_{n \times n}) = (\det(\boldsymbol{A}))^n (\det(\boldsymbol{B}))^m$ is considered in equation (a) [109]. Thus, the log-likelihood function of $\boldsymbol{y}$ can be further written as

$$\begin{aligned}\mathcal{L}(\boldsymbol{y};\sigma_{\rm n}^2) =& \ln\mathcal{P}(\boldsymbol{y};\sigma_{\rm n}^2)\\ =& - GMP\ln\left(\pi\sigma_{\rm n}^2\right) - P\ln\left(\det\left(\bar{\boldsymbol{W}}^{\rm H}\bar{\boldsymbol{W}}\right)\right)\\ &- \frac{1}{\sigma_{\rm n}^2}\sum_{p=1}^{P}(\boldsymbol{y}_p-\boldsymbol{x}_p)^{\rm H}\left(\bar{\boldsymbol{W}}^{\rm H}\bar{\boldsymbol{W}}\right)^{-1}(\boldsymbol{y}_p-\boldsymbol{x}_p). \end{aligned} \tag{5.18}$$

According to the ML criterion, by solving $\partial\mathcal{L}(\boldsymbol{y};\sigma_{\rm n}^2)/\partial\sigma_{\rm n}^2=0$, we can obtain the ML estimation $\widehat{\sigma}_{\rm n}^2$ of the noise variance $\sigma_{\rm n}^2$ as

$$\begin{aligned}\widehat{\sigma}_{\rm n}^2 =& \frac{1}{GMP}\sum_{p=1}^{P}(\boldsymbol{y}_p-\boldsymbol{x}_p)^{\rm H}\left(\bar{\boldsymbol{W}}^{\rm H}\bar{\boldsymbol{W}}\right)^{-1}(\boldsymbol{y}_p-\boldsymbol{x}_p)\\ \overset{(b)}{=}& \frac{1}{GMP}\sum_{p=1}^{P}(\boldsymbol{y}_p-\boldsymbol{x}_p)^{\rm H}(\boldsymbol{y}_p-\boldsymbol{x}_p). \end{aligned} \tag{5.19}$$

It should be noted that considering the hybrid combining matrix $\boldsymbol{W}^{(g)}, \forall g$ designed in Sect. 5.3.3 is a semi-unitary matrix in equation (b), the block diagonal matrix $\bar{\boldsymbol{W}}$ composed by $\{\boldsymbol{W}^{(g)}\}_{g=1}^{G}$ also satisfies the semi-unitary property, i.e., $\bar{\boldsymbol{W}}^{\rm H}\bar{\boldsymbol{W}}=\boldsymbol{I}_{GM}$. Then, according to the projection analysis theory, the estimation of the noise variance $\widehat{\sigma}_{\rm n}^2$ in Eq. (5.19) can be linked with the preset iteration termination threshold $\epsilon_{\rm th}$ in Algorithm 5.2. Specifically, in the ith iteration on the pth subcarrier, by using LS to the block channel vector in *line* 13, i.e., $\widetilde{\boldsymbol{h}}_{{\rm b},p}^{i}=\boldsymbol{Z}_{p\{:,\mathcal{I}^i\}}^{\dagger}\boldsymbol{y}_p$, the estimate $\widehat{\boldsymbol{x}}_p^i$ of the effective signal vector $\boldsymbol{x}_p$ in the ith iteration can be given by

$$\widehat{\boldsymbol{x}}_p^i = \boldsymbol{Z}_{p\{:,\mathcal{I}^i\}}\widetilde{\boldsymbol{h}}_{{\rm b},p}^{i} = \boldsymbol{Z}_{p\{:,\mathcal{I}^i\}}\boldsymbol{Z}_{p\{:,\mathcal{I}^i\}}^{\dagger}\boldsymbol{y}_p = \boldsymbol{P}_p^i\boldsymbol{y}_p, \tag{5.20}$$

where $\boldsymbol{P}_p^i=\boldsymbol{Z}_{p\{:,\mathcal{I}^i\}}\boldsymbol{Z}_{p\{:,\mathcal{I}^i\}}^{\dagger}\in\mathbb{C}^{GM\times GM}$ is the projection matrix of the selected sensing matrix $\boldsymbol{Z}_{p\{:,\mathcal{I}^i\}}$. Then, the estimated error vector $\boldsymbol{r}_p^i$ between the measurement vector $\boldsymbol{y}_p$ and $\widehat{\boldsymbol{x}}_p^i$ (residual vector in Algorithm 5.2) can be expressed as

$$\boldsymbol{r}_p^i = \boldsymbol{y}_p - \widehat{\boldsymbol{x}}_p^i = \left(\boldsymbol{I}_{GM}-\boldsymbol{P}_p^i\right)\boldsymbol{y}_p = \left(\boldsymbol{P}_p^i\right)^{\perp}\boldsymbol{y}_p, \tag{5.21}$$

where $\left(\boldsymbol{P}_p^i\right)^{\perp}=\boldsymbol{I}_{GM}-\boldsymbol{P}_p^i$ is the orthogonal projection matrix of $\boldsymbol{P}_p^i$. A physical interpretation of Eq. (5.20) and (5.21) is that the estimated effective signal vector $\widehat{\boldsymbol{x}}_p^i$ and the estimated error vector $\boldsymbol{r}_p^i$ are the projection and the orthogonal projection of the signal vector $\boldsymbol{y}_p$ on the subspace spanned by the sensing matrix $\boldsymbol{Z}_{p\{:,\mathcal{I}^i\}}$ [109, 136]. Therefore, iteratively searching for the orthogonal projection of the received signal vector on the subspace spanned by the selected sensing matrix is the essense of the joint AUD and CE of the proposed GMMV-BSOMP algorithm. In the meantime, the support set detection of active UEs and the channel vector arre estimated as well. After enough iterations (i.e., $i\geq K_{\rm a}$), $\boldsymbol{y}_p$ becomes the orthogonal projection on the subspace spanned by $\boldsymbol{Z}_{P\{:,\mathcal{I}^{K_a}\}}$, which is denoted as $\boldsymbol{r}_p^{K_a}$. $\boldsymbol{r}_p^{K_a}$ no longer contains useful signals from active UEs, but only noise. If the obtained active UE support set and channel vector are accurate enough, the estimated effective signal vector $\widehat{\boldsymbol{x}}_p^{K_a}$ can be approximated as the real signal vector $\boldsymbol{x}_p$, that is,

$$\boldsymbol{x}_p \approx \widehat{\boldsymbol{x}}_p^{K_a} = \boldsymbol{Z}_{P\{:,\mathcal{I}^{K_a}\}}\widetilde{\boldsymbol{h}}_{\mathrm{b},p}^{K_a}. \tag{5.22}$$

Substitute Eq. (5.22) into Eq. (5.19), we have

$$\begin{aligned}\widehat{\sigma}_{\mathrm{n}}^2 &\approx \frac{1}{GMP}\sum_{p=1}^{P}\left(\boldsymbol{y}_p-\boldsymbol{Z}_{P\{:,\mathcal{I}^{K_a}\}}\widetilde{\boldsymbol{h}}_{\mathrm{b},p}^{K_a}\right)^{\mathrm{H}}\left(\boldsymbol{y}_p-\boldsymbol{Z}_{P\{:,\mathcal{I}^{K_a}\}}\widetilde{\boldsymbol{h}}_{\mathrm{b},p}^{K_a}\right)\\ &= \frac{1}{GMP}\sum_{p=1}^{P}\left(\boldsymbol{r}_p^{K_a}\right)^{\mathrm{H}}\boldsymbol{r}_p^{K_a}.\end{aligned} \tag{5.23}$$

From *line* 18 in Algorithm5.2, the mean residual calculated by the last iteration (i.e. $i=K_a$) is $\epsilon=\frac{1}{GMP}\sum_{p=1}^{P}\left(\boldsymbol{r}_p^{K_a}\right)^{\mathrm{H}}\boldsymbol{r}_p^{K_a}$. Comparing the results in Eq. (5.23), $\widehat{\sigma}_{\mathrm{n}}^2\approx\epsilon$, which indicates that the residual mean ϵ eventually approaches the estimated noise variance $\widehat{\sigma}_{\mathrm{n}}^2$. From the ML perspective, it is optimal to regard the true noise variance as a preset threshold for iteration termination conditions. Therefore, the variance σ_{n}^2 of the real noise in the received signal can be set as the threshold ϵ_{th}, i.e., $\epsilon_{\mathrm{th}}=\sigma_{\mathrm{n}}^2$. More intuitively, when the active UE support set and the channel vector are estimated accurately, the useful signal components in the residual vector except noise will be iteratively eliminated. Then, the noise variance can be computed as the average of the all residual items.

5.3.5 Computational Complexity Analysis

The computational complexity of the MMV-BSOMP algorithm and GMMV-BSOMP algorithm proposed in Algorithm 5.1 and Algorithm 5.2 is mainly composed of the following three parts. Considering the computational complexity of the proposed algorithm in the ith iteration, namely

(1) Compute the (distributed) structured block correlation
Algorithm 5.1 (*line* 6) and Algorithm 5.2 (*line* 4): $\mathcal{O}(KN_{\mathrm{BS}}GMP)$;
(2) Use LS to estimate the block channel matrix or vector
Algorithm 5.1 (*line* 15): $\mathcal{O}(2i^2N_{\mathrm{BS}}^2GM+i^3N_{\mathrm{BS}}^3+iN_{\mathrm{BS}}GMP)$;
Algorithm 5.2 (*line* 13): $\mathcal{O}(2i^2N_{\mathrm{BS}}^2GMP+i^3N_{\mathrm{BS}}^3P+iN_{\mathrm{BS}}GMP)$;
(3) Update the residual matrix or vector
Algorithm 5.1 (*line* 17) and Algorithm 5.2 (*line* 15): $\mathcal{O}(GMiN_{\mathrm{BS}}P)$.

For Algorithm 5.2, there is an extra step namely "calculate residual mean" in *line* 18, whose computational complexity is $\mathcal{O}(GMP)$. In addition, in Algorithm 5.2,

Table 5.1 Comparisons of computational complexity for different joint AUD and CE algorithms

Joint AUD and CE algorithms	Computational complexity per iteration
BSAMP algorithm [135]	$\mathcal{O}\left(\left(KN_{\mathrm{BS}}+3jN_{\mathrm{BS}}+4j^2N_{\mathrm{BS}}^2/P\right)GMP+2j^3N_{\mathrm{BS}}^3\right)$
BSP algorithm [73]	$\mathcal{O}\left(\left(KN_{\mathrm{BS}}+3K_{\mathrm{a}}N_{\mathrm{BS}}+4K_{\mathrm{a}}^2N_{\mathrm{BS}}^2/P\right)GMP+2K_{\mathrm{a}}^3N_{\mathrm{BS}}^3\right)$
BOMP algorithm [71]	$\mathcal{O}\left(\left(KN_{\mathrm{BS}}+2iN_{\mathrm{BS}}+2i^2N_{\mathrm{BS}}^2\right)GMP+i^3N_{\mathrm{BS}}^3P\right)$
Proposed MMV-BSOMP algorithm	$\mathcal{O}\left(\left(KN_{\mathrm{BS}}+2iN_{\mathrm{BS}}+2i^2N_{\mathrm{BS}}^2/P\right)GMP+i^3N_{\mathrm{BS}}^3\right)$
Proposed GMMV-BSOMP algorithm	$\mathcal{O}\left(\left(KN_{\mathrm{BS}}+2iN_{\mathrm{BS}}+2i^2N_{\mathrm{BS}}^2+1\right)GMP+i^3N_{\mathrm{BS}}^3P\right)$

Note: i and j represent the iteration index and stage index, respectively.

the computational complexity of transforming the angular-domain channel to the spatial-domain channel after the iteration is $\mathcal{O}(\widehat{K}_{\mathrm{a}}N_{\mathrm{BS}}^2P)$. By comparing the computational complexity of the two algorithms mentioned above, it can be found that the computational complexity of LS estimation in Algorithm 5.2 is higher than that in Algorithm 5.1, This is because the proposed GMMV-BSOMP algorithm uses different sensing matrices on different subcarriers, there are P pseudo-inverse operations on the selected sensing matrix to obtain LS estimation of the block channel vector on P subcarriers. However, in Algorithm 5.1 there is only one pseudo-inverse operation.

Table 5.1 compares the computational complexity of the proposed MMV-BSOMP algorithm with GMMV-BSOMP algorithm, as well as of these traditional greedy CS recovery algorithms such as block sparsity adaptive matching pursuit (BSAMP) [135], block subspace pursuit (BSP) and block orthogonal matching pursuit (BOMP) in one iteration. It can be concluded that the matrix pseudo-inverse operation of LS channel estimation in all algorithms occupies most of the computational complexity. In addition, by comparing the computational complexity of these algorithms, proposed MMV-BSOMP algorithm has the lowest complexity, then followed by BSAMP and BSP algorithms. The complexity of BOMP algorithm and proposed GMMV-BSOMP algorithm are higher than other candidates due to multiple matrix pseudo-inverse operations. It should be pointed out that proposed GMMV-BSOMP algorithm leverages the common structured block sparsity in the angular-domain, which reduces the computational complexity of LS.

5.4 Simulation Results

For the indoor XL-MIMO-OFDM system, this section evaluates the performance of proposed joint AUD and CE schemes based on MMV-BSOMP algorithm and GMMV-BSOMP algorithm for uplink massive IoT access. To simplify the simulation scene, we consider the two-dimensional Cartesian coordinate as shown in Fig. 5.5, where the entire three-dimensional indoor space can be compressed into a two-dimensional plane with the size of $\overline{OC}\times\overline{OD}$. An mmWave XL-MIMO array is equipped with M subarrays embedded in the roof or wall, with the spacing between adjacent subarrays denoted by Δ. Randomly distributed area of UEs covers

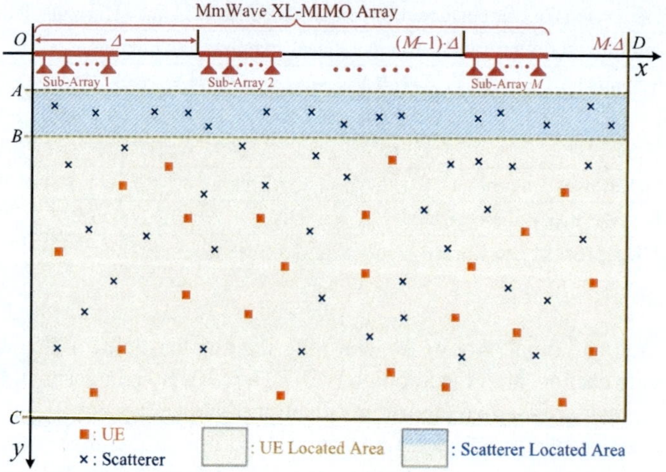

Fig. 5.5 Schematic diagram of indoor near-field massive IoT access simulation scenario based on mmWave XL-MIMO

$\{(x, y)|0 \leq x \leq \overline{OD} \cap \overline{OB} \leq y \leq \overline{OC}\}$, and the random distribution area of scatterers is $\{(x, y)|0 \leq x \leq \overline{OD} \cap \overline{OA} \leq y \leq \overline{OC}\}$. A common parameter setting is $\overline{OA} = 1$ m, $\overline{OB} = 3$ m, $\overline{OD} = M\Delta$ m, and $\overline{OC} = 3\overline{OD}/5$ m. The main simulation parameter settings are shown in Table 5.2. Considering that the maximum delay spread is affected by the propagation distance of the EM signal, the maximum transmission distance can be considered as twice the long side $\overline{OD}$ of the indoor space to generate $\tau_{\max}$, i.e., $\tau_{\max} = 2\overline{OD}/v_c$, and the number of required pilot subcarriers P can be determined by $\tau_{\max}$. In addition, the transmission distance and AoA defined in Sect. 5.2.2 can be accurately calculated from the randomly generated 2D coordinates of the UEs and scatterers' positions.

The following two performance evaluation metrics are considered here to evaluate the performance of the proposed joint AUD and CE schemes based on MMV-BSOMP algorithm and GMMV-BSOMP algorithm, which are the detection error probability P_e of AUD and the normalized mean square error (NMSE) of CE. The metrics can be expressed by

$$P_e = \left\| \zeta - \widehat{\zeta} \right\|_1 / K, \tag{5.24}$$

$$\text{NMSE} = 10 \log_{10} \left(\mathbb{E}\left(\left\| \boldsymbol{H} - \widehat{\mathbf{H}} \right\|_F^2 / \left\| \boldsymbol{H} \right\|_F^2 \right) \right). \tag{5.25}$$

In this section, to verify the effectiveness of the preset iteration termination threshold ϵ_{th}, the proposed joint AUD and CE schemes based on MMV-BSOMP algorithm and GMMV-BSOMP algorithm both consider the number of active UEs K_{a} is known and unknown. When K_{a} is unknown, it will be marked with "ϵ_{th}" in the legend. In addition, the comparative baseline schemes considered in the simulation include

Table 5.2 Main simulation parameter settings

Parameters	Values
Carrier frequency f_c	28 GHz
System bandwidth f_s	200 Megahertz (MHz)
The number of subarrays (Number of RF Links) M	10
The number of antennas in each subarray N_s	8
Adjacent subarray spacing Δ	5 m
Total number of subcarriers N_c	2048
Total UEs K	60
Active UEs K_a	6
Multipath number L_k	Equal probability to select an integer between 1–5
Rice factor γ_k	Equal probability to select integers between 10–20
The number of scatterers	Equal probability select an integer between 5–15
Power spectral density of background noise [137]	-174 dBm/Hz

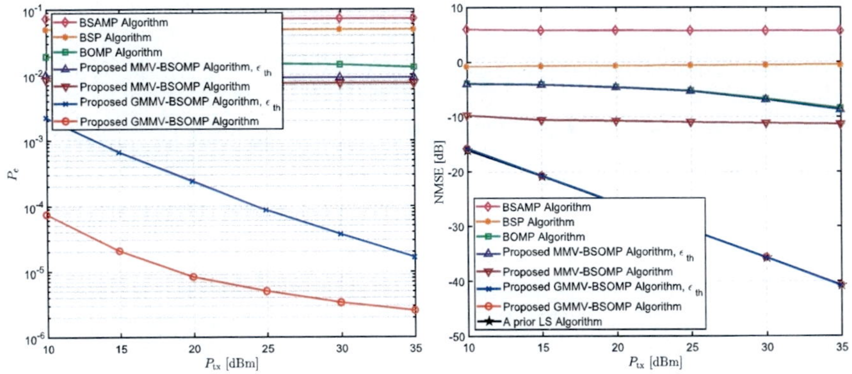

(a) AUD error probability P_e (b) NMSE

Fig. 5.6 Performance of different joint AUD and CE schemes for different transmit power P_{tx} with the time overhead $G = 50$

traditional CS algorithms such as BSAMP [135], BSP [73] and BOMP [71]. Meanwhile, prior LS algorithm with known active UE support set can also be considered as the lower bound in CE.

Figure 5.6 compares AUD detection error probability P_e and NMSE performance of different joint AUD and CE schemes as the transmit power P_{tx} varies, where the fixed time overhead is $G = 50$. Firstly, Fig. 5.6 shows that the P_e and NMSE performance of proposed GMMV-BSOMP algorithm are significantly improved with

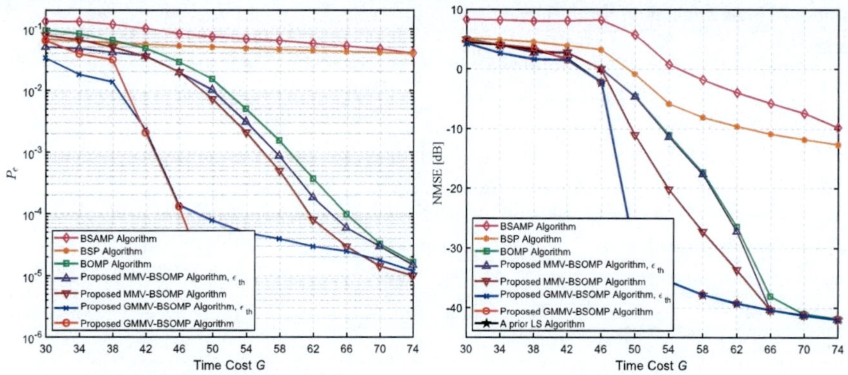

(a) AUD error probability P_e (b) NMSE

Fig. 5.7 Performance of different joint AUD and CE schemes for different time overhead G with the transmit power $P_{tx} = 25$ dBm

P_{tx} increasing. This demonstrates that in the indoor near-field massive IoT access scenario, the proposed GMMV-BSOMP algorithm can make full use of the common structured block sparsity and the diversity gain brought about by sufficiently diverse sensing matrices of the angular-domain mmWave XL-MIMO channel both when K_a is known and unknown, which results in a more accurate AUD and CE result. In the joint AUD and CE process, the better AUD performance is, the better CE performance achieves, and vice versa. Therefore, based on the excellent AUD performance in Fig. 5.6a, NMSE performance of proposed GMMV-BSOMP algorithm in Fig. 5.6b is very close to prior LS algorithm, and the NMSE curves show a linear downward trend with the increase of P_{tx}, which also verifies the effectiveness of the iteration termination threshold ϵ_{th} designed in Sect. 5.3.4.2. In contrast, due to the lack of sufficient diversity gain, AUD performance of the proposed MMV-BSOMP algorithm is limited compared with other three benchmark schemes based on greedy algorithms. Therefore, even when P_{tx} is relatively large, NMSE performance cannot be further improved with the increase of P_{tx}, which limits AUD performance. In Fig. 5.6, joint AUD and CE performance of proposed MMV-BSOMP algorithm is better than other three benchmark algorithms when K_a is known, and it is noted that although the proposed MMV-BSOMP algorithm has similar NMSE performance with BOMP algorithm when K_a is unknown, the algorithm still has the advantage of P_e. In addition, the BSAMP algorithm and the BSP algorithm cannot effectively utilize the common structural block sparsity of the mmWave XL-MIMO channel, so they cannot perform AUD well, resulting in poor CE performance as well.

Figure 5.7 compares the AUD detection error probability P_e and NMSE performance of different joint AUD and CE schemes versus time overhead G, where the transmit power is $P_{tx} = 25$ dBm. From Fig. 5.7, due to the inaccurate AUD under the spatially structured block sparsity and mmWave XL-MIMO channel conditions, there are clear performance gaps between BSAMP algorithm, BSP algorithm and

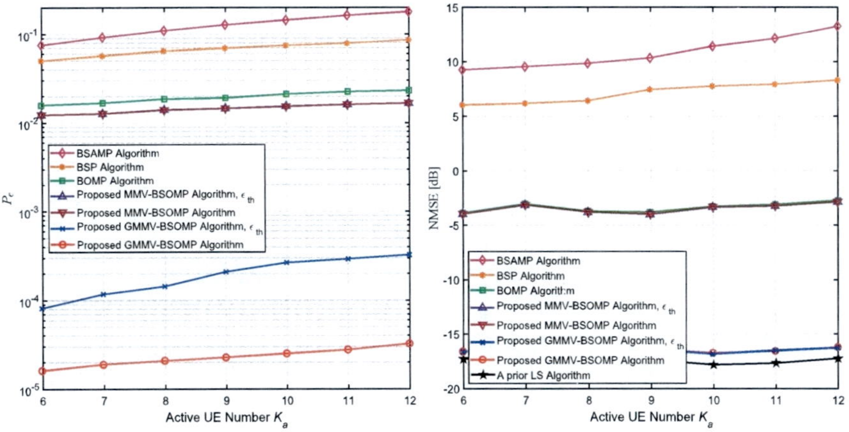

(a) AUD error probability P_e (b) NMSE

Fig. 5.8 Performance of different joint AUD and CE schemes for different active UE number $K_{\rm a}$ with the transmit power $P_{\rm tx} = 25\,{\rm dBm}$ and total UE number $K = 60$

other algorithms in P_e and NMSE. In addition, in Fig. 5.7a, if the number of OFDM symbols $G < 42$, P_e of proposed GMMV-BSOMP algorithm with unknown $K_{\rm a}$ is better than the case with known $K_{\rm a}$. When $G \geq 46$, the situation behaves oppositely, and there is a performance upper bound for P_e in proposed GMMV-BSOMP algorithm when the time cost G increases if $K_{\rm a}$ is unknown. In Fig. 5.7b, their NMSE performance almost overlaps with prior LS algorithm. This is because proposed GMMV-BSOMP algorithm takes the advantage of the diversity gain from diverse sensing matrices to reduce the probability of false alarms and missed detections. Therefore, the P_e and NMSE performance of GMMV-BSOMP algorithm proposed in this chapter is significantly better than other algorithms when $G < 66$. This demonstrates that GMMV-BSOMP algorithm can achieve indoor near-field massive IoT access with less time overhead, which is conducive to promoting ultra-reliable low-latency communications (uRLLC) in future B5G/6G wireless communication networks. In addition, P_e and NMSE performance of proposed MMV-BSOMP algorithm are also better than the other three benchmark schemes based on greedy algorithms. Since more time overhead can ensure better AUD and LS estimation in the iteration of proposed MMV-BSOMP algorithm and BOMP algorithm, when G is large enough, their NMSE performance in Fig. 5.7b can be close to proposed GMMV-BSOMP algorithm and prior LS algorithm.

To verify the effectiveness of different joint AUD and CE schemes under different sparsities, Fig. 5.8 compares the AUD error probability P_e and NMSE performance of these schemes with the number of active users $K_{\rm a}$ varying where transmit power is $P_{\rm tx} = 25$ dBm and total UEs $K = 60$. Since the LS estimator is used in estimating the channel gain based on the greedy CS schemes, the time cost G needs to satisfy the constraint of $G \geq K_{\rm a} N_{\rm s}$ to ensure that there is no rank deficiency during the

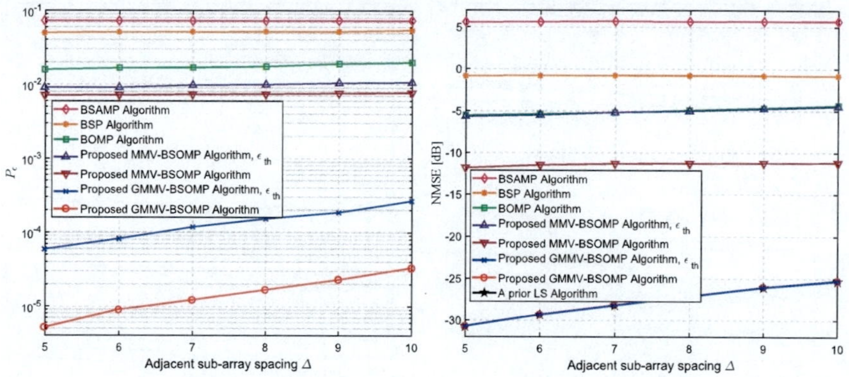

(a) AUD error probability P_e (b) NMSE

Fig. 5.9 Performance of different joint AUD and CE schemes for different adjacent subarray spacing Δ with transmitting power $P_{tx} = 25$ dBm, time overhead $G = 50$ and number of subarrays $M = 10$

matrix inversion in LS. To obtain reliable joint AUD and CE performance, the critical condition $G = K_a N_s$ can be directly considered here. From Fig. 5.8, although the AUD and CE performance of all algorithms decreases with the increase of sparsity, the P_e and NMSE performance of the proposed GMMV-BSOMP and MMV-BSOMP algorithms are still significantly better than other baseline algorithms. Combining Fig. 5.8a and b, it can be seen that although the proposed GMMV-BSOMP algorithm has better P_e performance when K_a is known than the case that K_a is unknown, but the NMSE performance of both cases is almost the same, which is close to the performance of prior LS algorithm.

In order to illustrate the impact of near-field environment on indoor massive IoT access, Fig. 5.9 compares AUD detection error probability P_e and NMSE performance of different joint AUD and CE schemes as the distance Δ between adjacent subarrays varies, given the fixed transmit power $P_{tx} = 25$ dBm, the time overhead $G = 50$ and the number of subarrays $M = 10$. In order to reflect the changes in the indoor near-field environment, we consider using different $\overline{OC} \times \overline{OE}$ to change the space size, and using the corresponding adjacent subarray spacing Δ to adapt to these changes. Specifically, since the number of subarrays is fixed at $M = 10$, the length of $\overline{OD} = M\Delta$ will increase as Δ increases. Therefore, the corresponding $\overline{OC} = 3\overline{OD}/5$, $\tau_{max} = 2\overline{OD}/v_c$, and the number of pilot subcarriers P can be computed. From Fig. 5.9, it can be found that with the increase of Δ, P_e and NMSE performance of proposed GMMV-BSOMP algorithm both decreases to some extent. This is because the larger the indoor space causes the more serious the path loss in free space propagation, which makes the signal strength received by the XL-MIMO-based mmWave BS weaker. According to the conclusion in Fig. 5.6, the performance of proposed GMMV-BSOMP will be significantly improved with the increase of the transmit power P_{tx}. It can be seen that proposed algorithms are sensitive to the strength of the received signal, while other candidates are not very sensitive to the received signal strength due to their limited detection and estimation capabilities. Furthermore, by comparing the P_e and NMSE curves of different joint AUD and CE schemes, similar conclusions can be drawn from Fig. 5.9 as observed in Fig. 5.6, that is, GMMV-BSOMP and MMV-BSOMP algorithms proposed in this chapter perform better than other algorithms. Therefore, the solutions proposed in this chapter are not only limited to massive IoT access scenarios in a small indoor area, but can also achieve good results in large spaces and high-density smart factory environments.

5.5 Summary

For indoor near-field uplink massive IoT access scenarios, this chapter proposes a joint AUD and CE scheme based on mmWave XL-MIMO, where the near-field SNS of mmWave XL-MIMO channels is considered. In the proposed scheme, a signal transmission model for uplink large-scale access and a mmWave XL-MIMO channel in the form of near-field spherical waves are established first, in which the channel exhibits both obvious SNS and coexistence of far-field conditions and near-field conditions. The joint AUD and CE problems are then formulated as MMV-CS and GMMV-CS problems according to the common support set of mmWave XL-MIMO channels on multiple pilot subcarriers. Then, the MMV-BSOMP and GMMV-BSOMP algorithms are designed using the common structured block sparsity of the mmWave XL-MIMO channel in the spatial and virtual angular domains, where a preset iteration termination threshold is designed to adaptively obtain the active user support set when the number of active users is unknown. Finally, the simulation results show that the proposed MMV-BSOMP algorithm and GMMV-BSOMP algorithm can achieve better joint AUD and CE performance compared with those of the existing CS algorithms.

Chapter 6
Channel Estimation and Data Transmission for THz UM-MIMO in Space-Air-Ground Integrated Network

Abstract This chapter explores the use of terahertz (THz) ultra-massive (UM)-MIMO for aeronautical communications in the context of the promising space-air-ground integrated network, aiming to provide efficient and reliable aeronautical communications. To this end, we focus on the challenges posed by the unique triple delay-beam-Doppler squint effects in aeronautical THz UM-MIMO channels, and propose a channel estimation (CE) and tracking scheme to address these issues. Specifically, the initial CE stage utilizes rough angle estimates obtained from navigation information and employs a grouping true-time delay unit (GTTDU) module to mitigate the delay-beam squint at the transceiver. Accurate azimuth and elevation angle estimates are then obtained through a prior-aided iterative angle estimation algorithm to achieve the beam alignment and refinement of the GTTDU module. Doppler shifts can also be estimated accurately to design the compensation matrix for the estimation of path delays and channel gains. In the data transmission stage, a data-aided decision-directed based channel tracking algorithm is developed to track fast time-varying channels. If tracking fails, the angles are re-estimated at the pilot-aided channel tracking stage using an equivalent sparse digital array, and angle ambiguity can be resolved based on previously estimated angles. The Cramèr-Rao lower bounds of the key channel parameters are derived. The numerical results demonstrate the effectiveness of the proposed solutions.

6.1 Introduction

As a key potential technology for the future sixth-generation (6G) wireless communication systems, terahertz (THz) communications can provide ubiquitous connection to realize a deeper and wider service range than the current network. The vast bandwidth is available within the range of THz frequency band from 0.1 to 10 THz, which is far beyond the millimeter-wave (mmWave) frequency band, supporting up to tens of gigahertz (GHz) ultra-high-bandwidth communications with the peak data rate of the order of Tbps [4, 14, 138]. Besides, THz communications can

Z. Gao et al., *Millimeter-Wave/Sub-Terahertz Ultra-Massive MIMO Transmission Technology*, https://doi.org/10.1007/978-981-99-2388-5_6

facilitate the realization of ultra-massive MIMO (UM-MIMO)[1] systems equipped with thousands of antenna elements (see, e.g., a uniform planar array (UPA) of size 1024×1024 in [139]), which effectively counteracts the severe path loss of THz signals and extends the distance of communication links by adopting the beamforming techniques [5, 19, 140]. Therefore, THz UM-MIMO technology is expected to become one of the potential candidates in the future 6G mobile communication systems [18]. However, due to the absorption of atmospheric molecules (such as typical atmospheric water molecules) and rain attenuation during the signal propagation [19, 141], the actual THz communications are limited to the short-range scenarios, such as indoor communications [142, 143]. Fortunately, atmospheric molecular absorption and rain attenuation mostly occur in the troposphere below 10 km, and these negative effects are almost negligible in the stratosphere and beyond [7, 144, 145].

On the other hand, future 6G wireless communication systems will seamlessly integrate air and space networks with terrestrial mobile cellular networks. In this context, the concept of the space-air-ground integrated network (SAGIN) has attracted extensive attention in academia and industry [147, 148]. As shown in Fig. 6.1, a typical SAGIN consists of a three-layer network structure, including the spaceborne network, the airborne network, and the terrestrial network [148]. Specifically, geostationary earth orbit (GEO), medium earth orbit (MEO), and low earth orbit (LEO) satellites operating at different altitudes constitute the spaceborne network, while the airborne platform network is composed of the aerial base stations (BSs), such as balloons, airships, etc. These aerial BSs can jointly serve a variety of aircraft and unmanned aerial vehicles (UAVs). Therefore, a large number of LEOs, aerial BSs, aircraft, and UAVs will build an aeronautical ad hoc network for the target of *Internet above the clouds* [149, 150]. For the outer space, the THz communication system is capable of achieving ultra-long-distance transmission with small transmit power since the propagation of THz signals is almost lossless [151]. This indicates that the aerial platforms can provide reliable and efficient space communication services with THz UM-MIMO technology. Besides, the demand for high-data-rate communications for passengers in civil aircraft cannot be ignored. Considering the civil aircraft is above the troposphere and near the bottom of the stratosphere in most of the flight time, the stable state facilitates the establishment of THz communication links. Therefore, this chapter mainly studies the aeronautical communications between the civil aircraft equipped with THz UM-MIMO in the stratosphere and the aerial BSs.

In order to ensure the quality of THz UM-MIMO-based aeronautical communications, it is essential for transceivers to obtain reliable channel state information (CSI) [152]. However, due to the high-speed mobility of the aircraft or UAVs in flight and the shaking of the aerial BSs, the air communication links present a severe fast time-varying fading characteristic, which makes it difficult to achieve accu-

[1] Again, we stress that we use the acronym *UM*-MIMO to narrowly refer to a kind of massive MIMO system which exhibits *squint effects* due to the large signal bandwidth, as shown in this chapter, while the full term *ultra-massive* MIMO (e.g., in the title of this book) is used to broadly refer to the evolution of massive MIMO.

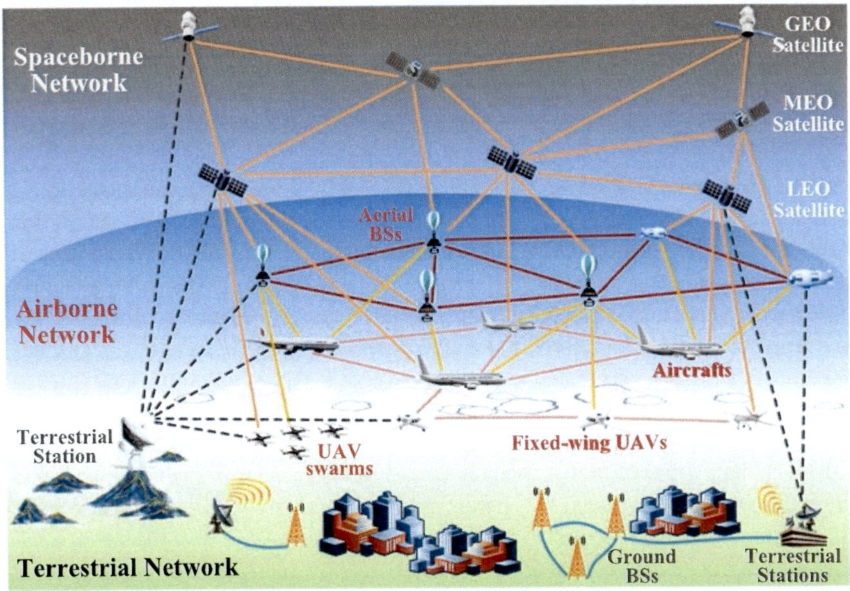

Fig. 6.1 Typical SAGIN scenario [146]

rate channel estimation (CE) and tracking. The existing solutions to this problem [87–93] consider the common mmWave massive MIMO systems or indoor short-range THz communications, while the UM-MIMO-based CE is not involved. Hence, these schemes cannot be directly applied to aeronautical THz UM-MIMO systems because of the severe challenges such as unprecedentedly large array apertures, large system bandwidths, and ultra-high mobile speeds.

6.2 CE and Data Transmission in Aeronautical Terahertz Ultra-Massive MIMO Systems

6.2.1 *Triple Delay-Beam-Doppler Squint Effects*

Compared to sub-6 GHz wireless communication systems or mmWave massive MIMO systems equipped with antenna arrays of limited aperture, the aeronautical THz UM-MIMO channels present the unique *triple-delay-beam-Doppler squint effects*. Specifically, THz UM-MIMO arrays in the form of UPA can integrate up to hundreds of antennas in a small physical size over the single horizontal or vertical dimension. If the arrival direction of the signal is not perpendicular to the array panel, it can be observed that the same received signals have different propagation delays across different antenna elements when passing through the entire array aperture. Meanwhile, due to the large system bandwidth used in THz communications, the delay gap between the first and last antennas can reach several symbol periods, resulting in high inter-symbol interference, even for the LoS-only scenario. The above phenomenon can be called the *delay squint effect* (also known as spatial-frequency wideband effects [153, 154] or aperture fill time effects in radar systems [155]), which is an inevitable challenge for THz UM-MIMO systems. Besides, this effect further leads to the *beam squint effect*, namely the beam direction becomes a function of the operating frequency. The main reason is that electromagnetic (EM) waves of different frequencies accumulate different phases at the same propagation distance, while the distance between adjacent antennas in the array is fixed according to the wavelength of the center carrier frequency. Therefore, the beam squint effect leads to the deviation of the generated beams from the desired direction for the signals at marginal carrier frequencies. Moreover, the high mobility of aircraft in aeronautical communications can cause serious Doppler shift. Due to the ultra-large bandwidth in the THz band, this shift also changes with different carrier frequencies, which is called the *Doppler squint effect*. Therefore, the aeronautical THz UM-MIMO systems suffer from triple delay-beam-Doppler squint effects.

However, recent researches [55, 156–159] mainly focus on the impact of the beam squint effect on mmWave or THz MIMO systems. Specifically, the authors in [55] studied the influence of the beam squint effect on the optimality of compressed subspace estimation and frequency-flat beamforming. Several hybrid transmit precoding (TPC) schemes were proposed in [156], which can project all frequencies to the center frequency to construct a common analog TPC matrix for all subcarriers to mitigate the beam squint effect. Besides, the authors in [157–159] proposed several CE schemes to estimate wideband massive MIMO channels by exploiting the mmWave channel characteristics under the beam squint effect, though the effect itself was not completely eliminated. To sum up, the triple delay-beam-Doppler squint effects that significantly reduce the performance of data transmission in THz UM-MIMO-based aeronautical communications are rarely considered in currently proposed CE and hybrid beamforming schemes [55, 87–93, 154, 156–159]. Therefore, it is urgent to propose an efficient signal processing paradigm of CE and data transmission dedicated for the aeronautical THz UM-MIMO systems.

6.2.2 The Overview of Proposed CE and Tracking Scheme

In general, this chapter mainly focuses on the aeronautical THz UM-MIMO-based communication link between the aircraft and the aerial BSs with the triple delay-beam-Doppler squint effects in the SAGIN.[2] To be specific, for the airborne network in the near space shown in Fig. 6.1, civil aircraft usually perform regular flight mis-

[2] The proposed schemes in this chapter can be also applied to the air-to-air links between the UAVs and aerial BSs, or the air-to-space links between the aircraft/UAVs and LEOs, as well as the links between the space-air networks and terrestrial BSs on the high-altitude mountains.

Fig. 6.2 Real-time flight tracking snapshot of civil airplanes in southern China [146].This snapshot is available on the website https://flightadsb.variflight.com/tracker/112.761836,29.084716/6

sions along the fixed route, as shown in Fig. 6.2. Based on this fact, a series of aerial BSs can be deployed near these routes to ensure that multiple aircraft can simultaneously establish reliable communication links with multiple aerial BSs, so as to build an aeronautical ad hoc network. Except for the high-altitude platform providing THz aeronautical communication services, there are few other scatterers in the upper stratosphere, hence the THz UM-MIMO channel with only one LoS path is considered in this chapter. Besides, multiple aerial BSs can jointly serve a high-speed moving aircraft through their respective THz LoS links, and their cooperation can be achieved through the THz air-to-air backbone links connecting different aerial BSs or air-to-ground backbone links.

In order to counteract the multi-path effect caused by multiple THz LoS links at the aircraft end, the orthogonal frequency division multiplexing (OFDM) technology can be applied to the aeronautical communication system.[3]

[3] In order to meet the high-quality service requirements of hundreds of people on the aircraft at the same time, the complicated high-order modulation methods, namely OFDM and quadrature amplitude modulation (QAM), can be adopted to improve the data transmission rate and throughput. Besides, due to the high peak-to-average power ratio (PAPR) of the OFDM system, the discrete

As mentioned before, the aeronautical communication channels exhibit fast time-varying fading characteristics because of the relative high-speed movement of the transceiver. Thus, it is very challenging to establish THz aeronautical communication links connecting the aircraft and the aerial BSs. On the one hand, the aeronautical communication system extracts several channel parameters (such as angle and Doppler shift, etc.) by exploiting the prior information acquired by the aerial BSs and the aircraft, including positioning, flight speed, direction and attitude, etc. On the other hand, due to the ultra-long transmission distance and extremely narrow beam width in aeronautical THz UM-MIMO communications, the positioning accuracy error and slight deviation caused by the posture rotation of the antenna array on the transceiver can lead to the undesired beam direction. Relying on these rough channel parameter estimates is not enough to establish an efficient and stable data transmission link. Therefore, how to effectively utilize the above prior information to establish and track fast time-varying communication links is crucial in the aeronautical THz UM-MIMO communication system.

6.2.2.1 The Procedure of Proposed Scheme

To support the THz UM-MIMO-based aeronautical communication systems, a low-overhead CE and tracking scheme is proposed in this chapter, of which the frame structure is depicted in Fig. 6.3. As can be seen, the scheme is divided into the following three stages: initial CE (to establish the THz communication links), data-aided channel tracking (to achieve data transmission while channel tracking), and pilot-aided channel tracking (to improve the angle estimate accuracy at the transceiver and ensure the stability and reliability of THz communication links). Specifically, the details are presented as follow.

- In the initial CE stage, the BSs and the aircraft can extract the rough angle estimates according to their respective positioning information and flight posture, so as to design rough transmit beamforming and receive combining and establish effective THz UM-MIMO communication links. Meanwhile, the transmitter and receiver can significantly mitigate the delay-beam squint effects by exploiting the grouping true-time delay unit (GTTDU) modules.
- After the establishment of THz communication links, the accurate estimates of azimuth and elevation angles can be obtained at the BSs and the aircraft, as well as the accurate Doppler and time delay estimates at the aircraft. During this stage, the rough estimate of the Doppler shift can be used to improve the accuracy of the parameter estimation. More specifically, the extra-large hybrid array can be equivalently regarded as the low-dimensional fully-digital array by utilizing the reconfigurable radio frequency (RF) selection network in the dedicated connection pattern of the sub-array selection scheme. Based on the equivalent model, the proposed prior-aided iterative angle estimation algorithm can obtain accurate

---

Fourier transformation-spread-OFDM (DFT-S-OFDM) technique is also a potential alternative for THz UM-MIMO-based aeronautical communication systems.

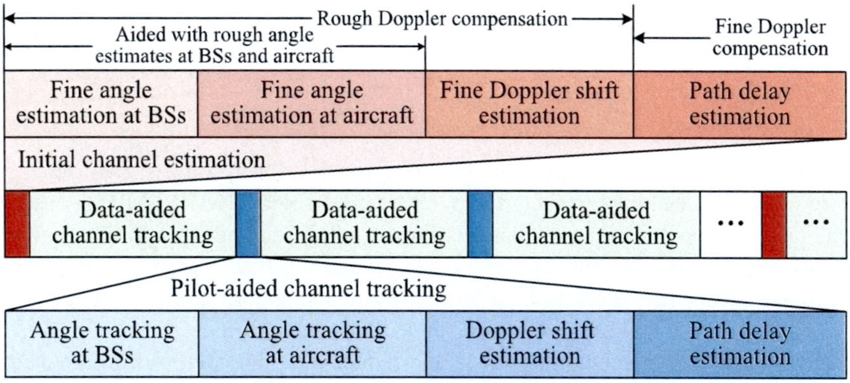

Fig. 6.3 The frame structure of the proposed CE and tracking scheme in this chapter [146]

estimates of azimuth and elevation angles at the BSs and the aircraft, respectively. These precise angle estimates can be used not only to achieve finer beam alignment, but also to optimize the GTTDU module designed on the transceiver to further eliminate the delay-beam squint effects. Meanwhile, due to the sufficiently high beamforming gain and signal-to-noise ratio (SNR) provided by the UM-MIMO array after beam alignment, the proposed prior-aided iterative Doppler shift estimation algorithm can be more efficient, where the Doppler squint effect can be mitigated by compensating the received signal with the initial rough Doppler shift estimates. After that, the path delays and channel gains can also be estimated.

- During the data transmission stage, the proposed data-aided decision-directed (DADD) channel tracking algorithm is used to track the equivalent channel after beam alignment. In this method, the detected data can be regarded as the equivalent pilots for subsequent CE, and the estimated channel can support the subsequent data detection. This process is performed alternately until the beginning of pilot-aided channel tracking stage.
- Finally, in the pilot-aided channel tracking stage, the extra-large hybrid array can be equivalently regarded as the fully-digital sparse array by reconstructing the connection pattern of the RF selection network in the sub-array selection scheme, in which the angle ambiguity created by sparse arrays can be eliminated via previous angle estimates at the receiver. Once accurate beam alignment is achieved, Doppler shift and path delay estimates can be obtained similar to the initial CE stage, and then the data transmission stage will begin. If the pilot-aided channel tracking stage fails to establish reliable THz communication links due to the inaccurate angle estimates, the angle and Doppler shift components will be re-estimated with subsequent operations.

6.2.2.2 The Contributions of Proposed Scheme

The contributions of the proposed CE and tracking scheme in this chapter can be concluded as follows.

- A low-overhead CE and tracking scheme is proposed for the huge spatial dimension and ultra-fast time-varying fading characteristics in THz UM-MIMO-based aeronautical communication channels. In the initial stage, the proposed scheme can obtain the azimuth and elevation angle, Doppler shift, and accurate path delay estimates by using the prior-aided iterative estimation algorithm, where the rough estimates of channel parameters (i.e., angle, Doppler shift, and etc.) can be leveraged to improve estimation accuracy and reduce the pilot overhead. In the data transmission stage, the proposed DADD-based channel tracking algorithm can reliably track the equivalent fast time-varying channel after beam alignment to further reduce the required pilot overhead.
- The proposed scheme can effectively overcome the unique triple delay-beam-Doppler squint effects in THz UM-MIMO aeronautical channels. It should be pointed out that this triple squint effects can hardly be observed in sub-6 GHz wireless communication systems with limited array aperture and system bandwidth, as well as in the mmWave massive MIMO systems. To deal with the delay-beam squint effect, the low-cost GTTDU module at the transceiver compensates the signal transmission delay of different antenna groups via navigation information. After that, the received SNR can be significantly increased, which ensures the establishment of reliable THz communication links before channel parameter estimation. The Rotman lens-based GTTDU module designed in Sect. 6.4 provides a feasible architecture of the tunable true-time delay unit (TTDU) module based phase shift network (PSN), which is also a potential direction for future related research. In addition, the proposed prior-aided iterative angle and Doppler shift estimation algorithm can further eliminate the residual beam and Doppler squint effects after the above processing, so as to obtain accurate angles and Doppler shifts used for subsequent data transmission.
- By designing the array selection scheme at the receiver, the ultra-large scale hybrid array can be equivalent to a low-dimensional fully-digital sparse array. Then, the robust array signal processing techniques such as the estimating signal parameters via rotational invariance techniques (ESPRIT) algorithm [101, 106] can be utilized to accurately estimate and track the azimuth and elevation angles at the transceiver. In addition, the obtained equivalent low-dimensional digital sparse array can significantly improve the angle estimation accuracy of the pilot-aided channel tracking stage, where the previous angle estimates are exploited to address the angle ambiguity caused by the sparse array.
- The Cramér-Rao lower bounds (CRLBs) of the key channel parameters, including angle, Doppler shift and path delay, are derived from the effective received signal model. According to the performance analysis, the CRLB of angle estimation derived in the pilot-aided channel tracking stage indicates that the sparse array can be leveraged to significantly improve the accuracy of angle estimation.

The numerical results show that the parameter estimation performance can coincide with the corresponding CRLBs, which demonstrates the effectiveness of the proposed scheme.

6.3 System Model

This section presents the formulation of signal transmission and channel modeling in the THz UM-MIMO-based aeronautical communications, where the azimuth and elevation angles are involved in full-dimensional (FD) UPA UM-MIMO channels [101, 160]. The considered scenario is depicted in Fig. 6.4, in which L aerial BSs jointly serve one aircraft through their respective THz LoS paths. The aircraft adopts the hybrid beamforming architecture with the partial sub-connected PSN [14, 141], while the analog beamforming architecture is adopted at each BS with only one RF chain. The total number of antennas at each BS is $N_{\rm BS} = N_{\rm BS}^{\rm h} N_{\rm BS}^{\rm v}$, where $N_{\rm BS}^{\rm h}$ and $N_{\rm BS}^{\rm v}$ are the numbers of antennas in the horizontal and vertical directions, respectively. Since the partial sub-connected PSN is adopted at the aircraft, we define $\widetilde{I}_{\rm AC}^{\rm h}$ and $\widetilde{I}_{\rm AC}^{\rm v}$ as the numbers of sub-arrays in the horizontal and vertical direction, respectively, and $M_{\rm AC}^{\rm h}$ and $M_{\rm AC}^{\rm v}$ as the numbers of antennas at each sub-array in the horizontal and vertical direction, respectively. Therefore, $N_{\rm AC}^{\rm h} = \widetilde{I}_{\rm AC}^{\rm h} M_{\rm AC}^{\rm h}$ and $N_{\rm AC}^{\rm v} = \widetilde{I}_{\rm AC}^{\rm v} M_{\rm AC}^{\rm v}$ are the total numbers of antennas in the horizontal and vertical directions, respectively. The total number of antennas in each sub-array (in the entire

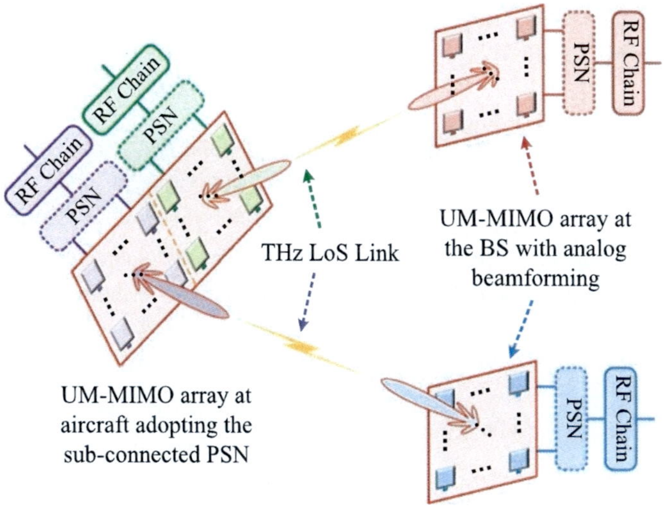

Fig. 6.4 The structure of the UM-MIMO arrays at the transceiver, where $L = 2$ BSs with analog beamforming architecture communicate with the aircraft adopting the sub-connected PSN via their respective THz LoS links [146]

array) is $M_{\rm AC}=M_{\rm AC}^{\rm h}M_{\rm AC}^{\rm v}$ ($N_{\rm AC}=N_{\rm AC}^{\rm h}N_{\rm AC}^{\rm v}$). Besides, the aircraft has $L=\widetilde{I}_{\rm AC}^{\rm h}\widetilde{I}_{\rm AC}^{\rm v}$ RF chains, and each RF chain (along with its corresponding sub-array) is designated to a certain BS.

6.3.1 Signal Transmission Model

According to the frame structure in Fig. 6.3, the aeronautical communication system estimates the channel angles at the BS and the aircraft in the uplink and downlink, respectively. For the lth BS in the uplink, the receive baseband signal at the kth ($1 \leqslant k \leqslant K$) subcarrier of the mth OFDM symbol is given as

$$y_{{\rm UL},l}^{[m]}[k] = \sqrt{P_l}\boldsymbol{q}_{{\rm RF},l}^{\rm H}\boldsymbol{H}_{{\rm UL},l}^{[m]}[k]\boldsymbol{P}_{\rm RF}\boldsymbol{P}_{\rm BB}^{[m]}[k]\boldsymbol{s}_{\rm UL}^{[m]}[k] + \boldsymbol{q}_{{\rm RF},l}^{\rm H}\boldsymbol{n}_{{\rm UL},l}^{[m]}[k], \tag{6.1}$$

where P_l denotes the transmit power, and K denotes the total number of subcarriers. In Eq. (6.1), $\boldsymbol{q}_{{\rm RF},l} \in \mathbb{C}^{N_{\rm BS}}$ denotes the analog combining vector at the lth BS, $\boldsymbol{P}_{\rm RF} \in \mathbb{C}^{N_{\rm AC}\times L}$ and $\boldsymbol{P}_{\rm BB}^{[m]}[k] \in \mathbb{C}^{L\times L}$ denote the analog and digital precoding matrix at the aircraft, respectively, $\boldsymbol{H}_{{\rm UL},l}^{[m]}[k] \in \mathbb{C}^{N_{\rm BS}\times N_{\rm AC}}$ denotes the uplink baseband channel matrix, $\boldsymbol{s}_{\rm UL}^{[m]}[k] \in \mathbb{C}^{L}$ denotes the transmit signal vector, $\boldsymbol{n}_{{\rm UL},l}^{[m]}[k] \in \mathbb{C}^{N_{\rm BS}}$ denotes the additive white Gaussian noise (AWGN) vector with covariance matrix $\sigma_n^2\boldsymbol{I}_{N_{\rm BS}}$, namely $\boldsymbol{n}_{{\rm UL},l}^{[m]}[k] \sim \mathcal{CN}\left(\boldsymbol{0}_{N_{\rm BS}}, \sigma_n^2\boldsymbol{I}_{N_{\rm BS}}\right)$. Similarly, for the aircraft in the downlink, the receive baseband signal at the kth subcarrier of the nth OFDM symbol can be written as

$$\boldsymbol{y}_{\rm DL}^{[n]}[k] = (\boldsymbol{W}_{\rm BB}^{[n]}[k])^{\rm H}\boldsymbol{W}_{\rm RF}^{\rm H}\left(\sum_{l=1}^{L}\sqrt{P_l}\boldsymbol{H}_{{\rm DL},l}^{[n]}[k]\boldsymbol{f}_{{\rm RF},l}s_{{\rm DL},l}^{[n]}[k] + \boldsymbol{n}_{\rm DL}^{[n]}[k]\right), \tag{6.2}$$

where $\boldsymbol{W}_{\rm RF} \in \mathbb{C}^{N_{\rm AC}\times L}$ and $\boldsymbol{W}_{\rm BB}^{[n]}[k] \in \mathbb{C}^{L\times L}$ denote the analog and digital combining matrix at the aircraft, respectively, $\boldsymbol{f}_{{\rm RF},l} \in \mathbb{C}^{N_{\rm BS}}$ denotes the lth analog precoding vector for the lth BS, $\boldsymbol{H}_{{\rm DL},l}^{[n]}[k] \in \mathbb{C}^{N_{\rm AC}\times N_{\rm BS}}$ denotes the downlink baseband channel matrix, $s_{{\rm DL},l}^{[n]}[k]$ denotes the transmit pilot signal (or the modulated/coded data), and $\boldsymbol{n}_{{\rm DL},l}^{[n]}[k] \in \mathbb{C}^{N_{\rm AC}}$ denotes the AWGN vector with the same distribution as $\boldsymbol{n}_{{\rm UL},l}^{[m]}[k] \in \mathbb{C}^{N_{\rm BS}}$.

6.3.2 THz UM MIMO Channel Model

Take the array at the BS as shown in Fig. 6.5 as an example, the delay squint effect of the THz UM-MIMO channel can be explained as follows. Select the $(1, 1)$th antenna as the reference point, and define $\boldsymbol{r}=\left(\sin(\theta_l^{\rm BS})\cos(\varphi_l^{\rm BS}), \sin(\varphi_l^{\rm BS}), \cos(\theta_l^{\rm BS})\cos(\varphi_l^{\rm BS})\right)$ as a unit direction vector, where $\theta_l^{\rm BS}$ and $\varphi_l^{\rm BS}$ denote the azimuth and elevation

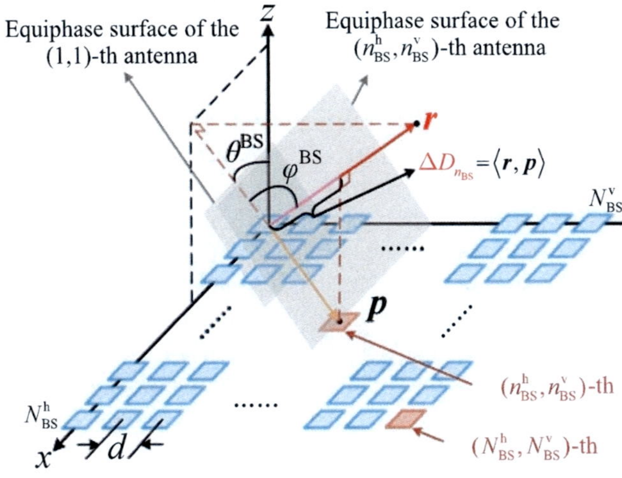

Fig. 6.5 The delay-beam squint effect of the THz UM-MIMO array, where the dimension of the array is $N_{\rm BS}^{\rm h} \times N_{\rm BS}^{\rm v}$ [146]

angle of the lth BS, respectively. Denote $(n_{\rm BS}^{\rm h}, n_{\rm BS}^{\rm v})$th antenna as the $n_{\rm BS}$th antenna with $n_{\rm BS} = (n_{\rm BS}^{\rm v} - 1)N_{\rm BS}^{\rm h} + n_{\rm BS}^{\rm h}$, the direction vector to the reference antenna can be expressed as $\boldsymbol{p} = \left((n_{\rm BS}^{\rm h} - 1)d, (n_{\rm BS}^{\rm v} - 1)d, 0\right)$, where d denotes the half-wavelength spacing between the adjacent antennas. Since the wave path-difference between the $n_{\rm BS}$th antennas and the first antenna, denoted as $\Delta D_{n_{\rm BS}}$, is equal to the distance between the equiphase surfaces of these the two antennas, it can be calculated as $\Delta D_{n_{\rm BS}} = \langle \boldsymbol{r}, \boldsymbol{p} \rangle = (n_{\rm BS}^{\rm h} - 1)d \sin(\theta_l^{\rm BS}) \cos(\varphi_l^{\rm BS}) + (n_{\rm BS}^{\rm v} - 1)d \sin(\varphi_l^{\rm BS})$. Then, the transmission delay caused by the path difference can be calculated as $\tau_l^{[n_{\rm BS}]} = \Delta D_{n_{\rm BS}}/c$, where c denotes the speed of light. It should be pointed out that $\tau_l^{[n_{\rm BS}]}$ is related to the antenna index and the azimuth and elevation angles. When the arrival direction of the signal is not perpendicular to the array plane and $n_{\rm BS}$ is large, $\tau_l^{[n_{\rm BS}]}$ can even be larger than the symbol period T_s,[4] which raises higher requirements on the signal processing of transceivers. Therefore, it is necessary to deal with the delay squint effect in the aeronautical THz UM-MIMO system.

Due to the channel reciprocity in the time division duplex (TDD) system, this subsection focuses on the modeling of the downlink channel matrix. According to the literature [158, 161], the space-delay domain downlink passband channel matrix of the lth BS at time t is defined as $\bar{\boldsymbol{H}}_{{\rm DL},l}^{(t)}(\tau) \in \mathbb{C}^{N_{\rm AC} \times N_{\rm BS}}$, where the $(n_{\rm AC}, n_{\rm BS})$th element ($1 \le n_{\rm AC} \le N_{\rm AC}$ and $1 \le n_{\rm BS} \le N_{\rm BS}$) can be further written as

4 Consider an extreme scenario that the direction-of-arrival (DoA) of the signal is the diagonal direction of the UPA (with dimension $n_{\rm BS} \times n_{\rm BS}$). In this case, the $n_{\rm BS}$ diagonal antennas of the UPA can be equivalent to a uniform linear array (ULA) with antenna spacing of $\sqrt{2}d$. For typical parameters such as $\theta_l^{\rm BS} = 60°$, carrier frequency $f_c = 0.1$ THz, bandwidth $f_s = 1$ GHz, and $n_{\rm BS} = 200$, the transmission delay of the entire array can be calculated as $\tau_l^{[n_{\rm BS}]} = \sqrt{2}(n_{\rm BS} - 1)\sin(\theta_l^{\rm BS})/(2f_c) \approx 1.219\, T_s$.

$$\{\bar{\boldsymbol{H}}_{\mathrm{DL},l}^{(t)}(\tau)\}_{n_{\mathrm{AC}},n_{\mathrm{BS}}} = \sqrt{G_l}\alpha_l e^{\mathrm{j}2\pi\psi_l t}\delta\big(\tau - \tau_l - \underbrace{(\tau_l^{[n_{\mathrm{AC}}]} + \tau_l^{[n_{\mathrm{BS}}]})}_{\textbf{Delay squint}}\big), \tag{6.3}$$

where G_l and $\alpha_l \sim \mathcal{CN}(0, \sigma_\alpha^2)$ denote the large-scale fading gain and the channel gain of the communication links,[5] respectively, $\psi_l = \underline{v}_l/\lambda_c$ denote the Doppler shift ($\underline{v}_l$ and λ_c denote the relative radial velocity and carrier wavelength, respectively), f_c denotes the carrier frequency, $\tau_l^{[n_{\mathrm{AC}}]}$ denotes the transmission delay between the n_{AC}th antenna ($n_{\mathrm{AC}} = (n_{\mathrm{AC}}^{\mathrm{v}} - 1)N_{\mathrm{AC}}^{\mathrm{h}} + n_{\mathrm{AC}}^{\mathrm{h}}$, which is also the $(n_{\mathrm{AC}}^{\mathrm{h}}, n_{\mathrm{AC}}^{\mathrm{v}})$th antenna of the UPA at the aircraft) and the selected reference point, $\delta(\cdot)$ denotes the Dirac pulse function, and τ_l denotes the path delay. In Eq. (6.3), the transmission delay $\tau_l^{[n_{\mathrm{AC}}]}$ at the aircraft and $\tau_l^{[n_{\mathrm{BS}}]}$ at the BS jointly constitute the delay squint.

After some manipulations, the downlink space-frequency domain channel matrix at the kth subcarrier of the nth OFDM symbol in (6.2) can be written as

$$\boldsymbol{H}_{\mathrm{DL},l}^{[n]}[k] = \sqrt{G_l}\alpha_l e^{\mathrm{j}2\pi\psi_{l,k}(n-1)T_{\mathrm{sym}}} e^{-\mathrm{j}2\pi\left(\frac{k-1}{K} - \frac{1}{2}\right)f_s\tau_l} \boldsymbol{A}_{\mathrm{DL},l}[k], \tag{6.4}$$

where T_{sym} and f_s denote the duration of OFDM symbol and system bandwidth, respectively, $\psi_{l,k} = \psi_{z,l} + \frac{\underline{v}_l}{c}(\frac{k-1}{K} - \frac{1}{2})f_s$ denotes the frequency-dependent Doppler shift at the kth subcarrier, $\psi_{z,l}$ denotes the Doppler shift at the central carrier frequency f_z (corresponding wavelength denoted as λ_z), and $\frac{\underline{v}_l}{c}(\frac{k-1}{K} - \frac{1}{2})f_s$ denotes the Doppler squint caused by ultra-large bandwidth in THz communications. Besides, $\boldsymbol{A}_{\mathrm{DL},l}[k] \in \mathbb{C}^{N_{\mathrm{AC}} \times N_{\mathrm{BS}}}$ is the downlink array response matrix associated with the array response vectors of the aircraft and the lth BS, which can be represented as

$$\boldsymbol{A}_{\mathrm{DL},l}[k] = \underbrace{\left(\boldsymbol{a}_{\mathrm{AC}}(\mu_l^{\mathrm{AC}}, \nu_l^{\mathrm{AC}})\boldsymbol{a}_{\mathrm{BS}}^{\mathrm{H}}(\mu_l^{\mathrm{BS}}, \nu_l^{\mathrm{BS}})\right)}_{\boldsymbol{A}_{\mathrm{DL},l}} \circ \underbrace{\left(\bar{\boldsymbol{a}}_{\mathrm{AC}}(\mu_l^{\mathrm{AC}}, \nu_l^{\mathrm{AC}}, k)\bar{\boldsymbol{a}}_{\mathrm{BS}}^{\mathrm{H}}(\mu_l^{\mathrm{BS}}, \nu_l^{\mathrm{BS}}, k)\right)}_{\bar{\boldsymbol{A}}_{\mathrm{DL},l}[k]\ \textbf{(Beam squint)}}, \tag{6.5}$$

where $\mu_l^{\mathrm{AC}} = \pi\sin(\theta_l^{\mathrm{AC}})\cos(\varphi_l^{\mathrm{AC}})$ ($\mu_l^{\mathrm{BS}} = \pi\sin(\theta_l^{\mathrm{BS}})\cos(\varphi_l^{\mathrm{BS}})$) and $\nu_l^{\mathrm{AC}} = \pi\sin(\varphi_l^{\mathrm{AC}})$ ($\nu_l^{\mathrm{BS}} = \pi\sin(\varphi_l^{\mathrm{BS}})$) denote the horizontal and vertical virtual angles at the aircraft (the lth BS), respectively, $\boldsymbol{A}_{\mathrm{DL},l}$ denotes the traditional downlink array response matrix when the transceivers are not affected by the squint effect, and $\bar{\boldsymbol{A}}_{\mathrm{DL},l}[k]$ denotes the array response squint matrix when considering beam squint effect. In addition, $\boldsymbol{a}_{\mathrm{AC}}(\mu_l^{\mathrm{AC}}, \nu_l^{\mathrm{AC}}) = \boldsymbol{a}_{\mathrm{v}}(\nu_l^{\mathrm{AC}}, N_{\mathrm{AC}}^{\mathrm{v}}) \otimes \boldsymbol{a}_{\mathrm{h}}(\mu_l^{\mathrm{AC}}, N_{\mathrm{AC}}^{\mathrm{h}})$ and $\boldsymbol{a}_{\mathrm{BS}}(\mu_l^{\mathrm{BS}}, \nu_l^{\mathrm{BS}}) = \boldsymbol{a}_{\mathrm{v}}(\nu_l^{\mathrm{BS}}, N_{\mathrm{BS}}^{\mathrm{v}}) \otimes \boldsymbol{a}_{\mathrm{h}}(\mu_l^{\mathrm{BS}}, N_{\mathrm{BS}}^{\mathrm{h}})$ denote the traditional array response vector at the aircraft and the lth BS [101, 105], respectively, while $\bar{\boldsymbol{a}}_{\mathrm{AC}}(\mu_l^{\mathrm{AC}}, \nu_l^{\mathrm{AC}}, k) = \bar{\boldsymbol{a}}_{\mathrm{v}}(\nu_l^{\mathrm{AC}}, N_{\mathrm{AC}}^{\mathrm{v}}, k) \otimes \bar{\boldsymbol{a}}_{\mathrm{h}}(\mu_l^{\mathrm{AC}}, N_{\mathrm{AC}}^{\mathrm{h}}, k)$ and $\bar{\boldsymbol{a}}_{\mathrm{BS}}(\mu_l^{\mathrm{BS}}, \nu_l^{\mathrm{BS}}, k) = \bar{\boldsymbol{a}}_{\mathrm{v}}(\nu_l^{\mathrm{BS}}, N_{\mathrm{BS}}^{\mathrm{v}}, k) \otimes \bar{\boldsymbol{a}}_{\mathrm{h}}(\mu_l^{\mathrm{BS}}, N_{\mathrm{BS}}^{\mathrm{h}}, k)$ denote the corresponding frequency-dependent array response squint

[5] For the THz communication links in the stratosphere and above, the frequency-dependent signal attenuation (such as atmospheric molecular absorption, rain attenuation, etc.) are negligible [144, 145], and the channel gain α_l can be modeled as the frequency-flat channel coefficient, which differs from the frequency-dependent channel coefficients considered in [161, 162].

vectors, respectively. Meanwhile, the horizontal (vertical) steering vector $\boldsymbol{a}_{\mathrm{h}}(\mu_l^{\mathrm{AC}}, N_{\mathrm{AC}}^{\mathrm{h}})$ ($\boldsymbol{a}_{\mathrm{v}}(\nu_l^{\mathrm{AC}}, N_{\mathrm{AC}}^{\mathrm{v}})$) can be further defined as

$$\boldsymbol{a}_{\mathrm{h}}(\mu_l^{\mathrm{AC}}, N_{\mathrm{AC}}^{\mathrm{h}}) = \left[1\ e^{\mathrm{j}\mu_l^{\mathrm{AC}}}\ \cdots\ e^{\mathrm{j}(N_{\mathrm{AC}}^{\mathrm{h}}-1)\mu_l^{\mathrm{AC}}}\right]^{\mathrm{T}}, \tag{6.6}$$

$$\boldsymbol{a}_{\mathrm{v}}(\nu_l^{\mathrm{AC}}, N_{\mathrm{AC}}^{\mathrm{v}}) = \left[1\ e^{\mathrm{j}\nu_l^{\mathrm{AC}}}\ \cdots\ e^{\mathrm{j}(N_{\mathrm{AC}}^{\mathrm{v}}-1)\nu_l^{\mathrm{AC}}}\right]^{\mathrm{T}}, \tag{6.7}$$

and the horizontal (vertical) steering squint vector can be defined as

$$\bar{\boldsymbol{a}}_{\mathrm{h}}(\mu_l^{\mathrm{AC}}, N_{\mathrm{AC}}^{\mathrm{h}}, k) = \left[1\ e^{\mathrm{j}\left(\frac{k-1}{K}-\frac{1}{2}\right)\frac{f_s}{f_z}\mu_l^{\mathrm{AC}}}\ \cdots\ e^{\mathrm{j}\left(\frac{k-1}{K}-\frac{1}{2}\right)\frac{f_s}{f_z}(N_{\mathrm{AC}}^{\mathrm{h}}-1)\mu_l^{\mathrm{AC}}}\right]^{\mathrm{T}}, \tag{6.8}$$

$$\bar{\boldsymbol{a}}_{\mathrm{v}}(\nu_l^{\mathrm{AC}}, N_{\mathrm{AC}}^{\mathrm{v}}, k) = \left[1\ e^{\mathrm{j}\left(\frac{k-1}{K}-\frac{1}{2}\right)\frac{f_s}{f_z}\nu_l^{\mathrm{AC}}}\ \cdots\ e^{\mathrm{j}\left(\frac{k-1}{K}-\frac{1}{2}\right)\frac{f_s}{f_z}(N_{\mathrm{AC}}^{\mathrm{v}}-1)\nu_l^{\mathrm{AC}}}\right]^{\mathrm{T}}. \tag{6.9}$$

Note that the associated vectors of the BS, such as $\boldsymbol{a}_{\mathrm{h}}(\mu_l^{\mathrm{BS}}, N_{\mathrm{BS}}^{\mathrm{h}})$, $\boldsymbol{a}_{\mathrm{v}}(\nu_l^{\mathrm{BS}}, N_{\mathrm{BS}}^{\mathrm{v}})$, $\bar{\boldsymbol{a}}_{\mathrm{h}}(\mu_l^{\mathrm{BS}}, N_{\mathrm{BS}}^{\mathrm{h}}, k)$, and $\bar{\boldsymbol{a}}_{\mathrm{v}}(\nu_l^{\mathrm{BS}}, N_{\mathrm{BS}}^{\mathrm{v}}, k)$, have similar definitions and expressions to Eqs. (6.6)–(6.9). The detailed derivation of donwlink space-frequency domain channel matrix $\boldsymbol{H}_{\mathrm{DL},l}^{[n]}[k]$ is given in Appendix E.

Similar to (6.4), for the lth BS, the uplink space-frequency domain baseband channel matrix $\boldsymbol{H}_{\mathrm{UL},l}^{[m]}[k]$ at the kth subcarrier of the mth OFDM symbol can be written as

$$\boldsymbol{H}_{\mathrm{UL},l}^{[m]}[k] = \sqrt{G_l}\alpha_l e^{\mathrm{j}2\pi\psi_{l,k}(m-1)T_{\mathrm{sym}}}\boldsymbol{A}_{\mathrm{UL},l}[k], \tag{6.10}$$

where the uplink array response matrix $\boldsymbol{A}_{\mathrm{UL},l}[k] \in \mathbb{C}^{N_{\mathrm{BS}}\times N_{\mathrm{AC}}}$ is given as follows

$$\boldsymbol{A}_{\mathrm{UL},l}[k] = \underbrace{\left(\boldsymbol{a}_{\mathrm{BS}}(\mu_l^{\mathrm{BS}}, \nu_l^{\mathrm{BS}})\boldsymbol{a}_{\mathrm{AC}}^{\mathrm{H}}(\mu_l^{\mathrm{AC}}, \nu_l^{\mathrm{AC}})\right)}_{\boldsymbol{A}_{\mathrm{UL},l}} \circ \underbrace{\left(\bar{\boldsymbol{a}}_{\mathrm{BS}}(\mu_l^{\mathrm{BS}}, \nu_l^{\mathrm{BS}}, k)\bar{\boldsymbol{a}}_{\mathrm{AC}}^{\mathrm{H}}(\mu_l^{\mathrm{AC}}, \nu_l^{\mathrm{AC}}, k)\right)}_{\bar{\boldsymbol{A}}_{\mathrm{UL},l}[k]\ \textbf{(Beam squint)}}. \tag{6.11}$$

6.4 Initial CE

As shown in Fig. 6.3, in the initial CE stage, the BS and the aircraft can sequentially estimate the precise azimuth/elevation angle, Doppler shift and path delay. The aeronautical system can extract some rough channel parameter estimates (such as angle and Doppler shift, etc.) according to the obtained positioning and flight attitude information, and further establish the initial THz UM-MIMO communication link. However, due to the positioning accuracy errors and attitude rotations of antenna arrays installed on aerial BSs and aircraft, these rough angle and Doppler shift estimates are not accurate enough for data transmission. Therefore, accurate estimation of the primary channel parameters remains an essential process.

The fully-digital MIMO architecture is the best approach to solve the delay-beam squint effect in the THz UM-MIMO system. However, since each antenna requires a dedicated RF chain, the hardware cost and power consumption are unacceptable.

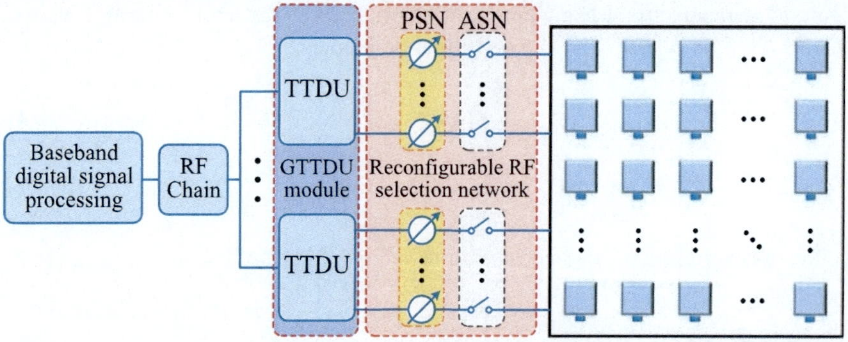

Fig. 6.6 The designed transceiver structure with GTTDU module, where each single RF link is connected to the array via the GTTDU module and the reconfigurable RF selection network consisting of the sub-connected PSNs and ASNs [146]

As for the delay-beam squint effects in the hybrid beamforming architecture, the CE schemes proposed in [55, 156–159] only employ some signal processing methods to reduce the influence of the squint effect on the CE results, rather than eliminate them in the signal transmission. These processing methods are only suitable for terrestrial mmWave or THz cellular networks with a large number of scatterers, because for the short-distance transmission (up to hundreds of meters), the receivers can still receive the corresponding signals on all carrier frequencies within the communication range, though the transmitted mmWave/THz signals are affected by the delay-beam squint effects. However, considering the long distance of the LoS link (up to hundreds of kilometers) and few redundant scatterers in the THz UM-MIMO-based aeronautical communication system, the receiver may only receive signals near the center carrier frequency because of the narrow radiating pencil beams and (even slight) delay-beam squint effects. Therefore, in addition to the essential signal processing, the aeronautical communication system also needs to carefully design the structure of the transceivers to eliminate the influence of the delay-beam squint effects, thereby ensuring that the reliable THz communication links can be established within all carrier frequencies.

Designing transceivers through TTDU modules is a common method to handle the delay-beam squint effects [163, 164]. The optimal TTDU module is to assign a dedicated real-time delay unit to each antenna [165], which leads to excessive hardware complexity and cost. Therefore, this section designs a sub-optimal transceiver structure based on the GTTDU module, as shown in Fig. 6.6. It can be observed that in addition to the UM-MIMO array, this transceiver also includes a GTTDU module and a reconfigurable RF selection network composed of some sub-connected PSNs and antenna switching networks (ASN) [166]. The ASN can control the activation state of the antenna elements, so as to ensure that the RF selection network can provide different connection modes in the angle estimation stage. In this GTTDU module, the hardware limitation caused by a group of antennas sharing the same TTDU can be solved in subsequent signal processing algorithms.

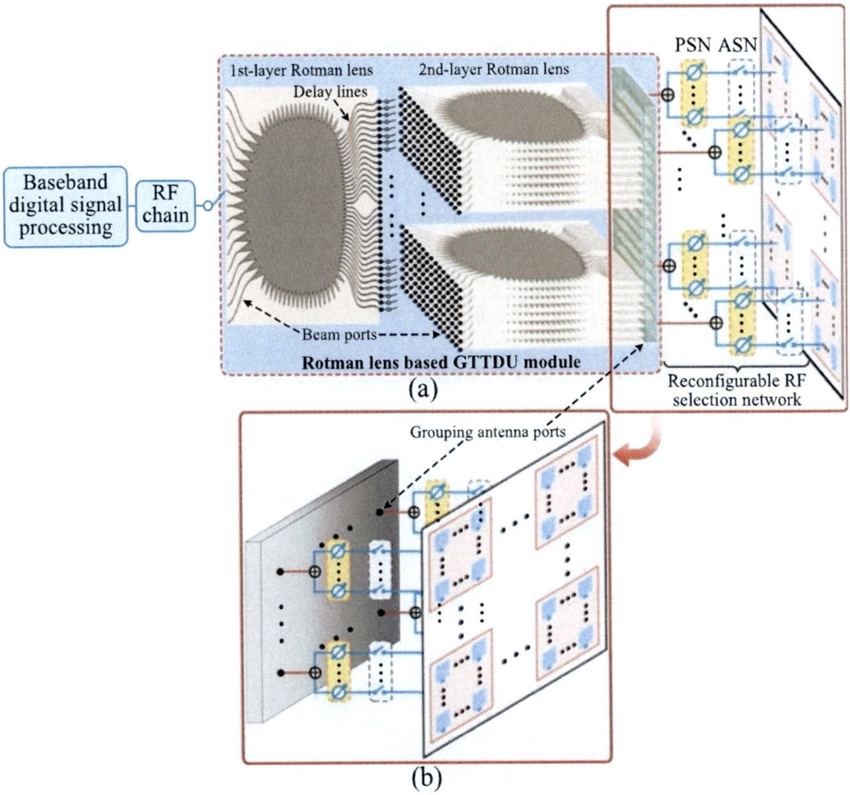

Fig. 6.7 **a** A feasible transceiver structure corresponding to Fig. 6.6 [146], where the Rotman lens based GTTDU module works as the practical tunable TTDU module [167]. **b** Two elevational views of the RF front-end (in solid boxes), including the grouped antenna ports, reconfigurable RF selection network, and THz UM-MIMO array [146]

It should be pointed out here that although the delay-beam squint effects of the entire UM-MIMO array cannot be ignored, it is relatively weak for an antenna within a group. Therefore, the designed GTTDU module can weaken the delay-beam squint effects between different antenna groups, and the residual phase deviation in each group of antennas can be further eliminated through the corresponding phase shifters in PSN.

In order to illustrate the feasibility of the transceiver structure designed in Fig. 6.6, a possible implementation including the GTTDU module based on the Rotman lens is given in Fig. 6.7. As a practical photonic implementation [168], this GTTDU module can exploit the optical properties of EM waves to realize the tunable TTDU module as a broadband phase shifter, which can adaptively adjust the phase shift for different carrier frequencies. Hence, it can precisely remove the influence caused by the beam squint effect [169, 170]. In Fig. 6.7, the total number of grouped antenna ports is the same as the number of antenna groups of the GTTDU module designed in Fig. 6.6, and the beam ports corresponding to the first and second layers of Rotman lens are employed to control the horizontal and vertical beam directions, respectively. Besides, this cascaded two-layer Rotman lens can achieve FD beamforming [167].

When the acquired angle is accurate enough, the GTTDU-based transceiver can significantly reduce the influence of delay-beam squint effects. Specifically, according to the prior information obtained from the navigation, the rough estimates of the azimuth and elevation angles at the BS (aircraft) can be defined as $\{\widetilde{\theta}_l^{\rm BS}\}_{l=1}^L$ ($\{\widetilde{\theta}_l^{\rm AC}\}_{l=1}^L$) and $\{\widetilde{\varphi}_l^{\rm BS}\}_{l=1}^L$($\{\widetilde{\varphi}_l^{\rm AC}\}_{l=1}^L$), respectively. Then, the corresponding horizontal and vertical virtual angles can be expressed as $\{\widetilde{\mu}_l^{\rm BS}\}_{l=1}^L$($\{\widetilde{\mu}_l^{\rm AC}\}_{l=1}^L$) and $\{\widetilde{\nu}_l^{\rm BS}\}_{l=1}^L$($\{\widetilde{\nu}_l^{\rm AC}\}_{l=1}^L$), respectively.

According to Eq. (6.4), the closed-form expression of the downlink channel matrix processed by the ideal TTDU module, denoted as $\widetilde{\boldsymbol{H}}_{{\rm DL},l}^{[n]}[k]$, is given in the following lemma. See Appendix F for derivation in detail.

Lemma 6.1 *According to the rough estimates of the above angle parameters, the transmission delay on the THz UM-MIMO array at the BS and the aircraft can be compensated through the ideal TTDU module. The compensated downlink space-frequency domain channel matrix* $\widetilde{\boldsymbol{H}}_{{\rm DL},l}^{[n]}[k]$ *can be expressed as*

$$\widetilde{\boldsymbol{H}}_{{\rm DL},l}^{[n]}[k] = \sqrt{G_l}\alpha_l e^{j2\pi\psi_{l,k}(n-1)T_{\rm sym}} e^{-j2\pi\left(\frac{k-1}{K}-\frac{1}{2}\right)f_s\tau_l}\widetilde{\boldsymbol{A}}_{{\rm DL},l}[k], \tag{6.12}$$

where the matrix $\widetilde{\boldsymbol{A}}_{{\rm DL},l}[k]$ *can be further expressed as*

$$\widetilde{\boldsymbol{A}}_{{\rm DL},l}[k] = \boldsymbol{A}_{{\rm DL},l}[k] \circ \underbrace{\left(\bar{\boldsymbol{a}}_{\rm AC}(\widetilde{\mu}_l^{\rm AC}, \widetilde{\nu}_l^{\rm AC}, k)\bar{\boldsymbol{a}}_{\rm BS}^{\rm H}(\widetilde{\mu}_l^{\rm BS}, \widetilde{\nu}_l^{\rm BS}, k)\right)^*}_{\widetilde{\bar{\boldsymbol{A}}}_{{\rm DL},l}[k]}. \tag{6.13}$$

By comparing $\widetilde{\bar{\boldsymbol{A}}}_{{\rm DL},l}[k]$ in Eq. (6.13) and $\bar{\boldsymbol{A}}_{{\rm DL},l}[k]$ in Eq. (6.5), it can be found that the beam squint component $\bar{\boldsymbol{A}}_{{\rm DL},l}[k]$ can be completely eliminated if the transceivers can obtain accurate angles. That is to say, when $\widetilde{\mu}_l^{\rm AC} = \mu_l^{\rm AC}$, $\widetilde{\nu}_l^{\rm AC} = \nu_l^{\rm AC}$, $\widetilde{\mu}_l^{\rm BS} = \mu_l^{\rm BS}$, and $\widetilde{\nu}_l^{\rm BS} = \nu_l^{\rm BS}$, we can obtain $\widetilde{\bar{\boldsymbol{A}}}_{{\rm DL},l}[k] = \bar{\boldsymbol{A}}_{{\rm DL},l}[k]$ and $\widetilde{\boldsymbol{A}}_{{\rm DL},l}[k] = \boldsymbol{A}_{{\rm DL},l}$. Besides, according to Eqs. (6.10) and (6.11), the compensated uplink space-frequency domain channel matrix $\widetilde{\boldsymbol{H}}_{{\rm UL},l}^{[m]}[k]$ has a similar expression to Eq. (6.12), so we omit it for simplicity.

In addition, the closed-form expression of the ideal TTDU module derived above can provide an upper bound for channel parameter estimation or data transmission. The solution proposed in this chapter can design the sub-optimal GTTDU module and the corresponding signal processing algorithm to approach the upper bound, and the actual downlink/uplink space-frequency domain channel matrix compensated by the GTTDU module can be derived from Eqs. (6.12) and (6.13). Specifically, assuming that all antenna groups in the GTTDU module have the same size, the sub-array dimensions allocated to any group at the BS and aircraft are denoted as $\widetilde{M}_{\rm BS}^{\rm h} \times \widetilde{M}_{\rm BS}^{\rm v}$ and $\widetilde{M}_{\rm AC}^{\rm h} \times \widetilde{M}_{\rm AC}^{\rm v}$, respectively. The central antenna in each sub-array can be employed as the reference for the transmission delay of this group of antennas when designing the corresponding TTDU module. At the same time, to minimize the beam squint effect caused by antenna grouping, the low-cost PSNs can be used to compensate the phase deviations of the antennas in each group except the central antenna one by one, where the phase deviations are calculated referring to the phase at the central carrier frequency. For convenience, the compensated uplink and downlink space-frequency domain channel matrices can also be expressed as $\widetilde{\boldsymbol{H}}^{[m]}_{\mathrm{UL},l}[k]$ and $\widetilde{\boldsymbol{H}}^{[n]}_{\mathrm{DL},l}[k]$, respectively.

It should be pointed out that the channel parameter estimation stage mentioned in this chapter (including the initial CE stage in this section and the subsequent pilot-aided channel tracking stage) adopts the orthogonal frequency division multiple access (OFDMA) to distinguish the pilot signals transmitted by different BSs for enhancing the estimation accuracy. Therefore, K subcarriers can be uniformly allocated to L BSs, which can be achieved by the alternate subcarrier index allocation scheme with equal intervals. Then, the sorted subcarrier index set allocated to the lth BS is $\mathcal{K}_l$ with $K_l = |\mathcal{K}_l|_c$. Furthermore, the proposed scheme can estimate the azimuth/elevation angle of the BS in the uplink, and obtain the rest of the main channel parameters in the downlink.

6.4.1 Accurate Angle Estimation Based on Reconfigurable RF Selection Network

6.4.1.1 Accurate Angle Estimation at the BS

Since the transceiver at the BS only adopts a single RF link, it is necessary to accumulate the observations during multiple OFDM symbols to estimate the angles. Meanwhile, to reduce the inter-carrier interference caused by large Doppler shift in an OFDM symbol, the rough estimate of Doppler shift can be exploited to compensate the received and transmitted signals, so that the compensated channel exhibits slow time-varying characteristics during multiple OFDM symbols. Considering the reconfigurable RF selection network designed in Figs. 6.6 and 6.7, if the transmitter transmits signals with the same configuration in consecutive OFDM symbols, and the receiver adopts designed sub-arrays to receive signals in different OFDM symbols, there will be regular phase differences between the received signals. By combining these regular phase differences, we can construct the array response vector corresponding to the equivalent low-dimensional fully-digital array. To this end, we design a sub-array selection scheme for the initial CE stage as shown in Fig. 6.8. In the proposed scheme, by exploiting the reconfigurable RF selection network to design specific connection modes during different periods, the receivers can not only reap the huge beamforming gain from the THz UM-MIMO array, but also make the ultra-large scale hybrid array equivalent to the low-dimensional fully-digital array. From the example of a 5×5 UPA given in Fig. 6.8, it can be observed that by controlling the reconfigurable RF selection network to select 4 sub-arrays with a dimension of 4×4 during 4 consecutive OFDM symbols, the array response vector corresponding to a 2×2 equivalent low-dimensional fully-digital array can be constructed.

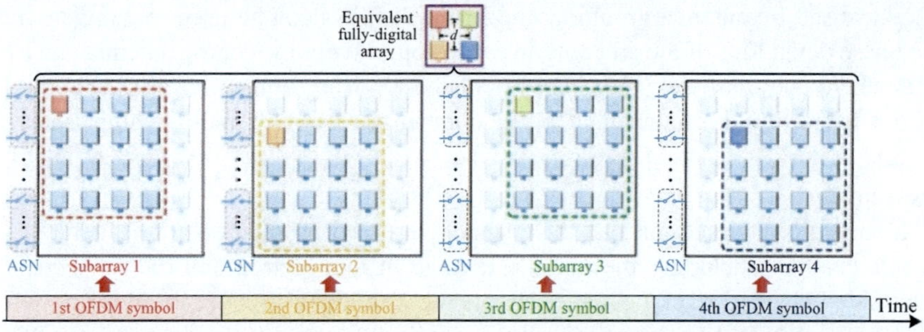

Fig. 6.8 The schematic diagram of the sub-array selection in the initial angle estimation stage [146], in which the different antenna connection modes are formed by controlling the switch state of the ASNs in the reconfigurable RF selection network. Taking a 5×5 UPA as an example, it can be divided into 4 sub-arrays with a size of 4×4, and the interval between each sub-array is the width of an antenna. The same RF link can sequentially select the designed sub-arrays in 4 consecutive OFDM symbols to receive signals, and these received signals are equivalent to that received by the low-dimensional fully-digital array with size being 2×2 and antenna spacing being d (i.e., a critical spacing)

Specifically, the angle estimation at the BS needs $I_{\rm BS}$ OFDM symbols, and the designed antenna connection pattern (i.e., the corresponding selected sub-array) is adopted in each OFDM symbol. First, by exploiting the rough angle estimates obtained at the aircraft and the BS, we can design the analog precoding and combining vectors at the transceiver, namely $\boldsymbol{p}_{{\rm RF},l}$ and $\boldsymbol{q}_{{\rm RF},l}^{[m]}(1\leq l\leq L,\ 1\leq m\leq I_{\rm BS})$, respectively. For $\boldsymbol{p}_{{\rm RF},l}$, we can initialize $\boldsymbol{p}_{{\rm RF},l}=\boldsymbol{0}_{N_{\rm AC}}$, then let $\boldsymbol{p}_{{\rm RF},l\{\mathcal{I}_{{\rm AC},l}\}}=\frac{1}{\sqrt{M_{\rm AC}}}\boldsymbol{a}_{\rm AC}(\widetilde{\mu}_l^{\rm AC},\widetilde{\nu}_l^{\rm AC})_{\{\mathcal{I}_{{\rm AC},l}\}}$, where $\mathcal{I}_{{\rm AC},l}$ with $M_{\rm AC}=|\mathcal{I}_{{\rm AC},l}|_c$ denotes the antenna index set of the sub-array corresponding to the lth BS in the entire aircraft array. To design $\{\boldsymbol{q}_{{\rm RF},l}^{[m]}\}_{m=1}^{I_{\rm BS}}$, we can divide the UM-MIMO array of the BS into $I_{\rm BS}=I_{\rm BS}^{\rm h}I_{\rm BS}^{\rm v}$ sub-arrays to generate the array response vectors for $I_{\rm BS}^{\rm h}\times I_{\rm BS}^{\rm v}$ equivalent low-dimensional fully-digital array, where the dimension of sub-array is $\bar{M}_{\rm BS}^{\rm h}\times\bar{M}_{\rm BS}^{\rm v}$ ($\bar{M}_{\rm BS}^{\rm h}=N_{\rm BS}^{\rm h}-I_{\rm BS}^{\rm h}+1$ and $\bar{M}_{\rm BS}^{\rm v}=N_{\rm BS}^{\rm v}-I_{\rm BS}^{\rm v}+1$) and the total number of antennas in sub-array is $\bar{M}_{\rm BS}=\bar{M}_{\rm BS}^{\rm h}\bar{M}_{\rm BS}^{\rm v}$. Define $m=(i_{\rm BS}^{\rm v}-1)I_{\rm BS}^{\rm h}+i_{\rm BS}^{\rm h}$ with $i_{\rm BS}^{\rm h}$ and $i_{\rm BS}^{\rm v}$ denoting the $(i_{\rm BS}^{\rm h},i_{\rm BS}^{\rm v})$th $(1\leq i_{\rm BS}^{\rm h}\leq I_{\rm BS}^{\rm h},\ 1\leq i_{\rm BS}^{\rm v}\leq I_{\rm BS}^{\rm v})$ sub-array, the antenna index set of the mth $(1\leq m\leq I_{\rm BS})$ sub-array (corresponding to the mth OFDM symbol) can be denoted as $\mathcal{I}_{\rm BS}^{[m]}$ with $\bar{M}_{\rm BS}=|\mathcal{I}_{\rm BS}^{[m]}|_c$. After that, we initialize $\boldsymbol{q}_{{\rm RF},l}^{[m]}=\boldsymbol{0}_{N_{\rm BS}}$ and let $\boldsymbol{q}_{{\rm RF},l\{\mathcal{I}_{\rm BS}^{[m]}\}}^{[m]}=\frac{1}{\sqrt{\bar{M}_{\rm BS}}}\boldsymbol{a}_{\rm BS}(\widetilde{\mu}_l^{\rm BS},\widetilde{\nu}_l^{\rm BS})_{\{\mathcal{I}_{\rm BS}^{[1]}\}}$.

According to Eq. (6.1), at the lth $(1\leq l\leq L)$ BS, the received signals $y_{{\rm UL},l}^{[m]}[k_l]$ at the k_lth subcarrier of the mth OFDM symbol can be expressed as

$$y_{{\rm UL},l}^{[m]}[k_l]=\sqrt{P_l}(\boldsymbol{q}_{{\rm RF},l}^{[m]})^{\rm H}\widetilde{\boldsymbol{H}}_{{\rm UL},l}^{'[m]}[k_l]\boldsymbol{p}_{{\rm RF},l}s_{{\rm UL},l}^{[m]}[k_l]+n_{{\rm UL},l}^{[m]}[k_l], \tag{6.14}$$

where $k_l\in\mathcal{K}_l$, $1\leq m\leq I_{\rm BS}$, $\widetilde{\boldsymbol{H}}_{{\rm UL},l}^{'[m]}[k_l]$ denotes the channel matrix through the GTTDU module and compensation of rough Doppler shift estimation, $s_{{\rm UL},l}^{[m]}[k_l]$ and $n_{{\rm UL},l}^{[m]}[k_l]$

denote the transmitted pilot and the noise, respectively. Stack the received signals at all K_l subcarriers as $\boldsymbol{y}_{\mathrm{UL},l}^{[m]} \in \mathbb{C}^{K_l}$, then substitute the uplink channel matrix in Eq. (6.10) into $\boldsymbol{y}_{\mathrm{UL},l}^{[m]} \in \mathbb{C}^{K_l}$, we can obtain

$$\begin{aligned}\boldsymbol{y}_{\mathrm{UL},l}^{[m]} &= \left[y_{\mathrm{UL},l}^{[m]}[\{\mathcal{K}_l\}_1] \cdots y_{\mathrm{UL},l}^{[m]}[\{\mathcal{K}_l\}_{K_l}]\right]^{\mathrm{T}} \\ &= \sqrt{P_l G_l}\alpha_l (\boldsymbol{q}_{\mathrm{RF},l}^{[m]})^{\mathrm{H}} \boldsymbol{A}_{\mathrm{UL},l} \boldsymbol{p}_{\mathrm{RF},l} \boldsymbol{s}_{\mathrm{UL},l}^{[m]} \circ \widetilde{\boldsymbol{y}}_{\mathrm{UL},l}^{[m]} + \boldsymbol{n}_{\mathrm{UL},l}^{[m]},\end{aligned} \tag{6.15}$$

where $\boldsymbol{s}_{\mathrm{UL},l}^{[m]} = \left[s_{\mathrm{UL},l}^{[m]}[\{\mathcal{K}_l\}_1] \cdots s_{\mathrm{UL},l}^{[m]}[\{\mathcal{K}_l\}_{K_l}]\right]^{\mathrm{T}} \in \mathbb{C}^{K_l}$, $\widetilde{\boldsymbol{y}}_{\mathrm{UL},l}^{[m]}$ denotes the error vector including residual beam squint caused by inaccurate prior information, and $\boldsymbol{n}_{\mathrm{UL},l}^{[m]}$ denotes the noise. Furthermore, we assume that the same pilot are transmitted in consecutive I_{BS} OFDM symbols, i.e., for $\forall m$, $s_{\mathrm{UL},l}[k_l] = s_{\mathrm{UL},l}^{[m]}[k_l]$ and $\boldsymbol{s}_{\mathrm{UL},l} = \boldsymbol{s}_{\mathrm{UL},l}^{[m]}$. Then, stack the transposition of $\{\boldsymbol{y}_{\mathrm{UL},l}^{[m]}\}_{m=1}^{I_{\mathrm{BS}}}$ from I_{BS} OFDM symbols, we have

$$\begin{aligned}\boldsymbol{Y}_{\mathrm{UL},l} &= \left[\boldsymbol{y}_{\mathrm{UL},l}^{[1]} \cdots \boldsymbol{y}_{\mathrm{UL},l}^{[I_{\mathrm{BS}}]}\right]^{\mathrm{T}} \\ &= \sqrt{P_l G_l}\alpha_l \left(\boldsymbol{Q}_{\mathrm{RF},l}^{\mathrm{H}} \boldsymbol{A}_{\mathrm{UL},l} \boldsymbol{p}_{\mathrm{RF},l} \boldsymbol{s}_{\mathrm{UL},l}^{\mathrm{T}}\right) \circ \widetilde{\boldsymbol{Y}}_{\mathrm{UL},l} + \boldsymbol{N}_{\mathrm{UL},l},\end{aligned} \tag{6.16}$$

where $\boldsymbol{Q}_{\mathrm{RF},l} = \left[\boldsymbol{q}_{\mathrm{RF},l}^{[1]} \cdots \boldsymbol{q}_{\mathrm{RF},l}^{[I_{\mathrm{BS}}]}\right] \in \mathbb{C}^{N_{\mathrm{BS}} \times I_{\mathrm{BS}}}$ and $\widetilde{\boldsymbol{Y}}_{\mathrm{UL},l} = \left[\widetilde{\boldsymbol{y}}_{\mathrm{UL},l}^{[1]} \cdots \widetilde{\boldsymbol{y}}_{\mathrm{UL},l}^{[I_{\mathrm{BS}}]}\right]$ denote the analog combining matrix and the residual beam squint matrix, respectively, and $\boldsymbol{N}_{\mathrm{UL},l}$ denotes the noise matrix.

By exploiting the analog combining matrix $\boldsymbol{Q}_{\mathrm{RF},l}$, we can obtain the array response vector of equivalent low-dimensional fully-digital array. Specifically, compared to $(\boldsymbol{q}_{\mathrm{RF},l}^{[1]})^{\mathrm{H}} \boldsymbol{a}_{\mathrm{BS}}(\mu_l^{\mathrm{BS}}, \nu_l^{\mathrm{BS}})$ with $m = 1$ in Eq. (6.15), when $2 \le m \le I_{\mathrm{BS}}$ and $m = (i_{\mathrm{BS}}^{\mathrm{v}} - 1)I_{\mathrm{BS}}^{\mathrm{h}} + i_{\mathrm{BS}}^{\mathrm{h}} (1 \le i_{\mathrm{BS}}^{\mathrm{h}} \le I_{\mathrm{BS}}^{\mathrm{h}}, 1 \le i_{\mathrm{BS}}^{\mathrm{v}} \le I_{\mathrm{BS}}^{\mathrm{v}})$, $(\boldsymbol{q}_{\mathrm{RF},l}^{[m]})^{\mathrm{H}} \boldsymbol{a}_{\mathrm{BS}}(\mu_l^{\mathrm{BS}}, \nu_l^{\mathrm{BS}})$ is equivalent to multiplying an additional phase shift $e^{\mathrm{j}\left((i_{\mathrm{BS}}^{\mathrm{h}}-1)\mu_l^{\mathrm{BS}} + (i_{\mathrm{BS}}^{\mathrm{v}}-1)\nu_l^{\mathrm{BS}}\right)}$. Then, we can combine these regular phase shifts to form the effective array response vector of an $I_{\mathrm{BS}}^{\mathrm{h}} \times I_{\mathrm{BS}}^{\mathrm{v}}$ equivalent fully-digital array, i.e., $\bar{\bar{\boldsymbol{a}}}_{\mathrm{BS}}(\mu_l^{\mathrm{BS}}, \nu_l^{\mathrm{BS}}) = \boldsymbol{a}_{\mathrm{v}}(\nu_l^{\mathrm{BS}}, I_{\mathrm{BS}}^{\mathrm{v}}) \otimes \boldsymbol{a}_{\mathrm{h}}(\mu_l^{\mathrm{BS}}, I_{\mathrm{BS}}^{\mathrm{h}}) \in \mathbb{C}^{I_{\mathrm{BS}}}$. Therefore, the uplink received signal matrix $\boldsymbol{Y}_{\mathrm{UL},l}$ in Eq. (6.16) can be rewritten as

$$\boldsymbol{Y}_{\mathrm{UL},l} = \gamma_{\mathrm{UL},l} \left(\bar{\bar{\boldsymbol{a}}}_{\mathrm{BS}}(\mu_l^{\mathrm{BS}}, \nu_l^{\mathrm{BS}}) \boldsymbol{s}_{\mathrm{UL},l}^{\mathrm{T}}\right) \circ \widetilde{\boldsymbol{Y}}_{\mathrm{UL},l} + \boldsymbol{N}_{\mathrm{UL},l}, \tag{6.17}$$

where $\gamma_{\mathrm{UL},l} = \sqrt{P_l G_l}\alpha_l (\boldsymbol{q}_{\mathrm{RF},l}^{[1]})^{\mathrm{H}} \boldsymbol{A}_{\mathrm{UL},l} \boldsymbol{p}_{\mathrm{RF},l}$ denotes the uplink equivalent channel gain after beam alignment.

For the received signal model in Eq. (6.17), we propose the following prior-aided iterative angle estimation algorithm in this subsection. In the first iteration, i.e., $i_{\mathrm{BS}} = 1$, by employing the two-dimensional unitary (TDU)-ESPRIT algorithm [101, 106] on $\boldsymbol{Y}_{\mathrm{UL},l}$, we can obtain the azimuth and elevation angle estimates at the lth BS as $\widehat{\theta}_l^{(i_{\mathrm{BS}})}$ and $\widehat{\varphi}_l^{(i_{\mathrm{BS}})}$, respectively, and the corresponding horizontal and vertical virtual angles as $\widehat{\mu}_l^{(i_{\mathrm{BS}})}$ and $\widehat{\nu}_l^{(i_{\mathrm{BS}})}$, respectively. To minimize the interference of $\widetilde{\boldsymbol{Y}}_{\mathrm{UL},l}$, the above estimated angles can be employed to compensate $\boldsymbol{Y}_{\mathrm{UL},l}$ in the subsequent iterations ($i_{\mathrm{BS}} \geq 2$) for enhanced performance. Specifically, for the i_{BS}th iteration, based on the initial rough angle estimates $\widetilde{\mu}_l^{\mathrm{BS}}$ and $\widetilde{\nu}_l^{\mathrm{BS}}$, as well as $\widehat{\mu}_l^{(i_{\mathrm{BS}}-1)}$ and $\widehat{\nu}_l^{(i_{\mathrm{BS}}-1)}$ obtained in ($i_{\mathrm{BS}}-1$)th iteration, we can design an angle compensation matrix $\widetilde{\boldsymbol{Y}}_{\mathrm{UL},l}^{(i_{\mathrm{BS}}-1)} = \left[\widetilde{\boldsymbol{y}}_{\mathrm{UL},l}^{(i_{\mathrm{BS}}-1)}[\{\mathcal{K}_l\}_1] \cdots \widetilde{\boldsymbol{y}}_{\mathrm{UL},l}^{(i_{\mathrm{BS}}-1)}[\{\mathcal{K}_l\}_{K_l}]\right]$, whose k_lth column vector $\widetilde{\boldsymbol{y}}_{\mathrm{UL},l}^{(i_{\mathrm{BS}}-1)}[k_l] \in \mathbb{C}^{I_{\mathrm{BS}}}$ is given as

$$\widetilde{\boldsymbol{y}}_{\mathrm{UL},l}^{(i_{\mathrm{BS}}-1)}[k_l] = \left(\bar{\boldsymbol{a}}_{\mathrm{v}}(\widetilde{\nu}_l^{\mathrm{BS}}, I_{\mathrm{BS}}^{\mathrm{v}}, k_l) \otimes \bar{\boldsymbol{a}}_{\mathrm{h}}(\widetilde{\mu}_l^{\mathrm{BS}}, I_{\mathrm{BS}}^{\mathrm{h}}, k_l)\right)^* \circ \left(\bar{\boldsymbol{a}}_{\mathrm{v}}(\widehat{\nu}_l^{(i_{\mathrm{BS}}-1)}, I_{\mathrm{BS}}^{\mathrm{v}}, k_l) \otimes \bar{\boldsymbol{a}}_{\mathrm{h}}(\widehat{\mu}_l^{(i_{\mathrm{BS}}-1)}, I_{\mathrm{BS}}^{\mathrm{h}}, k_l)\right). \quad (6.18)$$

Therefore, the signal matrix $\boldsymbol{Y}_{\mathrm{UL},l}^{(i_{\mathrm{BS}})} = \left(\widetilde{\boldsymbol{Y}}_{\mathrm{UL},l}^{(i_{\mathrm{BS}}-1)}\right)^* \circ \boldsymbol{Y}_{\mathrm{UL},l}$ after the compensation of $\widetilde{\boldsymbol{Y}}_{\mathrm{UL},l}^{(i_{\mathrm{BS}}-1)}$ can be written as

$$\boldsymbol{Y}_{\mathrm{UL},l}^{(i_{\mathrm{BS}})} = \gamma_{\mathrm{UL},l}\left(\bar{\bar{\boldsymbol{a}}}_{\mathrm{BS}}(\mu_l^{\mathrm{BS}}, \nu_l^{\mathrm{BS}})\boldsymbol{s}_{\mathrm{UL},l}^{\mathrm{T}}\right) \circ \left(\widetilde{\boldsymbol{Y}}_{\mathrm{UL},l} \circ \left(\widetilde{\boldsymbol{Y}}_{\mathrm{UL},l}^{(i_{\mathrm{BS}}-1)}\right)^*\right) + \boldsymbol{N}_{\mathrm{UL},l}^{(i_{\mathrm{BS}})}, \quad (6.19)$$

where $\boldsymbol{N}_{\mathrm{UL},l}^{(i_{\mathrm{BS}})}$ denotes the noise matrix after compensation. Then, the refined azimuth and elevation angles can be estimated by applying the TDU-ESPRIT algorithm on $\{\boldsymbol{Y}_{\mathrm{UL},l}^{(i_{\mathrm{BS}})}\}_{l=1}^{L}$. This process is iterated until reaching the maximum number of iterations $i_{\mathrm{BS}}^{\max}$. Finally, the azimuth and elevation angle estimates at the lth BS can be expressed as $\widehat{\theta}_l^{\mathrm{BS}} = \widehat{\theta}_l^{(i_{\mathrm{BS}}^{\max})}$ and $\widehat{\varphi}_l^{\mathrm{BS}} = \widehat{\varphi}_l^{(i_{\mathrm{BS}}^{\max})}$, respectively, and the corresponding virtual angels can be expressed as $\widehat{\mu}_l^{\mathrm{BS}} = \widehat{\mu}_l^{(i_{\mathrm{BS}}^{\max})}$ and $\widehat{\nu}_l^{\mathrm{BS}} = \widehat{\nu}_l^{(i_{\mathrm{BS}}^{\max})}$, respectively. The proposed prior-aided iterative angle estimation algorithm is summarized in Algorithm 6.1, which can effectively solve the beam squint effect by iteratively constructing the angle compensation matrix.

Remark 6.1 According to the analysis above, in the initial angle estimation stage, the specific antenna connection mode can be selected in the reconfigurable RF selection network for the desired sub-array to obtain equivalent low-dimensional fully-digital array. Then, we can acquire accurate angle estimates by robust array signal processing techniques. On the other hand, each selected $\bar{M}_{\mathrm{BS}}^{\mathrm{h}} \times \bar{M}_{\mathrm{BS}}^{\mathrm{v}}$ sub-array is large enough to reap sufficient FD beamforming gain through rough angle estimates, improving received SNR to combat the serious path loss in the long-distance THz communication links.

6.4.1.2 Accurate Angle Estimation at the Aircraft

According to the channel reciprocity between the downlink and uplink, the accurate angle estimation at the aircraft in the downlink is similar to that at the BS in the uplink. In this stage, the BS employs the accurate azimuth/elevation angle estimates to design the analog precoding vector, enhancing the received SNR at the aircraft through the

Algorithm 6.1: Pilot-Aided Iterative Angle Estimation Algorithm

Input: The rough virtual angle estimates $\{\widetilde{\mu}_l^{\mathrm{BS}}, \widetilde{\nu}_l^{\mathrm{BS}}, \widetilde{\mu}_l^{\mathrm{AC}}, \widetilde{\nu}_l^{\mathrm{AC}}\}$, the transmitted pilot signals $\boldsymbol{s}_{\mathrm{UL},l}$, the maximum number of iteration $i_{\mathrm{BS}}^{\max}$, and other parameters $\{N_{\mathrm{AC}}, M_{\mathrm{AC}}, \bar{M}_{\mathrm{BS}}, I_{\mathrm{BS}}, I_{\mathrm{BS}}^{\mathrm{h}}, I_{\mathrm{BS}}^{\mathrm{v}}, K_l\}$

Output: The accurate azimuth/elevation angle estimates $\{\widehat{\theta}_l^{\mathrm{BS}}, \widehat{\varphi}_l^{\mathrm{BS}}\}$, and the corresponding virtual angle estimates $\{\widehat{\mu}_l^{\mathrm{BS}}, \widehat{\nu}_l^{\mathrm{BS}}\}$

```
%% Sub-array selection and pilot transmission %%
Ensure the antenna indices of sub-array at the aircraft I_{AC,l};
Initialize the analog precoding vector p_{RF,l} = 0_{N_AC}, and let
  p_{RF,l{I_{AC,l}}} = (1/√M_AC) a_AC(μ̃_l^AC, ν̃_l^AC)_{I_{AC,l}};
for m = 1, ···, I_BS do
    Ensure the antenna indices of the selected sub-array at the BS I_BS^[m];
    Initialize the analog combing vector q_{RF,l}^[m] = 0_{N_BS}, and let
      q^[m]_{RF,l{I_BS^[m]}} = (1/√M̄_BS) a_BS(μ̃_l^BS, ν̃_l^BS)_{I_BS^[1]};
    Transmit the pilot signals s_{UL,l} and obtain y_{UL,l}^[m] in Eq. (6.15);
end
Stack {y_{UL,l}^[m]}_{m=1}^{I_BS} as Y_{UL,l} = [y_{UL,l}^[1] ··· y_{UL,l}^[I_BS]]^T in Eqs. (6.16) and (6.17);
%% Pilot-aided iterative angle estimation %%
for i_BS = 1, ···, i_BS^max do
    if i_BS = 1 then
        Apply the TDU-ESPRIT algorithm on the received signal matrix Y_{UL,l};
        Obtain {θ̂_l^(i_BS), φ̂_l^(i_BS)} and {μ̂_l^(i_BS), ν̂_l^(i_BS)} in the first iteration;
    else
        Construct the angle compensation matrix Ỹ_{UL,l}^(i_BS−1), whose k_l th column is
          ỹ_{UL,l}^(i_BS−1)[k_l] in Eq. (6.18);
        Obtain the compensated signal matrix Y_{UL,l}^(i_BS) = (Ỹ_{UL,l}^(i_BS−1))^* ∘ Y_{UL,l} in Eq.
          (6.19);
        Apply the TDU-ESPRIT algorithm on Y_{UL,l}^(i_BS);
        Obtain {θ̂_l^(i_BS), φ̂_l^(i_BS)} and {μ̂_l^(i_BS), ν̂_l^(i_BS)} in the i_BS th iteration;
    end
end
Return: θ̂_l^BS = θ̂_l^(i_BS^max), φ̂_l^BS = φ̂_l^(i_BS^max), μ̂_l^BS = μ̂_l^(i_BS^max), and ν̂_l^BS = ν̂_l^(i_BS^max)
```

beam alignment. Meanwhile, reducing the delay-beam squint effects by optimizing the GTTDU module at the BS. Specifically, the aircraft utilizes $I_{\mathrm{AC}}=I_{\mathrm{AC}}^{\mathrm{h}}I_{\mathrm{AC}}^{\mathrm{v}}$ OFDM symbols for accurate angle estimation, where the size of equivalent low-dimensional fully-digital array is $I_{\mathrm{AC}}^{\mathrm{h}}\times I_{\mathrm{AC}}^{\mathrm{v}}$. The analog precoding vector $\boldsymbol{f}_{\mathrm{RF},l}=\boldsymbol{a}_{\mathrm{BS}}(\widehat{\mu}_l^{\mathrm{BS}},\widehat{\nu}_l^{\mathrm{BS}})$ at the lth BS is designed by exploiting estimated $\{\widehat{\mu}_l^{\mathrm{BS}},\widehat{\nu}_l^{\mathrm{BS}}\}_{l=1}^{L}$. For the lth sub-array at the aircraft (corresponding to the lth BS), the selected antenna indices obtained from the reconfigurable RF selection network in the nth $(1\leq n\leq I_{\mathrm{AC}})$ OFDM symbol can be expressed as $\mathcal{I}_{\mathrm{AC},l}^{[n]}$ with $\bar{M}_{\mathrm{AC}}=|\mathcal{I}_{\mathrm{AC},l}^{[n]}|_c$. We can initialize the analog combing vector at the aircraft as $\boldsymbol{w}_{\mathrm{RF},l}^{[n]}=\boldsymbol{0}_{N_{\mathrm{AC}}}$, then let $\boldsymbol{w}_{\mathrm{RF},l\{\mathcal{I}_{\mathrm{AC},l}^{[n]}\}}^{[n]}=\frac{1}{\sqrt{\bar{M}_{\mathrm{AC}}}}\boldsymbol{a}_{\mathrm{AC}}(\widetilde{\mu}_l^{\mathrm{AC}},\widetilde{\nu}_l^{\mathrm{AC}})_{\{\mathcal{I}_{\mathrm{AC},l}^{[1]}\}}$.

Based on Eq. (6.2), for the lth RF chain of the aircraft, the received signal $y_{\mathrm{DL},l}^{[n]}[k_l]$ from the lth BS at the k_lth $(k_l\in\mathcal{K}_l)$ subcarrier of the nth $(1\leq n\leq I_{\mathrm{AC}})$ OFDM symbol can be written as

$$y_{\mathrm{DL},l}^{[n]}[k_l]=\sqrt{P_l}(\boldsymbol{w}_{\mathrm{RF},l}^{[n]})^{\mathrm{H}}\widetilde{\boldsymbol{H}}_{\mathrm{DL},l}^{'[n]}[k_l]\boldsymbol{f}_{\mathrm{RF},l}s_{\mathrm{DL},l}^{[n]}[k_l]+n_{\mathrm{DL},l}^{[n]}[k_l], \tag{6.20}$$

where $\widetilde{\boldsymbol{H}}_{\mathrm{DL},l}^{'[n]}[k_l]$ denotes the compensated downlink channel matrix, and $s_{\mathrm{DL},l}^{[n]}[k_l]$ and $n_{\mathrm{DL},l}^{[n]}[k_l]$ denote the transmitted pilot and the noise, respectively. By stacking the signals at all K_l subcarriers of I_{AC} OFDM symbols, the obtained downlink received signal matrix is given as

$$\boldsymbol{Y}_{\mathrm{DL},l}=\sqrt{P_lG_l}\alpha_le^{\mathrm{j}\pi f_s\tau_l}\left(\bar{\boldsymbol{W}}_{\mathrm{RF},l}^{\mathrm{H}}\boldsymbol{A}_{\mathrm{DL},l}\boldsymbol{f}_{\mathrm{RF},l}\underbrace{\left(\boldsymbol{a}_\tau(\mu_l^\tau,K_l)\circ\boldsymbol{s}_{\mathrm{DL},l}\right)^{\mathrm{T}}}_{\bar{\boldsymbol{s}}_{\mathrm{DL},l}}\right)\circ\widetilde{\boldsymbol{Y}}_{\mathrm{DL},l}+\boldsymbol{N}_{\mathrm{DL},l}, \tag{6.21}$$

where $\bar{\boldsymbol{W}}_{\mathrm{RF},l}=\left[\boldsymbol{w}_{\mathrm{RF},l}^{[1]}\cdots\boldsymbol{w}_{\mathrm{RF},l}^{[I_{\mathrm{AC}}]}\right]\in\mathbb{C}^{N_{\mathrm{AC}}\times I_{\mathrm{AC}}}$ and $\widetilde{\boldsymbol{Y}}_{\mathrm{DL},l}$ denote the analog combing matrix and the residual beam squint matrix, respectively, $\boldsymbol{s}_{\mathrm{DL},l}=\boldsymbol{s}_{\mathrm{DL},l}^{[n]}=\left[s_{\mathrm{DL},l}^{[n]}[\{\mathcal{K}_l\}_1]\cdots s_{\mathrm{DL},l}^{[n]}[\{\mathcal{K}_l\}_{K_l}]\right]^{\mathrm{T}}\in\mathbb{C}^{K_l}$ (for $\forall$ n), and $\boldsymbol{N}_{\mathrm{DL},l}$ denotes the noise matrix. Besides, the steering vector $\boldsymbol{a}_\tau(\mu_l^\tau,K_l)=\left[e^{\mathrm{j}(\{\mathcal{K}_l\}_1-1)\mu_l^\tau}\ e^{\mathrm{j}(\{\mathcal{K}_l\}_2-1)\mu_l^\tau}\ \ldots\ e^{\mathrm{j}(\{\mathcal{K}_l\}_{K_l}-1)\mu_l^\tau}\right]^{\mathrm{T}}$ corresponding to the path delay τ_l is defined in Eq. (6.21), and $\mu_l^\tau=-2\pi f_s\tau_l/K$ denotes the virtual delay. Then, similar to Eq. (6.17), $\boldsymbol{Y}_{\mathrm{DL},l}$ can be rewritten as

$$\boldsymbol{Y}_{\mathrm{DL},l}=\gamma_{\mathrm{DL},l}\left(\bar{\bar{\boldsymbol{a}}}_{\mathrm{AC}}(\mu_l^{\mathrm{AC}},\nu_l^{\mathrm{AC}})\bar{\boldsymbol{s}}_{\mathrm{DL},l}^{\mathrm{T}}\right)\circ\widetilde{\boldsymbol{Y}}_{\mathrm{DL},l}+\boldsymbol{N}_{\mathrm{DL},l}, \tag{6.22}$$

where $\gamma_{\mathrm{DL},l}=\sqrt{P_lG_l}\alpha_le^{\mathrm{j}\pi f_s\tau_l}(\boldsymbol{w}_{\mathrm{RF},l}^{[1]})^{\mathrm{H}}\boldsymbol{A}_{\mathrm{DL},l}\boldsymbol{f}_{\mathrm{RF},l}$ denotes the equivalent downlink channel gain after beam alignment, and $\bar{\bar{\boldsymbol{a}}}_{\mathrm{AC}}(\mu_l^{\mathrm{AC}},\nu_l^{\mathrm{AC}})=\boldsymbol{a}_{\mathrm{v}}(\nu_l^{\mathrm{AC}},I_{\mathrm{AC}}^{\mathrm{v}})\otimes\boldsymbol{a}_{\mathrm{h}}(\mu_l^{\mathrm{AC}},I_{\mathrm{AC}}^{\mathrm{h}})\in\mathbb{C}^{I_{\mathrm{AC}}}$ denotes the effective array response vector for the equivalent low-dimensional fully-digital array of the lth sub-array at the aircraft. For Eq. (6.22), we can apply the proposed pilot-aided iterative angle estimation algorithm again to acquire accurate angle estimates. By replacing the input parameters $\{\widetilde{\mu}_l^{\mathrm{BS}},\widetilde{\nu}_l^{\mathrm{BS}},\{\mathcal{I}_{\mathrm{BS}}^{[m]}\}_{m=1}^{I_{\mathrm{BS}}},\boldsymbol{s}_{\mathrm{UL},l},\bar{M}_{\mathrm{BS}},I_{\mathrm{BS}},I_{\mathrm{BS}}^{\mathrm{h}},I_{\mathrm{BS}}^{\mathrm{v}},i_{\mathrm{BS}},i_{\mathrm{BS}}^{\max}\}$ with $\{\widehat{\mu}_l^{\mathrm{BS}},\widehat{\nu}_l^{\mathrm{BS}},\{\mathcal{I}_{\mathrm{AC},l}^{[n]}\}_{n=1}^{I_{\mathrm{AC}}},\boldsymbol{s}_{\mathrm{DL},l},\bar{M}_{\mathrm{AC}},I_{\mathrm{AC}},I_{\mathrm{AC}}^{\mathrm{h}},I_{\mathrm{AC}}^{\mathrm{v}},i_{\mathrm{AC}},i_{\mathrm{AC}}^{\max}\}$, we can obtain the accurate azimuth and elevation angle estimates corresponding to the lth BS at the aircraft, as well as the virtual angles, denoted as $\widehat{\theta}_l^{\mathrm{AC}} = \widehat{\theta}_l^{(i_{\mathrm{AC}}^{\max})}$, $\widehat{\varphi}_l^{\mathrm{AC}} = \widehat{\varphi}_l^{(i_{\mathrm{AC}}^{\max})}$, $\widehat{\mu}_l^{\mathrm{AC}} = \widehat{\mu}_l^{(i_{\mathrm{AC}}^{\max})}$, and $\widehat{\upsilon}_l^{\mathrm{AC}} = \widehat{\upsilon}_l^{(i_{\mathrm{AC}}^{\max})}$.

6.4.2 *Accurate Doppler Shift Estimation Under Doppler-Squint Effect*

According to the acquired accurate angle estimates, the beam alignment can be achieved at the aircraft by designing the analog combing vectors of L sub-arrays as follows: For the lth vector, initialize $\boldsymbol{w}_{\mathrm{RF},l} = \boldsymbol{0}_{N_{\mathrm{AC}}}$, then let $\boldsymbol{w}_{\mathrm{RF},l\{\mathcal{I}_{\mathrm{AC},l}\}} = \frac{1}{\sqrt{M_{\mathrm{AC}}}} \boldsymbol{a}_{\mathrm{AC}}(\widehat{\mu}_l^{\mathrm{AC}}, \widehat{\upsilon}_l^{\mathrm{AC}})_{\{\mathcal{I}_{\mathrm{AC},l}\}}$. Meanwhile, the GTTDU module at the aircraft can be optimized based on the angle estimates to further reduce the delay-beam squint effect. Since the initial estimates of the Doppler shift is not precise enough for the data transmission, this subsection discusses about the accurate Doppler shift estimates by using N_{do} OFDM symbols in the downlink. To ensure that the aircraft can quasi-statically observe the effective channels within multiple OFDM symbols, the transmitter at the BS still needs to perform rough pre-compensation on the transmitted signals in this stage.

For the lth RF chain of the aircraft, the received signal $y_{\mathrm{do},l}^{[\bar{m}]}[k_l]$ at the k_lth ($k_l \in \mathcal{K}_l$) subcarrier of the $\bar{m}$th ($1 \le \bar{m} \le N_{\mathrm{do}}$) OFDM symbol can be written as

$$\begin{aligned} y_{\mathrm{do},l}^{[\bar{m}]}[k_l] &= \sqrt{P_l}\, \boldsymbol{w}_{\mathrm{RF},l}^{\mathrm{H}} \widetilde{\boldsymbol{H}}_{\mathrm{DL},l}^{'[\bar{m}]}[k_l] \boldsymbol{f}_{\mathrm{RF},l} s_{\mathrm{do},l}^{[\bar{m}]}[k_l] + n_{\mathrm{do},l}^{[\bar{m}]}[k_l] \\ &= \underbrace{\sqrt{P_l G_l}\, \alpha_l e^{\mathrm{j}\pi f_s \tau_l}}_{\gamma_{\mathrm{do},l}} e^{\mathrm{j}2\pi \Delta\widetilde{\psi}_{l,k_l} (\bar{m}-1) T_{\mathrm{sym}}} \underbrace{\boldsymbol{w}_{\mathrm{RF},l}^{\mathrm{H}} \widetilde{\boldsymbol{A}}_{\mathrm{DL},l}[k_l] \boldsymbol{f}_{\mathrm{RF},l} e^{\mathrm{j}(k_l-1)\mu_l^{\tau}} s_{\mathrm{do},l}[k_l]}_{\bar{s}_{\mathrm{do},l}[k_l]} \\ &\quad + n_{\mathrm{do},l}^{[\bar{m}]}[k_l], \end{aligned} \tag{6.23}$$

where $\Delta\widetilde{\psi}_{l,k_l} = \psi_{l,k_l} - \widetilde{\psi}_{l,k_l}$ denotes the residual Doppler shift at the k_lth subcarrier after the rough compensation of $\widetilde{\psi}_{l,k_l}$, and $s_{\mathrm{do},l}[k_l] = s_{\mathrm{do},l}^{[\bar{m}]}[k_l]$ (for $\forall \bar{m}$) and $n_{\mathrm{do},l}^{[\bar{m}]}[k_l]$ denote the transmitted pilot and the noise, respectively. Since $\Delta\widetilde{\psi}_{l,k_l}$ is too small to effectively estimates the Doppler shift through a small number of OFDM symbols, we can remove the compensated phase difference $e^{-\mathrm{j}2\pi \widetilde{\psi}_{l,k_l}(\bar{m}-1)T_{\mathrm{sym}}}$ from $y_{\mathrm{do},l}^{[\bar{m}]}[k_l]$ in Eq. (6.23) and obtain

$$\bar{y}_{\mathrm{do},l}^{[\bar{m}]}[k_l] = \gamma_{\mathrm{do},l} e^{\mathrm{j}(\bar{m}-1)\nu_l^{\psi}} \underbrace{\bar{s}_{\mathrm{do},l}[k_l]\, e^{\mathrm{j}\frac{2\pi f_s \upsilon_l}{c}\left(\frac{k_l-1}{K} - \frac{1}{2}\right)(\bar{m}-1)T_{\mathrm{sym}}}}_{\widetilde{y}_{\mathrm{do},l}^{[\bar{m}]}[k_l](\upsilon_l)} + \bar{n}_{\mathrm{do},l}^{[\bar{m}]}[k_l], \tag{6.24}$$

where $\nu_l^{\psi} = 2\pi \psi_{z,l} T_{\mathrm{sym}}$ denotes the virtual Doppler shift, and $\widetilde{y}_{\mathrm{do},l}^{[\bar{m}]}[k_l]$ and $\bar{n}_{\mathrm{do},l}^{[\bar{m}]}[k_l]$ denote the Doppler shift and the noise, respectively. By stacking the signals at all K_l subcarriers of N_{do} OFDM symbols, the received signal matrix $\boldsymbol{Y}_{\mathrm{do},l} \in \mathbb{C}^{N_{\mathrm{do}} \times K_l}$ can be expressed as

$$\boldsymbol{Y}_{\mathrm{do},l} = \gamma_{\mathrm{do},l}\left(\boldsymbol{a}_{\psi}(\nu_l^{\psi}, N_{\mathrm{do}})\bar{\boldsymbol{s}}_{\mathrm{do},l}^{\mathrm{T}}\right) \circ \widetilde{\boldsymbol{Y}}_{\mathrm{do},l}(\underline{v}_l) + \boldsymbol{N}_{\mathrm{do},l}, \tag{6.25}$$

where $\boldsymbol{a}_{\psi}(\nu_l^{\psi}, N_{\mathrm{do}}) = \left[1\ e^{\mathrm{j}\nu_l^{\psi}} \cdots e^{\mathrm{j}(N_{\mathrm{do}}-1)\nu_l^{\psi}}\right]^{\mathrm{T}} \in \mathbb{C}^{N_{\mathrm{do}}}$ denotes the steering vector associated with the Doppler shift, $\widetilde{\boldsymbol{Y}}_{\mathrm{do},l}(\underline{v}_l)$ denotes the Doppler shift matrix whose $(\bar{m}, k_l)$th element is $\{\widetilde{\boldsymbol{Y}}_{\mathrm{do},l}(\underline{v}_l)\}_{\bar{m},k_l} = \widetilde{y}_{\mathrm{do},l}^{[\bar{m}]}[k_l](\underline{v}_l)$, and $\boldsymbol{N}_{\mathrm{do},l}$ denotes the noise matrix.

Based on Eq. (6.25), we propose the pilot-aided iterative Doppler shift estimation algorithm in this subsection. First, by defining the rough Doppler shift estimate at the central carrier frequency as $\widetilde{\psi}_{z,l}$, the initial relative radial velocity can be calculated as $\widehat{\underline{v}}_l^{(0)} = \widetilde{\psi}_{z,l}\lambda_z$. In the i_{do}th iteration, $\widehat{\underline{v}}_l^{(i_{\mathrm{do}}-1)}$ obtained in the last iteration can be employed to construct the Doppler compensation matrix $\widetilde{\boldsymbol{Y}}_{\mathrm{do},l}(\widehat{\underline{v}}_l^{(i_{\mathrm{do}}-1)})$, whose $(\bar{m}, k_l)$th element is $\widetilde{y}_{\mathrm{do},l}^{[\bar{m}]}[k_l](\widehat{\underline{v}}_l^{(i_{\mathrm{do}}-1)})$. Note that this value can be acquired by replacing $\underline{v}_l$ with $\widehat{\underline{v}}_l^{(i_{\mathrm{do}}-1)}$ in $\widetilde{y}_{\mathrm{do},l}^{[\bar{m}]}[k_l](\underline{v}_l)$ of Eq. (6.24). The compensated matrix $\boldsymbol{Y}_{\mathrm{do},l}^{(i_{\mathrm{do}})} = \widetilde{\boldsymbol{Y}}_{\mathrm{do},l}^{*}(\widehat{\underline{v}}_l^{(i_{\mathrm{do}}-1)}) \circ \boldsymbol{Y}_{\mathrm{do},l}$ can be rewritten as

$$\boldsymbol{Y}_{\mathrm{do},l}^{(i_{\mathrm{do}})} = \gamma_{\mathrm{do},l}\left(\boldsymbol{a}_{\psi}(\nu_l^{\psi}, \bar{m})\bar{\boldsymbol{s}}_{\mathrm{do},l}^{\mathrm{T}}\right) \circ \left(\widetilde{\boldsymbol{Y}}_{\mathrm{do},l}(\underline{v}_l) \circ \widetilde{\boldsymbol{Y}}_{\mathrm{do},l}^{*}(\widehat{\underline{v}}_l^{(i_{\mathrm{do}}-1)})\right) + \boldsymbol{N}_{\mathrm{do},l}^{(i_{\mathrm{do}})}, \tag{6.26}$$

where $\boldsymbol{N}_{\mathrm{do},l}^{(i_{\mathrm{do}})}$ denotes the corresponding noise matrix. In the i_{do}th iteration, the Doppler shift at the central carrier frequency can be estimated by applying the total Least Squares (TLS)-ESPRIT algorithm [171] on $\boldsymbol{Y}_{\mathrm{do},l}^{(i_{\mathrm{do}})}$. Then, the accurate relative radial velocity $\widehat{\underline{v}}_l^{(i_{\mathrm{do}})} = \widehat{\psi}_{z,l}^{(i_{\mathrm{do}})}\lambda_z$ is obtained, and the Doppler compensation matrix is updated. When arriving at the maximum number of iteration $i_{\mathrm{do}}^{\max}$, the proposed algorithm outputs the accurate Doppler shift estimates $\widehat{\psi}_{z,l} = \widehat{\psi}_{z,l}^{(i_{\mathrm{do}}^{\max})}$ corresponding to the lth BS, and the estimates can be extended to all subcarriers $\{\widehat{\psi}_{l,k}\}_{k=1}^{K}$. The whole steps of the proposed pilot-aided iterative Doppler shift estimation algorithm is concluded in Algorithm 6.2, which can effectively solve the Doppler squint effect by iteratively constructing the Doppler compensation matrix.

6.4.3 Estimation of Path Delay and Channel Gain

In this stage, the acquired accurate Doppler shift is utilized in the compensation, and the path delay is estimated through N_{de} OFDM symbols in the downlink. Review the steering vector associated with the path delay $\boldsymbol{a}_{\tau}(\mu_l^{\tau}, K_l) = \left[e^{\mathrm{j}(\{\mathcal{K}_l\}_1-1)\mu_l^{\tau}}\ e^{\mathrm{j}(\{\mathcal{K}_l\}_2-1)\mu_l^{\tau}} \cdots e^{\mathrm{j}(\{\mathcal{K}_l\}_{K_l}-1)\mu_l^{\tau}}\right]^{\mathrm{T}}$ and $\mu_l^{\tau} = -2\pi f_s \tau_l / K$, the downlink received signal $y_{\mathrm{de},l}^{[\bar{n}]}[k_l]$ at the k_lth ($k_l \in \mathcal{K}_l$) subcarrier of the $\bar{n}$th ($1 \le \bar{n} \le N_{\mathrm{de}}$) OFDM symbol can be expressed as

Algorithm 6.2: Pilot-Aided Iterative Doppler Shift Estimation Algorithm

Input: The estimated virtual angles $\{\widehat{\mu}_l^{\mathrm{BS}}, \widehat{\nu}_l^{\mathrm{BS}}, \widehat{\mu}_l^{\mathrm{AC}}, \widehat{\nu}_l^{\mathrm{AC}}\}$, the rough estimates of the Doppler shift $\widetilde{\psi}_{z,l}$ and its extensions on different subcarriers $\{\widetilde{\psi}_{l,k_l}\}_{k_l=1}^{K_l}$, the pilot signals $\{s_{\mathrm{do},l}[k_l]\}_{k_l=1}^{K_l}$, the maximum number of iterations $i_{\mathrm{do}}^{\max}$, the wavelength at the central carrier frequency λ_z, and other parameters $\{N_{\mathrm{AC}}, M_{\mathrm{AC}}, N_{\mathrm{do}}, K_l\}$

Output: The accurate Doppler shift estimates at the central carrier freqency $\widehat{\psi}_{z,l}$ and its extensions on different subcarriers $\{\widehat{\psi}_{l,k}\}_{k=1}^{K}$

%% Pilot signal transmission and pre-processing %%
Ensure the antenna indices at the aircraft $\mathcal{I}_{\mathrm{AC},l}$, and initialize $\boldsymbol{w}_{\mathrm{RF},l} = \boldsymbol{0}_{N_{\mathrm{AC}}}$;
Set $\boldsymbol{w}_{\mathrm{RF},l\{\mathcal{I}_{\mathrm{AC},l}\}} = \frac{1}{\sqrt{M_{\mathrm{AC}}}}\boldsymbol{a}_{\mathrm{AC}}(\widehat{\mu}_l^{\mathrm{AC}}, \widehat{\nu}_l^{\mathrm{AC}})_{\{\mathcal{I}_{\mathrm{AC},l}\}}$ and $\boldsymbol{f}_{\mathrm{RF},l} = \boldsymbol{a}_{\mathrm{BS}}(\widehat{\mu}_l^{\mathrm{BS}}, \widehat{\nu}_l^{\mathrm{BS}})$;
for $\bar{m} = 1, \cdots, N_{\mathrm{do}}$ **do**
  for $k_l = 1, \cdots, K_l$ **do**
    Transmit the pilot signals $s_{\mathrm{do},l}[k_l]$ and obtain $y_{\mathrm{do},l}^{[\bar{m}]}[k_l]$;
    Remove the compensated phase $e^{-\mathrm{j}2\pi\widetilde{\psi}_{l,k_l}(\bar{m}-1)T_{\mathrm{sym}}}$ from $y_{\mathrm{do},l}^{[\bar{m}]}[k_l]$ to obtain $\bar{y}_{\mathrm{do},l}^{[\bar{m}]}[k_l]$;
  end
end
Obtain $\boldsymbol{Y}_{\mathrm{do},l}$ by stacking $\{\{\bar{y}_{\mathrm{do},l}^{[\bar{m}]}[k_l]\}_{k_l=1}^{K_l}\}_{\bar{m}=1}^{N_{\mathrm{do}}}$;
%% Pilot-aided iterative Doppler shift estimation %%
Initialize $i_{\mathrm{do}} = 0$ and $\widehat{\underline{\upsilon}}_l^{(0)} = \widetilde{\psi}_{z,l}\lambda_z$;
while $i_{\mathrm{do}} \leq i_{\mathrm{do}}^{\max}$ **do**
  if $i_{\mathrm{do}} = 0$ **then**
    Apply the TLS-ESPRIT algoirhtm on $\boldsymbol{Y}_{\mathrm{do},l}$ to obtain $\widehat{\psi}_{z,l}^{(0)}$;
  else
    Construct the Doppler compensation matrix $\widetilde{\boldsymbol{Y}}_{\mathrm{do},l}(\widehat{\underline{\upsilon}}_l^{(i_{\mathrm{do}}-1)})$, whose $(\bar{m}, k_l)$th element is $\widetilde{y}_{\mathrm{do},l}^{[\bar{m}]}[k_l](\widehat{\underline{\upsilon}}_l^{(i_{\mathrm{do}}-1)})$;
    Apply the TLS-ESPRIT algorithm on the compensated signal matrix $\boldsymbol{Y}_{\mathrm{do},l}^{(i_{\mathrm{do}})} = \widetilde{\boldsymbol{Y}}_{\mathrm{do},l}^{*}(\widehat{\underline{\upsilon}}_l^{(i_{\mathrm{do}}-1)}) \circ \boldsymbol{Y}_{\mathrm{do},l}$;
    Obtain the Doppler shift estimates $\widehat{\psi}_{z,l}^{(i_{\mathrm{do}})}$ in the i_{do}th iteration with $\widehat{\underline{\upsilon}}_l^{(i_{\mathrm{do}})} = \widehat{\psi}_{z,l}^{(i_{\mathrm{do}})}\lambda_z$;
  end
  $i_{\mathrm{do}} = i_{\mathrm{do}} + 1$;
end
Return: $\widehat{\psi}_{z,l} = \widehat{\psi}_{z,l}^{(i_{\mathrm{do}}^{\max})}$ with its extensions on different subcarriers $\{\widehat{\psi}_{l,k}\}_{k=1}^{K}$

$$
\begin{aligned}
y_{\mathrm{de},l}^{[\bar{n}]}[k_l] &= \sqrt{P_l}\boldsymbol{w}_{\mathrm{RF},l}^{\mathrm{H}}\widetilde{\boldsymbol{H}}_{\mathrm{DL},l}^{'[\bar{n}]}[k_l]\boldsymbol{f}_{\mathrm{RF},l}s_{\mathrm{de},l}^{[\bar{n}]}[k_l] + n_{\mathrm{de},l}^{[\bar{n}]}[k_l] \\
&= \underbrace{\sqrt{P_l G_l}\alpha_l e^{\mathrm{j}\pi f_s \tau_l}\boldsymbol{w}_{\mathrm{RF},l}^{\mathrm{H}}\boldsymbol{A}_{\mathrm{DL},l}\boldsymbol{f}_{\mathrm{RF},l}}_{\gamma_{\mathrm{de},l}} e^{\mathrm{j}(k_l-1)\mu_l^{\tau}} \underbrace{e^{\mathrm{j}2\pi(\psi_{z,l}-\widehat{\psi}_{z,l})(\bar{n}-1)T_{\mathrm{sym}}} s_{\mathrm{de},l}^{[\bar{n}]}}_{\bar{s}_{\mathrm{de},l}^{[\bar{n}]}} \cdot \widetilde{y}_{\mathrm{de},l}^{[\bar{n}]}[k_l] \\
&\quad + n_{\mathrm{de},l}^{[\bar{n}]}[k_l],
\end{aligned} \tag{6.27}
$$

where $s_{\mathrm{de},l}^{[\bar{n}]} = s_{\mathrm{de},l}^{[\bar{n}]}[k_l]$ (for $\forall k_l \in \mathcal{K}_l$) denotes the transmitted pilot,[6] $\widetilde{y}_{\mathrm{de},l}^{[\bar{n}]}[k_l]$ denotes the error term including the residual beam-Doppler squint errors caused by the CE error, and $n_{\mathrm{de},l}^{[\bar{n}]}[k_l]$ denotes the noise. Stack the received signals at all K_l subcarriers into $\boldsymbol{y}_{\mathrm{de},l}^{[\bar{n}]} \in \mathbb{C}^{K_l}$, we have

$$
\begin{aligned}
\boldsymbol{y}_{\mathrm{de},l}^{[\bar{n}]} &= \left[y_{\mathrm{de},l}^{[\bar{n}]}[\{\mathcal{K}_l\}_1] \cdots y_{\mathrm{de},l}^{[\bar{n}]}[\{\mathcal{K}_l\}_{K_l}]\right]^{\mathrm{T}} \\
&= \gamma_{\mathrm{de},l}\boldsymbol{a}_{\tau}(\mu_l^{\tau}, K_l)\bar{s}_{\mathrm{de},l}^{[\bar{n}]} \circ \widetilde{\boldsymbol{y}}_{\mathrm{de},l}^{[\bar{n}]} + \boldsymbol{n}_{\mathrm{de},l}^{[\bar{n}]},
\end{aligned} \tag{6.28}
$$

where $\widetilde{\boldsymbol{y}}_{\mathrm{de},l}^{[\bar{n}]} = \left[\widetilde{y}_{\mathrm{de},l}^{[\bar{n}]}[\{\mathcal{K}_l\}_1] \cdots \widetilde{y}_{\mathrm{de},l}^{[\bar{n}]}[\{\mathcal{K}_l\}_{K_l}]\right]^{\mathrm{T}}$ and $\boldsymbol{n}_{\mathrm{de},l}^{[\bar{n}]}$ denote the corresponding error vector and the noise vector, respectively. Stack the signals during N_{de} OFDM symbols, we can obtain $\boldsymbol{Y}_{\mathrm{de},l} \in \mathbb{C}^{K_l \times N_{\mathrm{de}}}$ as follows

$$
\begin{aligned}
\boldsymbol{Y}_{\mathrm{de},l} &= \left[\boldsymbol{y}_{\mathrm{de},l}^{[1]} \cdots \boldsymbol{y}_{\mathrm{de},l}^{[N_{\mathrm{de}}]}\right] \\
&= \gamma_{\mathrm{de},l}\left(\boldsymbol{a}_{\tau}(\mu_l^{\tau}, K_l)\bar{\boldsymbol{s}}_{\mathrm{de},l}^{\mathrm{T}}\right) \circ \widetilde{\boldsymbol{Y}}_{\mathrm{de},l} + \boldsymbol{N}_{\mathrm{de},l},
\end{aligned} \tag{6.29}
$$

where $\bar{\boldsymbol{s}}_{\mathrm{de},l} = \left[\bar{s}_{\mathrm{de},l}^{[1]} \cdots \bar{s}_{\mathrm{de},l}^{[N_{\mathrm{de}}]}\right]^{\mathrm{T}} \in \mathbb{C}^{N_{\mathrm{de}}}$, and $\widetilde{\boldsymbol{Y}}_{\mathrm{de},l} = \left[\widetilde{\boldsymbol{y}}_{\mathrm{de},l}^{[1]} \cdots \widetilde{\boldsymbol{y}}_{\mathrm{de},l}^{[N_{\mathrm{de}}]}\right]$ and $\boldsymbol{N}_{\mathrm{de},l}$ denote the residual beam-Doppler squint matrix and the noise matrix, respectively. Then, the path delay $\{\widehat{\tau}_l\}_{l=1}^{L}$ can be estimated by applying the TLS-ESPRIT algorithm on $\boldsymbol{Y}_{\mathrm{de},l}$. From Eq. (6.29), it can be observed that the accuracy of path delay estimations depends on the precision of angle and Doppler shift estimations, which can be verified in the numerical experiments.

Finally, the received signal matrix $\boldsymbol{Y}_{\mathrm{de},l}$ can be exploited to estimate the channel gain. Specifically, set $\boldsymbol{Y}_{\mathrm{de},l} = \bar{\alpha}_l \bar{\boldsymbol{Y}}_{\mathrm{de},l}$, where $\bar{\alpha}_l = \sqrt{P_l G_l}\alpha_l$ denotes the effective channel gain. Then, $\bar{\boldsymbol{Y}}_{\mathrm{de},l}$ can be reconstructed using the estimated main channel parameters (i.e., azimuth/elevation angles at the transceivers, Doppler shift, and path delay), defined as $\widehat{\bar{\boldsymbol{Y}}}_{\mathrm{de},l}$. Here, the residual beam-Doppler squint and the noise matrix in $\boldsymbol{Y}_{\mathrm{de},l}$ are omitted. Therefore, the estimated $\bar{\alpha}_l$ can be calculated as

[6] In the delay estimation stage, it is assumed that the same pilot signals are transmitted for all K_l subcarriers, which may lead to a high peak-to-average power ratio (PAPR) in the OFDM system. Therefore, we adopt the predefined scrambling sequence designed in [101] and spread it on all subcarriers to reduce the excessive PAPR effectively.

$$\widehat{\alpha}_l = \frac{1}{N_{\text{de}}K_l} \sum_{\bar{n}=1}^{N_{\text{de}}} \sum_{k_l=1}^{K_l} \left[\boldsymbol{Y}_{\text{de},l}\right]_{k_l,\bar{n}} / \left[\widehat{\boldsymbol{Y}}_{\text{de},l}\right]_{k_l,\bar{n}}. \tag{6.30}$$

6.5 Data-Aided Channel Tracking

The estimates of main channel parameters obtained in last section are used in the following data transmission and channel tracking stages. Although the THz UM-MIMO aeronautical communication channel exhibits fast time-varying fading characteristics due to its large Doppler shift, the main channel parameters including angle, delay, Doppler shift, and channel gain can be considered to be relatively quasi-statical within the short duration of symbol T_{sym}. Hence, it is reasonable to select N_{C} OFDM symbols as a *time interval* and assume the unchanged channel parameters during this interval. It is worth noting that after rough or accurate Doppler compensation, the channel associated with each OFDM symbol in a time interval still changes slowly due to the error of Doppler compensation. Therefore, after a long period of accumulation, the channel will change significantly if the CSI information cannot be updated in time, deteriorating the accuracy of the data detection.

To improve the reliability and efficiency of data transmission, in this section we propose a DADD-based channel tracking algorithm to track the equivalent channel after beam alignment in real-time, which can avoid frequent estimation of time-varying channels, thus saving pilot overheads. The proposed tracking algorithm exploits the channel correlation between two adjacent OFDM symbols, in which the equivalent channel estimated in the previous symbol can be approximately considered as the real-time channel of the next symbol, and data detection is performed sequentially. Simultaneously, the proposed method also adopts the channel coding such as Turbo or Low-Density-Parity-Check (LDPC) codes to reduce the error propagation during decision feedback. Note that in the data transmission stage, to improve the spectral efficiency and throughput of the system, L BSs need to jointly serve the aircraft based on the same time-frequency resources instead of adopting the OFDMA access method. Specifically, considering the pth OFDM symbol in the qth time interval, the downlink channel matrix $\boldsymbol{H}_{\text{DL},l}^{[n]}[k]$ can be rewritten as $\boldsymbol{H}_{\text{DL},l}^{[r]}[k]$, where $r=(q-1)N_{\text{C}}+p$. Define the data sequence from the lth BS in the rth OFDM symbol as $\boldsymbol{x}_l^{[r]}$. After the channel coding and modulation, the data sequence is mapped into the transmitted signal vector $\boldsymbol{s}_l^{[r]}=\left[s_l^{[r]}[1]\cdots s_l^{[r]}[K]\right]^{\text{T}} \in \mathbb{C}^K$. At the aircraft, the received downlink baseband signal at the kth subcarrier of the rth OFDM symbol $\boldsymbol{y}^{[r]}[k] \in \mathbb{C}^L$ can be expressed as

$$\begin{aligned} \mathbf{y}^{[r]}[k] &= \left[y_1^{[r]}[k] \cdots y_L^{[r]}[k]\right]^{\mathrm{T}} \\ &= \boldsymbol{W}_{\mathrm{RF}}^{\mathrm{H}} \left(\sum_{l=1}^{L} \sqrt{P_l} \widetilde{\boldsymbol{H}}_{\mathrm{DL},l}^{'[r]}[k] \boldsymbol{f}_{\mathrm{RF},l} s_l^{[r]}[k] + \boldsymbol{n}^{[r]}[k] \right), \end{aligned} \tag{6.31}$$

where $\boldsymbol{W}_{\mathrm{RF}} = \left[\boldsymbol{w}_{\mathrm{RF},1} \cdots \boldsymbol{w}_{\mathrm{RF},L}\right]$ and $\boldsymbol{n}^{[r]}[k]$ denotes the noise vector. In Eq. (6.31), the lth element of $\mathbf{y}^{[r]}[k]$ (the transmitted signal from the lth BS) is given as

$$y_l^{[r]}[k] = \underbrace{\sqrt{P_l} \boldsymbol{w}_{\mathrm{RF},l}^{\mathrm{H}} \widetilde{\boldsymbol{H}}_{\mathrm{DL},l}^{'[r]}[k] \boldsymbol{f}_{\mathrm{RF},l}}_{h_l^{[r]}[k]} s_l^{[r]}[k] + \underbrace{\boldsymbol{w}_{\mathrm{RF},l}^{\mathrm{H}} \sum_{\substack{l'=1 \\ l' \neq l}}^{L} \sqrt{P_{l'}} \widetilde{\boldsymbol{H}}_{\mathrm{DL},l'}^{'[r]}[k] \boldsymbol{f}_{\mathrm{RF},l'} s_{l'}^{[r]}[k] + n_l^{[r]}[k]}_{z_l^{[r]}[k]}, \tag{6.32}$$

where the second term denotes the interference from other BSs, $n_l^{[r]}[k]$ denotes the noise, $h_l^{[r]}[k]$ denotes the equivalent channel coefficient after beam alignment, and $z_l^{[r]}[k]$ denotes the interference plus noise. Due to the large angle difference between different BSs, the narrow beam from THz UM-MIMO array can reduce the interference, thus Eq. (6.32) can be simplified as $y_l^{[r]}[k] = h_l^{[r]}[k] s_l^{[r]}[k] + z_l^{[r]}[k]$, where the interference $z_l^{[r]}[k]$ is regarded as the minor noise. For the rth OFDM symbol from the lth BS, we define the true equivalent channel vector $\boldsymbol{h}_l^{[r]} = \left[h_l^{[r]}[1] \cdots h_l^{[r]}[K]\right]^{\mathrm{T}} \in \mathbb{C}^K$.

According to the estimated equivalent channel coefficients $\{\widehat{h}_l^{[r-1]}[k]\}_{l=1}^{L}$ in the $(r-1)$th OFDM symbol, the digital combining matrix can be designed as $\boldsymbol{W}_{\mathrm{BB}}^{[r]}[k] = \mathrm{diag}(\widehat{h}_1^{[r-1]}[k] \cdots \widehat{h}_L^{[r-1]}[k])$. By applying it on the basedband signal $\mathbf{y}^{[r]}[k]$, we obtain the the estimates of $\boldsymbol{s}^{[r]}[k] = \left[s_1^{[r]}[k] \cdots s_L^{[r]}[k]\right] \in \mathbb{C}^L$ as

$$\widehat{\boldsymbol{s}}^{[r]}[k] = \left[\widehat{s}_1^{[r]}[k] \cdots \widehat{s}_L^{[r]}[k]\right]^{\mathrm{T}} = \left(\boldsymbol{W}_{\mathrm{BB}}^{[r]}[k]\right)^{\mathrm{H}} \mathbf{y}^{[r]}[k], \tag{6.33}$$

where the lth element can be further expressed as

$$\widehat{s}_l^{[r]}[k] = \frac{\widehat{h}_l^{[r]}[k]}{\widehat{h}_l^{[r-1]}[k]} s_l^{[r]}[k] + \frac{z_l^{[r]}[k]}{\widehat{h}_l^{[r-1]}[k]}. \tag{6.34}$$

Define $\widehat{\boldsymbol{s}}_l^{[r]} \in \mathbb{C}^K$ as the estimates of the received signals from all K subcarriers on the lth RF chain. To track the equivalent channel, i.e., $\boldsymbol{h}_l^{[r]}$, in the rth OFDM symbol, we first demodulate and decode $\boldsymbol{s}_l^{[r]}$ to obtain the detected data sequence $\widehat{\boldsymbol{x}}_l^{[r]}$. Then, the reconstructed signal vector $\widetilde{\boldsymbol{s}}_l^{[r]}$ is obtained by encoding and modulating $\widehat{\boldsymbol{x}}_l^{[r]}$, in which the error correction capability of channel coding is exploited to improve the accuracy of $\widetilde{\boldsymbol{s}}_l^{[r]}$. By using $\widetilde{\boldsymbol{s}}_l^{[r]}$ as the known pilots to estimate the equivalent channels in the current OFDM symbol, we have $\widehat{h}_l^{[r]}[k] = y_l^{[r]}[k] / \widetilde{s}_l^{[r]}[k]$, based on which the digital combining matrix of next OFDM symbol is given as $\widehat{\boldsymbol{h}}_l^{[r]} \in \mathbb{C}^K$ for the following the channel equalization. Besides, the channel parameters estimated in the initial CE stage can be employed to reconstruct the estimates of the equivalent channel $\{\widehat{\boldsymbol{h}}_l^{[0]}\}_{l=1}^{L}$ after beam alignment, where the kth element of $\widehat{\boldsymbol{h}}_l^{[0]}$ can be written as

$$\widehat{h}_l^{[0]}[k] = \widehat{\alpha}_l e^{-\mathrm{j}2\pi\left(\frac{k-1}{K}-\frac{1}{2}\right)f_s\widehat{\tau}_l} \boldsymbol{w}_{\mathrm{RF},l}^{\mathrm{H}} \widehat{\boldsymbol{A}}_{\mathrm{DL},l} \boldsymbol{f}_{\mathrm{RF},l}, \tag{6.35}$$

where $\widehat{\boldsymbol{A}}_{\mathrm{DL},l}$ is the reconstructed downlink array response matrix relying on the accurate angle estimates.

After a period of time, the previously estimated channel parameters will deviate from the current equivalent channel. Therefore, the time correlation of two adjacent OFDM symbols is employed to monitor the quality of the equivalent channel vector tracked at this stage in real-time, to decide whether to terminate the tracking. Specifically, for estimated $\widehat{\boldsymbol{h}}_l^{[r]}$ in the rth OFDM symbol, its kth coefficient $\widehat{h}_l^{[r]}[k]$ can be considered wrong if the following inequality holds,

$$\left|\widehat{h}_l^{[r]}[k] - \widehat{h}_l^{[r-1]}[k]\right| > \frac{\varepsilon}{K}\sum_{k=1}^{K}\left|\widehat{h}_l^{[r-1]}[k]\right|, \tag{6.36}$$

where ε is the determinate threshold. We define $\widetilde{\mathcal{K}}_l^{[r]}$ as a set that collects the indices of subcarriers in the wrong estimates satisfying the above inequality, and select $\widetilde{K}$ as the number of tolerable errors. This tracking stage is terminated when $|\widetilde{\mathcal{K}}_l^{[r]}|_c > \widetilde{K}$ for $\forall l$. The proposed DADD-based channel tracking algorithm is summarized in Algorithm 6.3.

6.6 Pilot-Aided Channel Tracking

According to the previous analysis, the channel parameters including angle and Doppler shift usually change slowly, thus they can be considered as prior information in this stage. The difference between pilot-aided channel tracking and the initial CE lies in the array response vector. In this section, the azimuth/elevation angles at the BS or the aircraft are estimated by constructing the response vector of a low-dimensional equivalent fully-digital sparse array. However, the initial CE only considers the equivalent fully-digital array with the critical antenna spacing (i.e., half wavelength). Existing results show that exploiting sparse arrays can significantly enhance the accuracy of angle estimation at the expense of angle ambiguity [172, 173]. Fortunately, this problem can be solved by previously estimated angles. For the sake of brevity, this section focuses on pilot-aided tracking at the BS, which can be easily extended to the aircraft.

Specifically, the BS can construct an $I_{\mathrm{BS}}^{'\mathrm{h}} \times I_{\mathrm{BS}}^{'\mathrm{v}}$ equivalent fully-digital sparse array with $I'_{\mathrm{BS}} = I_{\mathrm{BS}}^{'\mathrm{h}} I_{\mathrm{BS}}^{'\mathrm{v}}$ OFDM symbols, where the I'_{BS} sub-arrays are selected accord-

Algorithm 6.3: DADD-Based Channel Tracking Algorithm

Input: The estimated channel parameters $\{\widehat{\theta}_l^{\rm BS}, \widehat{\varphi}_l^{\rm BS}, \widehat{\theta}_l^{\rm AC}, \widehat{\varphi}_l^{\rm AC}, \widehat{\psi}_{z,l}, \widehat{\tau}_l, \widehat{\alpha}_l\}_{l=1}^L$, the determinate threshold ε, and other parameters $\{K, L, N_{\rm C}, \widetilde{K}\}$

Output: The estimated equivalent channel vectors $\{\widehat{\boldsymbol{h}}_l^{[r]}\}_{l=1}^L$ and the detected data sequences $\{\widehat{\boldsymbol{x}}_l^{[r]}\}_{l=1}^L$, $r=1, 2, 3, \cdots$

Initialization: Set $\widetilde{\mathcal{K}}_l^{[0]}=\emptyset$ and obtain $\{\widehat{\boldsymbol{h}}_l^{[0]}\}_{l=1}^L$ according to Eq. (6.35);
for $q=1, 2, 3, \cdots$ **do**
  for $p=1, \cdots, N_{\rm C}$ **do**
    if $\left|\widetilde{\mathcal{K}}_l^{[r-1]}\right|_c \leq \widetilde{K}, \forall l$ **then**
      Encode and modulate the initial data sequences $\{\boldsymbol{x}_l^{[r]}\}_{l=1}^L$, then map it to transmitted signal vectors $\{\boldsymbol{s}_l^{[r]}\}_{l=1}^L$;
      Obtain the baseband received signal $\{\boldsymbol{y}^{[r]}[k]\}_{k=1}^K$ as Eq. (6.31);
      Design the digital combining matrix $\boldsymbol{W}_{\rm BB}^{[r]}[k]=\text{diag}(\widehat{h}_1^{[r-1]}[k]\cdots\widehat{h}_L^{[r-1]}[k])$, $1\leq k\leq K$;
      Obtain $\{\widehat{\boldsymbol{s}}^{[r]}[k]\}_{k=1}^K$ according to Eq. (6.33), then demodulate and decode the extracted $\{\widehat{\boldsymbol{s}}_l^{[r]}\}_{l=1}^L$ to obtain $\{\widehat{\boldsymbol{x}}_l^{[r]}\}_{l=1}^L$;
      Encode and modulate $\{\widehat{\boldsymbol{x}}_l^{[r]}\}_{l=1}^L$ again to reconstruct the transmitted signals $\{\widetilde{\boldsymbol{s}}_l^{[r]}\}_{l=1}^L$;
      Substitute $\widetilde{s}_l^{[r]}[k]$ to Eq. (6.32) to estimate the current equivalent channels $\widehat{h}_l^{[r]}[k]=y_l^{[r]}[k]/\widetilde{s}_l^{[r]}[k]$, $1\leq k\leq K$, $1\leq l\leq L$;
      Initialize $\widetilde{\mathcal{K}}_l^{[r]}=\emptyset$, $1\leq l\leq L$;
      for $k=1, \cdots, K$ and $l=1, \cdots, L$ **do**
        if $\left|\widehat{h}_l^{[r]}[k]-\widehat{h}_l^{[r-1]}[k]\right| > \frac{\varepsilon}{K}\sum_{k=1}^K \left|\widehat{h}_l^{[r-1]}[k]\right|$ **then**
          Set $\widetilde{\mathcal{K}}_l^{[r]}=\widetilde{\mathcal{K}}_l^{[r]}\cup k$;
        end
      end
    else
      Return: $\{\widehat{\boldsymbol{h}}_l^{[r]}\}_{l=1}^L$ and $\{\widehat{\boldsymbol{x}}_l^{[r]}\}_{l=1}^L$, $r=1, 2, 3, \cdots$;
      Terminate the current stage and enter the pilot-aided channel tracking stage
    end
  end
end

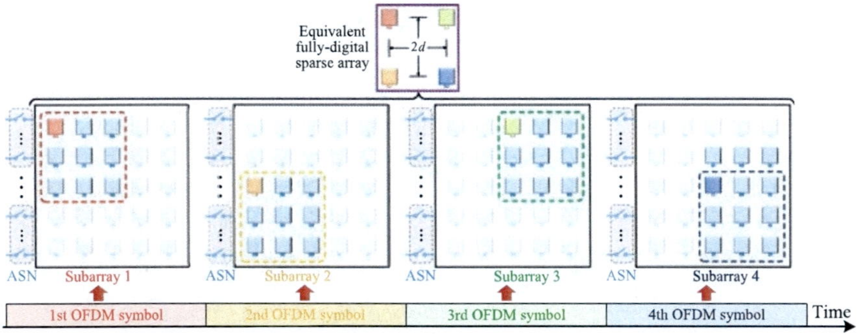

Fig. 6.9 The schematic diagram of the sub-array selection in the pilot-aided angle tracking stage [146], where the different antenna connection modes are formed by controlling the switch state of the ASNs in the reconfigurable RF selection network. Taking a 5×5 UPA as an example, it can be divided into 5 sub-arrays with a size of 3×3, and the interval between each sub-array is the width of two antennas. The same RF link can sequentially select the designed sub-arrays in 4 consecutive OFDM symbols to receive signals, and these received signals are equivalent to that received by the low-dimensional fully-digital array with size being 2×2 and antenna spacing being $2d$ (i.e., a sparse spacing)

ing to the antenna connection mode of the reconfigurable RF selection network. Define Ωd as the sparse array spacing, where d is the critical antenna spacing. The size of the selected sub-array is $\bar{M}'^{\text{h}}_{\text{BS}} \times \bar{M}'^{\text{v}}_{\text{BS}}$, and the total number of antennas is $\bar{M}'_{\text{BS}} = \bar{M}'^{\text{h}}_{\text{BS}} \bar{M}'^{\text{v}}_{\text{BS}}$, where $\bar{M}'^{\text{h}}_{\text{BS}} = N^{\text{h}}_{\text{BS}} - \Omega(I'^{\text{h}}_{\text{BS}} - 1)$ and $\bar{M}'^{\text{v}}_{\text{BS}} = N^{\text{v}}_{\text{BS}} - \Omega(I'^{\text{v}}_{\text{BS}} - 1)$. For the example in Fig. 6.9, a 5×5 UPA is divided into 4 sub-arrays of size 3×3. These sub-arrays can be employed to construct the array response vector of a 2×2 low-dimensional equivalent fully-digital sparse array with $\Omega = 2$.

Similar to Eq. (6.17), the uplink received signal matrix is denoted as $\bar{\boldsymbol{Y}}_{\text{UL},l} \in \mathbb{C}^{I'_{\text{BS}} \times K_l}$, and the effective array response vector of the sparse array can be expressed as

$$\bar{\bar{\boldsymbol{a}}}_{\text{BS}}(\bar{\mu}_l^{\text{BS}}, \bar{\nu}_l^{\text{BS}}) = \boldsymbol{a}_{\text{v}}(\bar{\nu}_l^{\text{BS}}, I'^{\text{v}}_{\text{BS}}) \otimes \boldsymbol{a}_{\text{h}}(\bar{\mu}_l^{\text{BS}}, I'^{\text{h}}_{\text{BS}}) \in \mathbb{C}^{I'_{\text{BS}}}, \tag{6.37}$$

where $\bar{\mu}_l^{\text{BS}} = \Omega \mu_l^{\text{BS}}$ and $\bar{\nu}_l^{\text{BS}} = \Omega \nu_l^{\text{BS}}$.

The estimates of $\bar{\mu}_l^{\text{BS}}$ and $\bar{\nu}_l^{\text{BS}}$, denoted as $\widehat{\bar{\mu}}_l^{\text{BS}}$ and $\widehat{\bar{\nu}}_l^{\text{BS}}$, respectively, can be obtained through the proposed prior-aided iterative angle estimation in Algorithm 6.1, although the angle ambiguity exists. To solve this problem, we first define a sorted index set $\mathcal{B} = \{-1, -1+\frac{1}{\Omega}, -1+\frac{2}{\Omega}, \cdots, 1\}$ with $|\mathcal{B}|_c = 2\Omega + 1$, and set $\tilde{\bar{\mu}}_l^{\text{BS}} = \widehat{\bar{\mu}}_l^{\text{BS}} / \Omega$, $\tilde{\bar{\nu}}_l^{\text{BS}} = \widehat{\bar{\nu}}_l^{\text{BS}} / \Omega$. Their estimates can be obtained as $\widehat{\mu}'^{\text{BS}}_l = \tilde{\bar{\mu}}_l^{\text{BS}} + b^\star_\mu \pi$ and $\widehat{\mu}'^{\text{BS}}_l = \tilde{\bar{\nu}}_l^{\text{BS}} + b^\star_\nu \pi$, respectively, where $b^\star_\mu \in \mathcal{B}$ and $b^\star_\nu \in \mathcal{B}$ denotes the indices of optimal angles. Since the number of elements in $\mathcal{B}$ is limited, we adopt the exhaustive method to find $b^\star_\mu$ and $b^\star_\nu$. Specifically, the prior information is exploited by setting $\tilde{\mu}_l^{\text{BS}} = \widehat{\mu}_l^{\text{BS}}$ and $\tilde{\nu}_l^{\text{BS}} = \widehat{\nu}_l^{\text{BS}}$, then $b^\star_\mu$ and $b^\star_\nu$ can be obtained by solving

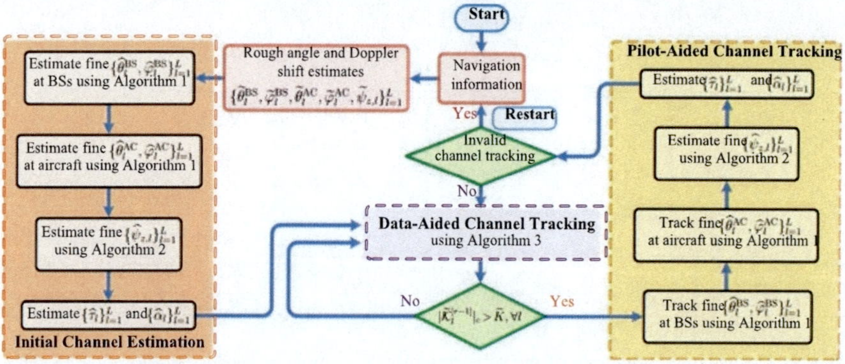

Fig. 6.10 Flow diagram of the proposed CE and tracking scheme [146]

$$b_\mu^\star = \arg\min_{b_\mu \in \mathcal{B}} \left| \tilde{\bar{\mu}}_l^{\mathrm{BS}} + b_\mu \pi - \tilde{\mu}_l^{\mathrm{BS}} \right|, \tag{6.38}$$

$$b_\nu^\star = \arg\min_{b_\nu \in \mathcal{B}} \left| \tilde{\bar{\nu}}_l^{\mathrm{BS}} + b_\nu \pi - \tilde{\nu}_l^{\mathrm{BS}} \right|. \tag{6.39}$$

Except for the exhaustive search in Eqs. (6.38) and (6.39), other steps are the same as the steps in Sect. 6.4.1.1. Then, the BS can obtain the accurate estimates of the azimuth/elevation angles, denoted as $\{\widehat{\theta}_l^{\mathrm{BS}}, \widehat{\varphi}_l^{\mathrm{BS}}\}_{l=1}^L$. Also, other channel parameters estimated in the pilot-aided channel tracking stage, including Doppler shift, path delay, and channel gain, are more accurate than those estimated in the initial CE stage. The equivalent channel after update can be used in the following data transmission, as shown in Fig. 6.3.

To explain the relationship between different CE and tracking stage more clearly, we provide the block diagram of the proposed scheme in Fig. 6.10. First, the rough estimates of angle and Doppler shift are obtained from aeronautical navigation information as the prior knowledge for the initial CE stage. Next, the accurate estimates of azimuth/elevation angles and the Doppler shift can be calculated from Algorithms 6.1 and 6.2, respectively, based on which the path delays and the channel gains can be estimated. Then, the data-aided channel tracking stage starts and terminates when inequality (6.36) holds. In the following pilot-aided channel tracking stage, the channel parameters are estimated again through the similar methods used in the initial stage. Besides, the equivalent sparse array is exploited to enhance the estimation accuracy. Finally, it is necessary to judge whether the channel tracking results are valid (i.e., whether the BS and the aircraft can establish a stable THz communication link). If so, the data-aid channel tracking is continued, otherwise the aeronautical navigation information is reacquired, and a new round of CE and tracking is started.

6.7 Performance Analysis

6.7.1 CRLB of Channel Parameter Estimation

The CRLB is an important analytical tool when evaluating the performance of channel parameter estimation, as it can provide a theoretical lower bound on the variance of any unbiased estimates [174]. Based on the effective received signal model in Sect. 6.4, this subsection investigates the CRLB for the main channel parameters, i.e., the azimuth/elevation angles, Doppler shift, and path delay at the BS and aircraft. It should be pointed out that the influence of the actual triple delay-beam-Doppler squint effects on the channel parameter estimation in the aeronautical THz UM-MIMO channel is not considered when deriving the CRLB.

6.7.1.1 CRLB of Angle Estimation

To investigate the estimation performance in the initial angle estimation stage and the following angel tracking stage, we consider the received signal model for an $\bar{I}_{\rm BS}^{\rm h} \times \bar{I}_{\rm BS}^{\rm v}$equivalent fully-digital sparse array, where $\Omega \geq 1$. According to Eq. (6.17), the effective received signal without triple squint effects, denoted as $\bar{\boldsymbol{Y}}_{{\rm UL},l} \in \mathbb{C}^{\bar{I}_{\rm BS}\times K_l}$, can be written as

$$\begin{aligned}\bar{\boldsymbol{Y}}_{{\rm UL},l} &= \left[\bar{\boldsymbol{y}}_{{\rm UL},l}[\{\mathcal{K}_l\}_1]\cdots\bar{\boldsymbol{y}}_{{\rm UL},l}[\{\mathcal{K}_l\}_{K_l}]\right] \\ &= \gamma_{{\rm UL},l}\bar{\bar{\boldsymbol{a}}}_{\rm BS}(\bar{\mu}_l^{\rm BS},\bar{\nu}_l^{\rm BS})\boldsymbol{s}_{{\rm UL},l}^{\rm T} + \bar{\boldsymbol{N}}_{{\rm UL},l},\end{aligned} \tag{6.40}$$

where $1 \leq l \leq L$, $\bar{I}_{\rm BS} = \bar{I}_{\rm BS}^{\rm h}\bar{I}_{\rm BS}^{\rm v}$, $\bar{\bar{\boldsymbol{a}}}_{\rm BS}(\bar{\mu}_l^{\rm BS},\bar{\nu}_l^{\rm BS}) = \boldsymbol{a}_{\rm v}(\bar{\nu}_l^{\rm BS},\bar{I}_{\rm BS}^{\rm v}) \otimes \boldsymbol{a}_{\rm h}(\bar{\mu}_l^{\rm BS},\bar{I}_{\rm BS}^{\rm h}) \in \mathbb{C}^{\bar{I}_{\rm BS}}$, $\bar{\mu}_l^{\rm BS} = \Omega\mu_l^{\rm BS} = \Omega\pi\sin(\theta_l^{\rm BS})\cos(\varphi_l^{\rm BS})$, $\bar{\nu}_l^{\rm BS} = \Omega\nu_l^{\rm BS} = \Omega\pi\sin(\varphi_l^{\rm BS})$, and $\bar{\boldsymbol{N}}_{{\rm UL},l} \sim \mathcal{CN}(0,\sigma_n^2)$ denotes the noise matrix. By defining $\boldsymbol{\eta}_l = [\alpha_l, (\boldsymbol{\xi}_l^{\rm BS})^{\rm T}]^{\rm T}$ and $\boldsymbol{\xi}_l^{\rm BS} = [\bar{\nu}_l^{\rm BS},\bar{\mu}_l^{\rm BS}]^{\rm T}$, its log-likelihood function can be written as

$$\begin{aligned}&\ln p(\bar{\boldsymbol{Y}}_{{\rm UL},l};\boldsymbol{\eta}_l) \\ &= -\bar{I}_{\rm BS}K_l\ln(\pi\sigma_n^2) - \frac{1}{\sigma_n^2}\sum_{k_l=1}^{K_l}\Big(\left[\bar{\boldsymbol{y}}_{{\rm UL},l}[\{\mathcal{K}_l\}_{k_l}] - \gamma_{{\rm UL},l}\bar{\bar{\boldsymbol{a}}}_{\rm BS}(\bar{\mu}_l^{\rm BS},\bar{\nu}_l^{\rm BS})s_{{\rm UL},l}[\{\mathcal{K}_l\}_{k_l}]\right]^{\rm H} \\ &\times \left[\bar{\boldsymbol{y}}_{{\rm UL},l}[\{\mathcal{K}_l\}_{k_l}] - \gamma_{{\rm UL},l}\bar{\bar{\boldsymbol{a}}}_{\rm BS}(\bar{\mu}_l^{\rm BS},\bar{\nu}_l^{\rm BS})s_{{\rm UL},l}[\{\mathcal{K}_l\}_{k_l}]\right]\Big).\end{aligned} \tag{6.41}$$

Then, the Fisher information matrix (FIM) $\boldsymbol{G}(\boldsymbol{\eta}_l)$ is given as follows

$$\{\boldsymbol{G}(\boldsymbol{\eta}_l)\}_{i,j} = -\mathbb{E}\left(\frac{\partial^2\ln p(\bar{\boldsymbol{Y}}_{{\rm UL},l};\boldsymbol{\eta}_l)}{\partial\{\boldsymbol{\eta}_l\}_i\partial\{\boldsymbol{\eta}_l\}_j}\right). \tag{6.42}$$

Therefore, the CRLB of $\boldsymbol{\xi}_l^{\mathrm{BS}}$ that consists of $\bar{\mu}_l^{\mathrm{BS}}$ and $\bar{\nu}_l^{\mathrm{BS}}$ can be expressed as [107, 175]

$$\mathrm{CRLB}_{\boldsymbol{\xi}_l^{\mathrm{BS}}} = \boldsymbol{G}^{-1}(\eta_l) = \frac{\sigma_n^2}{2\left|\gamma_{\mathrm{UL},l}\right|^2} \left\{ \sum_{k_l=1}^{K_l} \Re\left\{ \boldsymbol{B}_{\mathrm{BS},k_l}^{\mathrm{H}} \boldsymbol{\Gamma}_{\mathrm{BS}}^{\mathrm{H}} \left(\boldsymbol{I}_{\bar{I}_{\mathrm{BS}}} - \boldsymbol{\Phi}_{\mathrm{BS}} \right) \boldsymbol{\Gamma}_{\mathrm{BS}} \boldsymbol{B}_{\mathrm{BS},k_l} \right\} \right\}^{-1}, \quad (6.43)$$

where $\boldsymbol{B}_{\mathrm{BS},k_l} = \boldsymbol{I}_2 \otimes s_{\mathrm{UL},l}[\{\mathcal{K}_l\}_{k_l}]$, $\boldsymbol{\Gamma}_{\mathrm{BS}} = \left[\boldsymbol{a}_{\mathrm{v}}(\bar{\nu}_l^{\mathrm{BS}}, \bar{I}_{\mathrm{BS}}^{\mathrm{v}}) \otimes \frac{\partial \boldsymbol{a}_{\mathrm{h}}(\bar{\mu}_l^{\mathrm{BS}}, \bar{I}_{\mathrm{BS}}^{\mathrm{h}})}{\partial \bar{\mu}_l^{\mathrm{BS}}}, \frac{\partial \boldsymbol{a}_{\mathrm{v}}(\bar{\nu}_l^{\mathrm{BS}}, \bar{I}_{\mathrm{BS}}^{\mathrm{v}})}{\partial \bar{\nu}_l^{\mathrm{BS}}} \otimes \boldsymbol{a}_{\mathrm{h}} (\bar{\mu}_l^{\mathrm{BS}}, \bar{I}_{\mathrm{BS}}^{\mathrm{h}})\right]$, and the projection operator $\boldsymbol{\Phi}_{\mathrm{BS}} = \bar{\bar{\boldsymbol{a}}}_{\mathrm{BS}}(\bar{\mu}_l^{\mathrm{BS}}, \bar{\nu}_l^{\mathrm{BS}})\left(\bar{\bar{\boldsymbol{a}}}_{\mathrm{BS}}^{\mathrm{H}}(\bar{\mu}_l^{\mathrm{BS}}, \bar{\nu}_l^{\mathrm{BS}})\bar{\bar{\boldsymbol{a}}}_{\mathrm{BS}} (\bar{\mu}_l^{\mathrm{BS}}, \bar{\nu}_l^{\mathrm{BS}})\right)^{-1} \bar{\bar{\boldsymbol{a}}}_{\mathrm{BS}}^{\mathrm{H}}(\bar{\mu}_l^{\mathrm{BS}}, \bar{\nu}_l^{\mathrm{BS}})$.

Furthermore, to obtain the CRLB for the azimuth/elevation angles, we first define the relationship between the virtual angle and the actual physical angle as

$$\boldsymbol{J}(\boldsymbol{\xi}_l^{\mathrm{BS}}) = \begin{bmatrix} \varphi_l^{\mathrm{BS}} \\ \theta_l^{\mathrm{BS}} \end{bmatrix} = \begin{bmatrix} \arcsin\left(\frac{\bar{\nu}_l^{\mathrm{BS}}}{\Omega\pi}\right) \\ \arcsin\left(\frac{\bar{\mu}_l^{\mathrm{BS}}}{\Omega\pi\cos(\varphi_l^{\mathrm{BS}})}\right) \end{bmatrix}. \quad (6.44)$$

According to the CRLB with vector parameter transformation [174], by defining $\partial \boldsymbol{J}(\boldsymbol{\xi}_l^{\mathrm{BS}})/\partial \boldsymbol{\xi}_l^{\mathrm{BS}}$ as the Jacobian matrix, the CRLB of the azimuth angle θ_l^{BS} and the elevation angle φ_l^{BS} can be written as

$$\mathrm{CRLB}_{\varphi_l^{\mathrm{BS}}}(\Omega) = \left[\frac{\partial \boldsymbol{J}(\boldsymbol{\xi}_l^{\mathrm{BS}})}{\partial \boldsymbol{\xi}_l^{\mathrm{BS}}} \mathrm{CRLB}_{\boldsymbol{\xi}_l^{\mathrm{BS}}} \frac{\partial \boldsymbol{J}(\boldsymbol{\xi}_l^{\mathrm{BS}})^{\mathrm{T}}}{\partial \boldsymbol{\xi}_l^{\mathrm{BS}}}\right]_{1,1} = \frac{\left[\mathrm{CRLB}_{\boldsymbol{\xi}_l^{\mathrm{BS}}}\right]_{1,1}}{\Omega^2\left(\pi^2 - (\nu_l^{\mathrm{BS}})^2\right)}, \quad (6.45)$$

$$\mathrm{CRLB}_{\theta_l^{\mathrm{BS}}}(\Omega) = \left[\frac{\partial \boldsymbol{J}(\boldsymbol{\xi}_l^{\mathrm{BS}})}{\partial \boldsymbol{\xi}_l^{\mathrm{BS}}} \mathrm{CRLB}_{\boldsymbol{\xi}_l^{\mathrm{BS}}} \frac{\partial \boldsymbol{J}(\boldsymbol{\xi}_l^{\mathrm{BS}})^{\mathrm{T}}}{\partial \boldsymbol{\xi}_l^{\mathrm{BS}}}\right]_{2,2} = \frac{\left[\mathrm{CRLB}_{\boldsymbol{\xi}_l^{\mathrm{BS}}}\right]_{2,2}}{\Omega^2\left(\pi^2\cos^2(\varphi_l^{\mathrm{BS}}) - (\mu_l^{\mathrm{BS}})^2\right)}, \quad (6.46)$$

respectively. Since there are L aerial BSs, the final CRLB can be calculated as $\mathrm{CRLB}_{\theta^{\mathrm{BS}}}(\Omega) = \frac{1}{L}\sum_{l=1}^{L} \mathrm{CRLB}_{\theta_l^{\mathrm{BS}}}(\Omega)$ amd $\mathrm{CRLB}_{\varphi^{\mathrm{BS}}}(\Omega) = \frac{1}{L}\sum_{l=1}^{L} \mathrm{CRLB}_{\varphi_l^{\mathrm{BS}}}(\Omega)$. With the similar steps, we can obtain the CRLB for the azimuth and elevation angles at the aircraft, denoted as $\mathrm{CRLB}_{\theta^{\mathrm{AC}}}(\Omega)$ and $\mathrm{CRLB}_{\varphi^{\mathrm{AC}}}(\Omega)$, respectively. The derivation in detail is omitted for simplicity.

Remark 6.2 According to Eq. (6.43), if the system parameter configuration at the transceivers is the same except for Ω, $\boldsymbol{\xi}_l^{\mathrm{BS}}$ will be a constant. Therefore, $\mathrm{CRLB}_{\theta_l^{\mathrm{BS}}}(1) = \Omega^2 \cdot \mathrm{CRLB}_{\theta_l^{\mathrm{BS}}}(\Omega)$ and $\mathrm{CRLB}_{\varphi_l^{\mathrm{BS}}}(1) = \Omega^2 \cdot \mathrm{CRLB}_{\varphi_l^{\mathrm{BS}}}(\Omega)$. That is to say, compared to the array with critical antenna spacing (i.e., half-wavelength), the sparse array with larger spacing $\Omega > 1$ can achieve about $20\lg\Omega$ dB mean square error (MSE) gain on the CRLB, verifying that the sparse array can enhance the angle estimation accuracy.

6.7.1.2 CRLB for Doppler Shift and Path Delay Estimation

Similar to the above derivation, we can provide the CRLB of virtual Doppler ν_l^{ψ} and virtual delay μ_l^{τ} as follows

$$\mathrm{CRLB}_{\nu_l^{\psi}} = \frac{\sigma_n^2}{2\left|\gamma_{\mathrm{do},l}\right|^2}\left\{\sum_{k_l=1}^{K_l}\Re\left\{\left|\bar{s}_{\mathrm{do},l}[\{\mathcal{K}_l\}_{k_l}]\right|^2\left(\frac{\partial \boldsymbol{a}_{\psi}(\nu_l^{\psi},N_{\mathrm{do}})}{\partial \nu_l^{\psi}}\right)^{\mathrm{H}}\left(\boldsymbol{I}_{N_{\mathrm{do}}}-\boldsymbol{\Phi}_{\mathrm{Do}}\right)\frac{\partial \boldsymbol{a}_{\psi}(\nu_l^{\psi},N_{\mathrm{do}})}{\partial \nu_l^{\psi}}\right\}\right\}^{-1}, \tag{6.47}$$

$$\mathrm{CRLB}_{\mu_l^{\tau}} = \frac{\sigma_n^2}{2\left|\gamma_{\mathrm{de},l}\right|^2}\left\{\sum_{n=1}^{N_{\mathrm{de}}}\Re\left\{\left|\bar{s}_{\mathrm{de},l}^{[n]}\right|^2\left(\frac{\partial \boldsymbol{a}_{\tau}(\mu_l^{\tau},K_l)}{\partial \mu_l^{\tau}}\right)^{\mathrm{H}}\left(\boldsymbol{I}_{N_{\mathrm{de}}}-\boldsymbol{\Phi}_{\mathrm{De}}\right)\frac{\partial \boldsymbol{a}_{\tau}(\mu_l^{\tau},K_l)}{\partial \mu_l^{\tau}}\right\}\right\}^{-1}, \tag{6.48}$$

where the projection operators $\boldsymbol{\Phi}_{\mathrm{Do}}$ and $\boldsymbol{\Phi}_{\mathrm{De}}$ are similar to $\boldsymbol{\Phi}_{\mathrm{BS}}$ in Eq. (6.43). Considering the vector parameter transformation and L aerial BSs, the CRLB of Doppler shift $\psi_{z,l}$ and normalized delay $\bar{\tau}_l = f_s\tau_l$ can be obtain as $\mathrm{CRLB}_{\psi_z} = \frac{1}{L}\sum_{l=1}^{L}\mathrm{CRLB}_{\psi_{z,l}} = \frac{1}{L}\sum_{l=1}^{L}\frac{\mathrm{CRLB}_{\nu_l^{\psi}}}{(2\pi T_{\mathrm{sym}})^2}$ and $\mathrm{CRLB}_{\bar{\tau}} = \frac{1}{L}\sum_{l=1}^{L}\mathrm{CRLB}_{\bar{\tau}_l} = \frac{1}{L}\sum_{l=1}^{L}\frac{K^2\mathrm{CRLB}_{\mu_l^{\tau}}}{(2\pi)^2}$, respectively.

6.7.2 *Computational Complexity Analysis*

The computational complexity of the CE and tracking scheme proposed in this chapter mainly consists of two parts. The first part is to estimate and track channel parameters, including applying the TDU-ESPRIT algorithm and the TLS-ESPRIT algorithm to estimate the azimuth/elevation angle, Doppler shift and path delay at the BS and the aircraft. Since most of the trivial calculations with small computational complexity can be ignored, we only focus on the steps involving a large number of complex multiplication operations. Specifically, for the angle estimation and tracking stage at the BS and the aircraft, the total computational complexity is $\mathcal{O}\left(2LI_{\mathrm{BS}}K_l+2LI_{\mathrm{AC}}K_l+2LI'_{\mathrm{BS}}K_l+2LI'_{\mathrm{AC}}K_l\right)$. Besides, the computational complexity of the Doppler shift and path delay estimation is $\mathcal{O}\left(8LN_{\mathrm{Do}}^2K_l+8LK_l^2N_{\mathrm{De}}\right)$. In the second part of data-aided channel tracking, the computational complexities of the reconstruction of the equivalent channel after the initial beam alignment and the tracking of the subsequent equivalent channel are $\mathcal{O}\left(L(N_{\mathrm{AC}}+N_{\mathrm{BS}}+3K)\right)$ and $\mathcal{O}\left(LK\right)$, respectively.

According to the above analysis, it can be seen that although both the BS and the aircraft are equipped with THz UM-MIMO arrays consisting of tens of thousands of antenna elements, the proposed CE and tracking solution only performs low-dimensional effective signal processing at the level of subcarrier numbers. Therefore, the computational complexity of the proposed scheme is in polynomial time.

6.8 Numerical Experiments

6.8.1 *Settings*

The subsection evaluates the performance of the proposed estimation and tracking scheme in THz UM-MIMO-based aeronautical communication, where the simulation scenario is considered as Fig. 6.11. Without loss of generality, in Fig. 6.11a, the reference altitude of $L=2$ aerial BSs and one aircraft is 20 km and $D_{\rm AC}=10$ km (at the top of the troposphere or the bottom of the stratosphere), respectively. The vertical distance between the BSs and the aircraft is $D_{\rm AB}=10$ km, and the distance between two BSs is $D_{\rm BS}=200$ km. The abstract space coordinate system is provided in Fig. 6.11b, where point O is the origin of coordinates, and the coordinates of points A, B and C are $(0{,}0{,}D_{\rm AB})$, $(0{,}D_{\rm BS}{,}D_{\rm AB})$ and $(D_{\rm BS}/2{,}D_{\rm BS}/2,\,0)$, respectively. The aircraft is randomly distributed in a circle with point C as the center and $R_{\rm a}=50$ km as the radius. It is flying horizontally at the speed of $v_{\rm AC}=200$ m/s, and its direction falls within the intersection angle $\angle OCD$. To simplify the simulation scenario, the altitude of the aerial BSs and the aircraft is considered to change with angle over time.

In the simulations, we consider the central carrier frequency is $f_z=0.1$ THz, the system bandwidth $f_s=1$ GHz, the number of horizontal/vertical antennas of all sub-arrays at the BS and the aircraft is $N_{\rm BS}^{\rm h}=N_{\rm BS}^{\rm v}=M_{\rm AC}^{\rm h}=M_{\rm AC}^{\rm v}=200$, respectively, the number of horizontal/vertical sub-arrays at the aircraft is $\widetilde{I}_{\rm AC}^{\rm h}=1$ and $\widetilde{I}_{\rm AC}^{\rm v}=2$, respectively, and the dimension of the selected equivalent fully-digital (sparse) array is $I_{\rm BS}^{\rm h}=I_{\rm BS}^{\rm v}=I_{\rm AC}^{\rm h}=I_{\rm AC}^{\rm v}=5(I_{\rm BS}^{'\rm h}=I_{\rm BS}^{'\rm v}=I_{\rm AC}^{'\rm h}=I_{\rm AC}^{'\rm v}=5)$. Besides, the number of OFDM symbols used to estimate and track Doppler shift and path delay is $N_{\rm do}=6$ and $N_{\rm de}=10$, respectively. The number of subcarriers is $K=2048$, and the length of cyclic prefix is $N_{\rm cp}=128$. The channel model is considered as follows. The azimuth and elevation angles at the BS and the aircraft $\{\theta_l^{\rm BS},\varphi_l^{\rm BS},\theta_l^{\rm AC},\varphi_l^{\rm AC}\}_{l=1}^{L}$ are randomly generated from $[-\pi/3,\pi/3]$. Note that the long distance between adjacent BSs leads to large differences between the angles $\{\theta_l^{\rm AC},\varphi_l^{\rm AC}\}_{l=1}^{L}$, where the angles can be calculated according to locations of the aircraft in Fig. 6.11b. Moreover, the Doppler shift $\{\psi_{z,l}\}_{l=1}^{L}$ is calculated relying on the velocity $\boldsymbol{v}_d$ and the coordinates of the

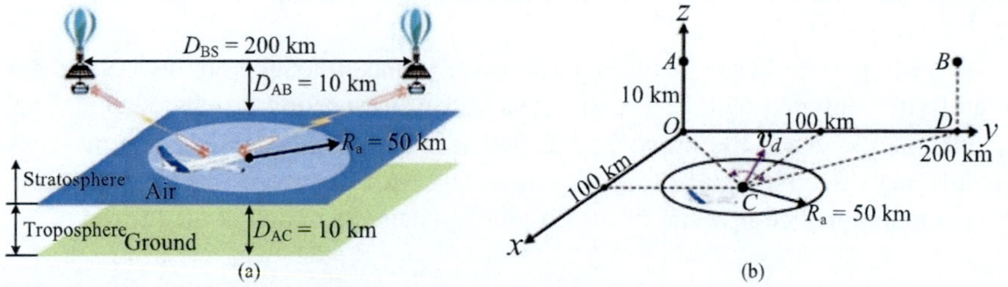

Fig. 6.11 The simulation scenario of the THz UM-MIMO-based aeronautical communication system [146]: **a** Schematic diagram. **b** Space coordinates

BSs and the aircraft. The path delay τ_l obeys the uniform distribution $\mathcal{U}(0, N_{\text{cp}}T_s)$, and the lth channel gain α_l obeys the complex Gaussian distribution $\mathcal{CN}(0, 1)$, i.e., $\sigma_\alpha^2 = 1$. Considering the positioning error in the aeronautical communication system and the shaking of the platform on which the array mounts, the rough estimates of azimuth and elevation angles $\{\widetilde{\theta}_l^{\text{BS}}, \widetilde{\varphi}_l^{\text{BS}}, \widetilde{\theta}_l^{\text{AC}}, \widetilde{\varphi}_l^{\text{AC}}\}_{l=1}^L$ can be randomly selected from the range of offset $\pm 5°$ on the true value, and the rough estimates of the Doppler shift $\{\widetilde{\psi}_{z,l}\}_{l=1}^L$ is chosen randomly from the range of offset $\pm 0.01\psi_{z,l}$ on the true value. In the fast time-varying fading channel, the correlation of the parameters between the qth and $(q+1)$th time interval is defined as $x^{[q+1]} = x^{[q]} + s_{\text{pm}}\rho_x N_{\text{C}} T_{\text{sym}}$, where x represents any one of the parameters $\{\alpha_l, \tau_l, \psi_{z,l}, \theta_l^{\text{AC}}, \varphi_l^{\text{AC}}, \theta_l^{\text{BS}}, \varphi_l^{\text{BS}}\}$, ρ_x is the changing rate associated with x, s_{pm} is the binary random variable chosen from $\{-1, 1\}$, $N_{\text{C}} = 70$, $T_{\text{sym}} = (N_{\text{cp}} + K)T_s = 2.176\,\mu\text{s}$, and the duration of a time interval is $T_{\text{TI}} = N_{\text{C}} T_{\text{sym}} = 152.32\,\mu\text{s}$. For each parameter, ρ_x can be set as $\rho_{\alpha_l} = \alpha_l^{(1)}/2$, $\rho_{\tau_l} = \tau_l^{(1)}/2$, $\rho_{\psi_{z,l}} = 0.01\psi_{z,l}^{(1)}$, $\rho_\theta^{\text{AC}} = \rho_\varphi^{\text{AC}} = \pi/4$, and $\rho_\theta^{\text{BS}} = \rho_\varphi^{\text{BS}} = \pi/12$. Based on these settings, the maximum angle change in a time interval is approximated as $\frac{\pi}{4} \times T_{\text{TI}} \approx 0.0069°$. Since this value is small enough, the assumption about the time interval in Sect. 6.5 is reasonable. For the data-aided channel tracking, we set $\varepsilon = 0.2$ and $\widetilde{K} = K/2$. Without loss of generality, the transmitted power P_l is considered to be the reciprocal of the large-scale fading gain G_l, i.e., $P_l G_l = 1$. Furthermore, we define $\sigma_\alpha^2/\sigma_n^2$ as the transmitted SNR in uplink/downlink throughout the simulations, where σ_n^2 is the noise variance.

6.8.2 Numerical Results

To evaluate the accuracy of channel parameter estimation at the transceivers, the root mean square error (RMSE) is defined as $\text{RMSE}_x = \sqrt{\mathbb{E}\left(\frac{1}{L}\|\boldsymbol{x} - \widehat{\boldsymbol{x}}\|_2^2\right)}$, where $\boldsymbol{x} \in \mathbb{R}^L$ and $\widehat{\boldsymbol{x}}$ denote the true and estimated channel parameter vector, respectively, and $\{\boldsymbol{x}\}_l$ is one of parameters $\{\theta_l^{\text{BS}}, \varphi_l^{\text{BS}}, \theta_l^{\text{AC}}, \varphi_l^{\text{AC}}, \psi_{z,l}, \tau_l\}$. The existing CE and tracking schemes [87–92, 158] are not applicable for the THz UM-MIMO-based aeronautical communication channels with fast time-varying fading characteristics. Therefore, we select the beam sweeping method (considering the triple delay-beam-Doppler squint effects) in IEEE standards 802.11ad [33] as one of the benchmarks in our simulations, where the sweeping range is $\pm 5°$ around the rough angle estimates obtained at the BS and the aircraft.

The RMSE performance for accurate angle estimation at the BS during the initial CE stage is evaluated in Fig. 6.12. Specifically, the curves labeled "no TTDU module" and "ideal TTDU module" represent the cases of transceivers without and with ideal TTDU modules, respectively, and the curves labeled "conventional scheme" represent the case of schemes directly applying the TDU-ESPRIT algorithm to estimate angles in existing mmWave systems [101], while $i_{\text{BS}}^{\max} = 1$ and $i_{\text{BS}}^{\max} = 2$ denotes the maximum number of iteration in the proposed pilot-aided iterative angle estimation algorithm. From Fig. 6.12, it can be seen that the performance of the proposed

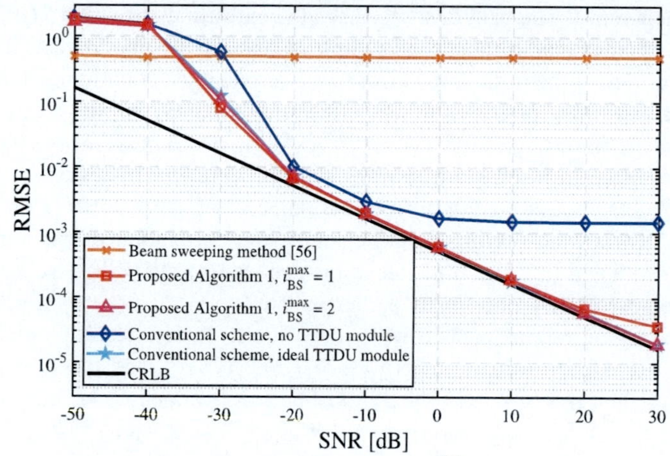

(a) azimuth angle θ^{BS}

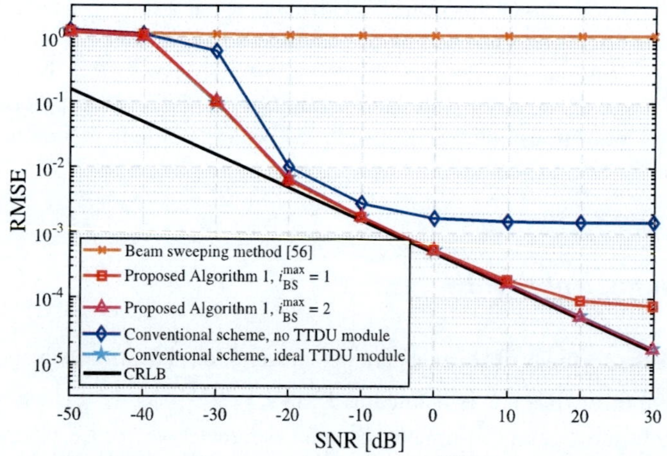

(b) elevation angle φ^{BS}

Fig. 6.12 The RMSE performance of the azimuth and elevation angles at the BS $\{\theta^{\mathrm{BS}}, \varphi^{\mathrm{BS}}\}$ in the initial angle estimation stage [146]

algorithm nearly overlaps with that of the conventional scheme considering the ideal TTDU module when $i_{\mathrm{BS}}^{\max} = 2$, and their CRLBs of the azimuth and elevation angles are very close at high SNR. Hence, the proposed algorithm only needs $i_{\mathrm{BS}}^{\max} = 2$ iterations to reach the upper performance bound of applying the ideal TTDU module to eliminate the delay-beam squint effects.

Figure 6.13 depicts the RMSE performance of the accurate angle estimation at the aircraft in the initial CE stage, where the estimation accuracy of $\{\theta_l^{\mathrm{AC}}, \varphi_l^{\mathrm{AC}}\}_{l=1}^{L}$ relies on the RMSE performance of $\{\theta_l^{\mathrm{BS}}, \varphi_l^{\mathrm{BS}}\}_{l=1}^{L}$ in Fig. 6.12. In this case, we consider "method 1" and "method 2" in our simulations, where the former adopts the estimated $\{\theta_l^{\mathrm{BS}}, \varphi_l^{\mathrm{BS}}\}_{l=1}^L$ with SNR $= -20$ dB fixed at the BS and the latter adopts the same SNR at the BS and the aircraft to estimate angles.[7] From Fig. 6.13, it can be observed that "method 2" can obtain better angle estimation performance than "method 1" when SNR is larger than -20 dB, and their RMSE gap can reach 12 dB after SNR ≥ -10 dB. This is because that "method 2" exploits the accurate angle estimates at the BS at high SNRs, thus obtaining a larger beam alignment gain than "method 1".

The RMSE performance of the accurate Doppler shift estimation in the initial channel parameter estimation stage is investigated in Fig. 6.14, where the angle estimates at the BS and the aircraft with SNR $= -20$ dB are used for the beam alignment. Here, the curve labeled "no Doppler squint" represents the case of channel model without Doppler squint effect, while the curve labeled "conventional scheme, $i_{\mathrm{do}}^{\max} = 0$" represents the case of directly applying the TLS-ESPRIT algorithm on $\boldsymbol{Y}_{\mathrm{do},l}$ in Eq. (6.25), to obtain $\widehat{\psi}_{z,l}^{(0)}$ in the proposed pilot-aided iterative Doppler shift estimation algorithm. Since the high beam alignment gain of the THz UM-MIMO array leads to high received SNR in the Doppler shift estimation, the RMSE performance can approach the CRLB at low-SNR region (even when SNR $= -100$ dB), as shown in Fig. 6.14. Moreover, the RMSE floor exists at high SNRs in the case of $i_{\mathrm{do}}^{\max} = 0$ and $i_{\mathrm{do}}^{\max} = 1$. By exploiting the designed Doppler compensation matrix, the RMSE of the proposed algorithm after $i_{\mathrm{do}}^{\max} = 2$ iterations is close to that of the conventional scheme without Doppler squint when SNR > -80 dB.

In Fig. 6.15, we provide the RMSE performance of the proposed path delay estimation scheme in the initial CE stage, where the SNR in estimation of angles and Doppler shifts is fixed with SNR $= -20$ dB and SNR $= 20$ dB, respectively, and the curves labeled "triple squint" and "no triple squint" represent the cases of channel model with and without triple squint effects, respectively. As expected, the RMSE performance of the proposed scheme under triple squint effects at SNR $= 20$ dB is better than that at SNR $= -20$ dB because of the accurate angle and Doppler estimates. Besides, due to the estimation error of the channel parameters $\{\widehat{\theta}_l^{\mathrm{BS}}, \widehat{\varphi}_l^{\mathrm{BS}}, \widehat{\theta}_l^{\mathrm{AC}}, \widehat{\varphi}_l^{\mathrm{AC}}, \widehat{\psi}_{z,l}\}_{l=1}^L$, the RMSE floor of the normalized delay $\bar{\tau}$ is unavoidable at high-SNR region.

Based on these estimated parameters, the whole initial CE stage can be measured with normalized mean square error (NMSE) [101], i.e.,

$$\mathrm{NMSE}_{\boldsymbol{H}_{\mathrm{DL}}^{[2]}} = \mathbb{E}\left(\frac{1}{L}\sum_{l=1}^{L}\left(\sum_{k=1}^{K}\left\|\boldsymbol{H}_{\mathrm{DL},l}^{[2]}[k] - \widehat{\boldsymbol{H}}_{\mathrm{DL},l}^{[2]}[k]\right\|_F^2 \Big/ \sum_{k=1}^{K}\left\|\boldsymbol{H}_{\mathrm{DL},l}^{[2]}[k]\right\|_F^2\right)\right), \tag{6.49}$$

where $\boldsymbol{H}_{\mathrm{DL},l}^{[2]}[k]$ and $\widehat{\boldsymbol{H}}_{\mathrm{DL},l}^{[2]}[k]$ denote the true channel matrix and the reconstructed channel matrix at the kth subcarrier of the second OFDM symbol in Eq. (6.4), respectively. Figure 6.16 shows the NMSE performance of the proposed scheme with different system bandwidths. It can be seen that the performance of the proposed scheme

[7] Note that the angles estimated at the BS at low SNRs are even worse than the prior rough angle estimates. To ensure the feasibility of "method 2" at low SNRs, we consider to replace the estimated $\{\widehat{\theta}_l^{\mathrm{BS}}, \widehat{\varphi}_l^{\mathrm{BS}}\}_{l=1}^L$ with the rough angle estimates $\{\widetilde{\theta}_l^{\mathrm{BS}}, \widetilde{\varphi}_l^{\mathrm{BS}}\}_{l=1}^L$ for beam alignment at the BS when SNR ≤ -20 dB.

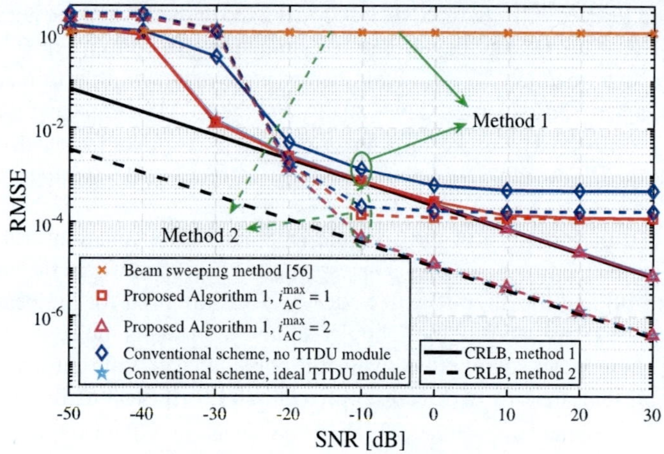

(a) azimuth angle θ^{AC}

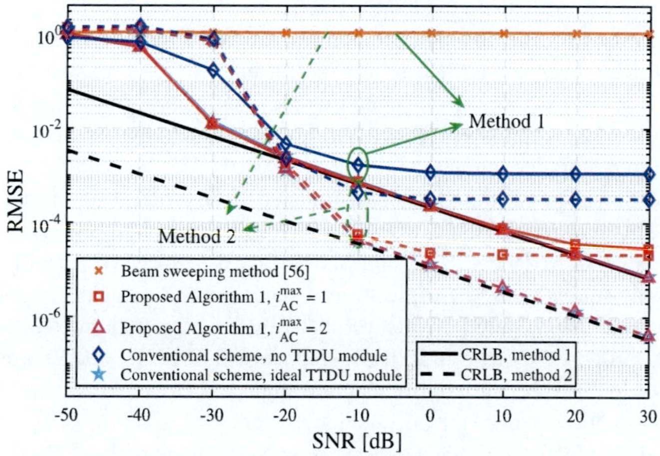

(b) elevation angle φ^{AC}

Fig. 6.13 The RMSE performance of the azimuth and elevation angles at the aircraft $\{\theta^{\mathrm{AC}}, \varphi^{\mathrm{AC}}\}$ in the initial angle estimation stage [146]

is very close to that without triple squint effects and does not deteriorate with the system bandwidth increasing from $f_s = 1$ GHz to 51 GHz.

Furthermore, the average spectrum efficiency (ASE) performance of the data transmission stage is also considered in our simulations, which is defined as

$$\mathrm{ASE} = \sum_{l=1}^{L} \left(\frac{1}{K} \sum_{k=1}^{K} \log_2 \left(1 + |h_l^{[2]}[k]|^2 / |\mathbb{E}(z_l^{[2]}[k])|^2 \right) \right), \tag{6.50}$$

where $h_l^{[2]}[k]$ and $z_l^{[2]}[k]$ denote the equivalent channel coefficient and interference plus noise at the kth subcarrier of the second OFDM symbol after beam alignment.

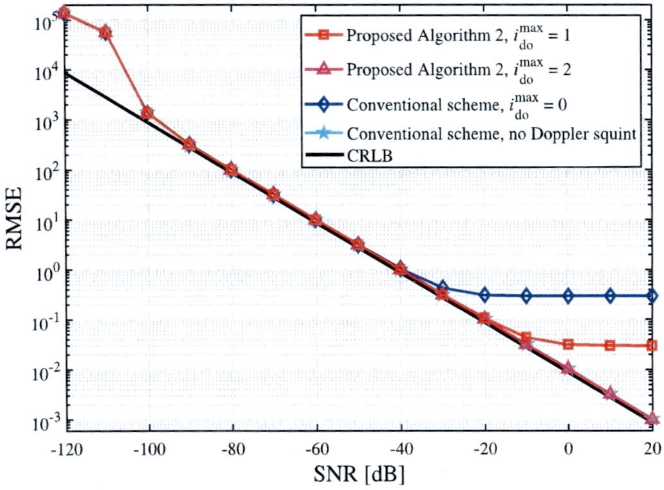

Fig. 6.14 The RMSE performance of the Doppler shift ψ_z in the initial channel parameter estimation stage [146]

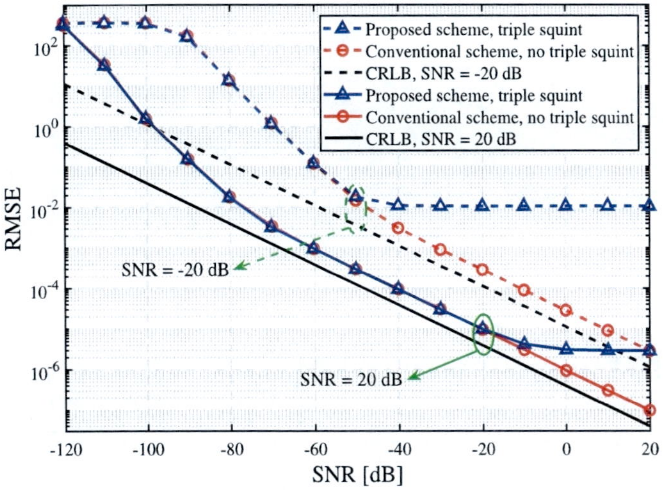

Fig. 6.15 The RMSE performance of the normalized delay $\bar{\tau}$ in the initial channel parameter estimation stage [146]

The ASE performance of the proposed scheme in different CSI conditions is compared in Fig. 6.17, where the case of the perfect CSI at the BS and the aircraft can be regarded as the upper bound of ASE performance. Regardless of the presence of the triple squint effects, the ASE performance with estimated CSI can almost reach the upper bound when SNR ≥ -14 dB. In addition, since the residual beam alignment errors caused by the delay-beam squint effect still exist in the practical GTTDU

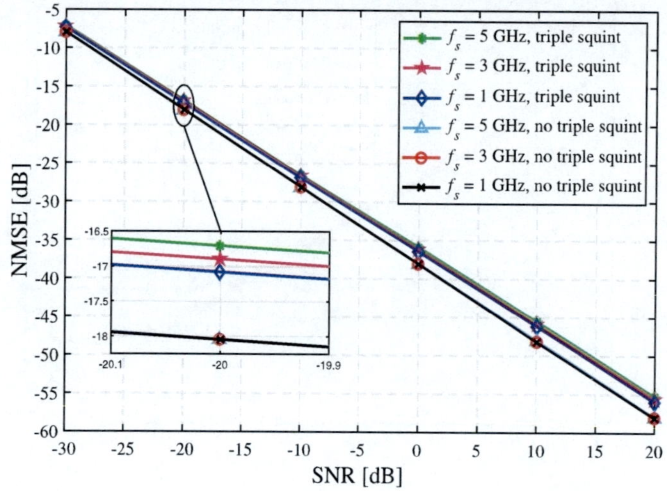

Fig. 6.16 The NMSE performance at different bandwidth $f_s = \{1, 3, 5\}$ GHz in the initial CE stage [146]

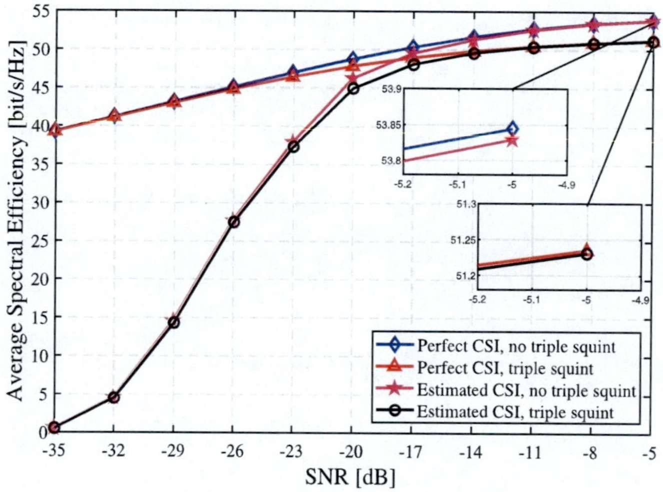

Fig. 6.17 The average spectrum efficiency at different CSI in the data transmission stage [146]

module, the ASE performance of the proposed scheme with the triple squint effects is around 2.5 bit/s/Hz lower than that of the proposed scheme without this effect at high SNRs.

The throughput of the THz UM-MIMO-based system adopting different TTDU modules is compared in Fig. 6.18, where Δf denotes the frequency spacing between adjacent subcarriers. Typically, we set $\Delta f \approx 0.488$ MHz when $f_s = 1$ GHz and $K = 2048$. In Fig. 6.18a, the throughput of the beam sweeping method and the conven-

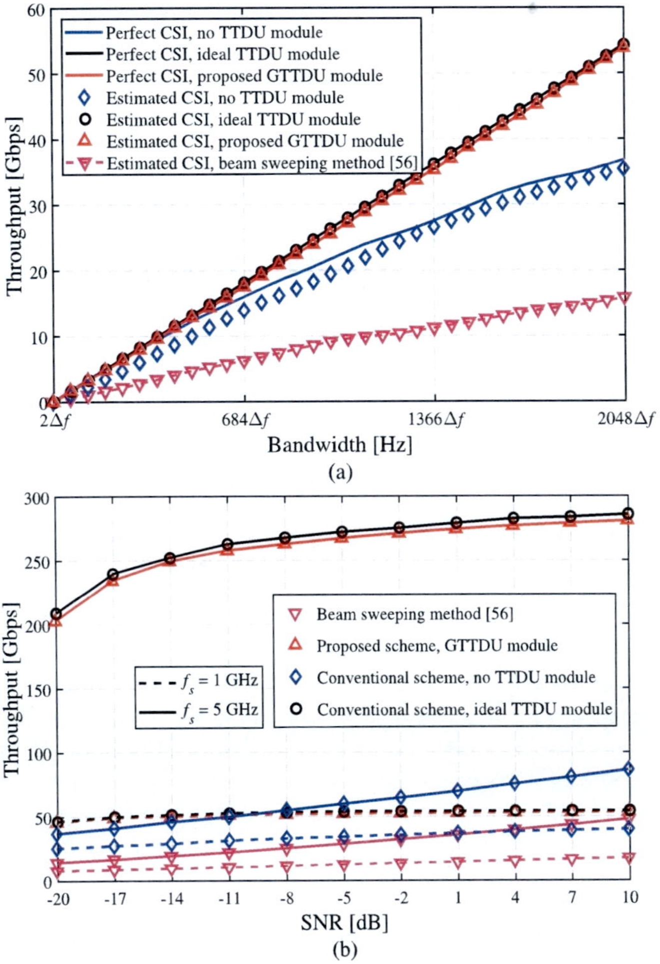

Fig. 6.18 The throughput performance of the THz UM-MIMO system with different TTDU modules [146]

tional scheme without the TTDU module is much lower than that of the other cases due to the delay-beam squint effects. In contrast, the proposed GTTDU module and the ideal TTDU module can effectively improve the throughput. Then, we consider the larger bandwidth in Fig. 6.18b, i.e., $f_s = 5\,\text{GHz}$, while the beam sweeping method and the conventional scheme without the TTDU module has limited throughput improvement. This is because no matter how large the bandwidth is, only the signals at the subcarriers near the center carrier frequency can be received with the delay-beam squint effects.

Figure 6.19 studies the throughput of THz UM-MIMO-based transceivers with different sizes of UPA, where $\text{SNR} = 10\,\text{dB}$, $f_s = 1$ GHz, and the transmitted power is

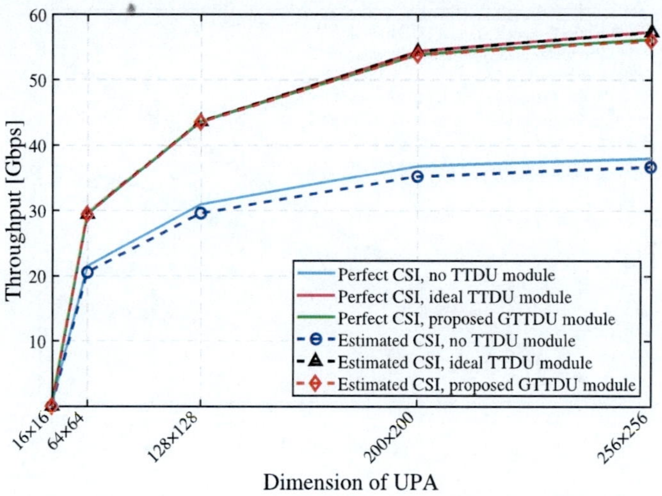

Fig. 6.19 The throughput performance of the THz UM-MIMO system with different TTDU modules [146].

the same. It shows that the 16×16 common array cannot establish effective communication links in THz aeronautical communications of ultra-long-distance. Since the THz UM-MIMO array can radiate a pencil-shaped beam with less interference, the system throughput is significantly enhanced as the size of the antenna array increases. However, according to the trend of the curve labeled "no TTDU module", increasing the array size causes a stronger delay-beam squint effects, which limits the improvement of the throughput. For the transceivers equipped with 256×256 UM-MIMO array, the performance of adopting the proposed GTTDU module is close to that of adopting the ideal TTDU module. Compared to the transceivers with 16×16 UM-MIMO array, it can achieve more than 55 Gbps throughput improvement. Therefore, it is essential to adopt UM-MIMO arrays in THz aeronautical communications to simultaneously meet the high data rate requirements of hundreds of users in the cabin.

To evaluate the performance of the proposed DADD-based channel tracking algorithm in the data transmission stage, we consider the magnitude and the NMSE of the equivalent channel, where the NMSE in the rth OFDM symbol is given by

$$\mathrm{NMSE}_{\boldsymbol{h}^{[r]}} = \mathbb{E}\left(\frac{1}{L}\sum_{l=1}^{L}\left(\left\|\boldsymbol{h}_l^{[r]} - \widehat{\boldsymbol{h}}_l^{[r]}\right\|_2^2 \Big/ \left\|\boldsymbol{h}_l^{[r]}\right\|_2^2\right)\right). \tag{6.51}$$

Figure 6.20 provides the magnitude (SNR $= -20$ dB) and the NMSE (SNR $= \{-20, -10, 0\}$ dB) of the equivalent channel tracked by the proposed DADD-based channel tracking algorithm with different time intervals, where the Turbo channel coding and quadrature phase-shift keying (QPSK) modulation are considered. It can

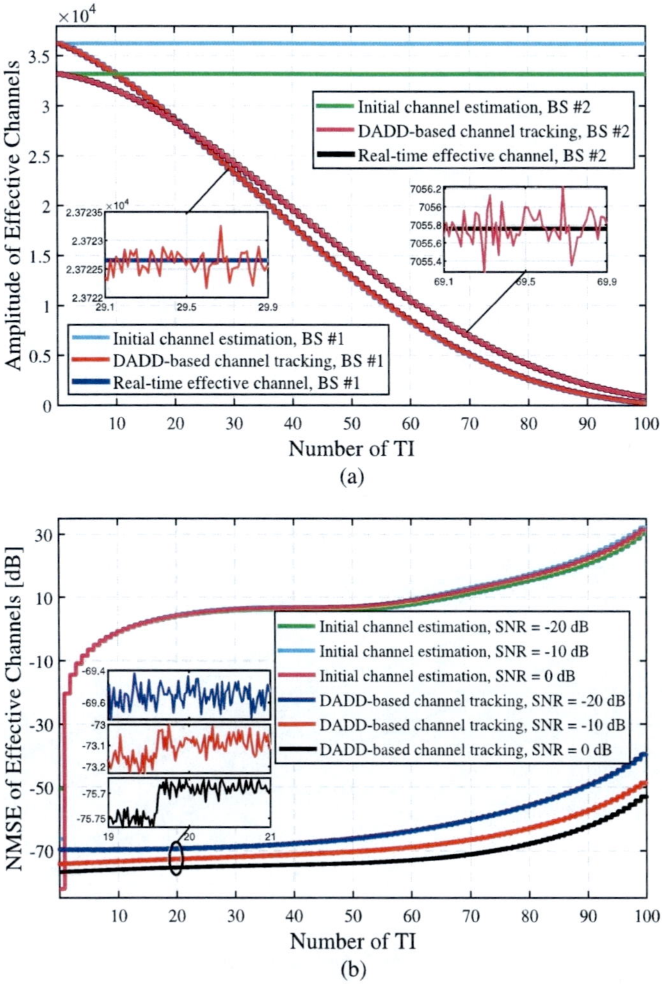

Fig. 6.20 The performance of the proposed DADD-based channel tracking algorithm in the data-aided channel tracking stage [146].

be observed from Fig. 6.20a that the magnitude of the equivalent channel decreases rapidly with time, indicating that the beam alignment gain is decreasing, while the proposed DADD-based channel tracking algorithm can keep up with the changes in real time. In Fig. 6.20b, the NMSE performance of the proposed tracking algorithm slowly deteriorates as the number of time intervals increases, but the NMSE of the benchmark without tracking process decreases rapidly within several time intervals. This demonstrates that the proposed tracking algorithm can ensure reliable THz communication link in data transmission with long duration.

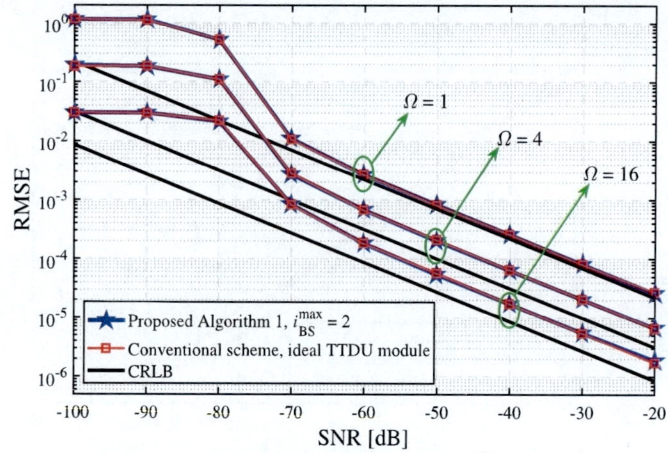

(a) azimuth angle at BS: θ^{BS}

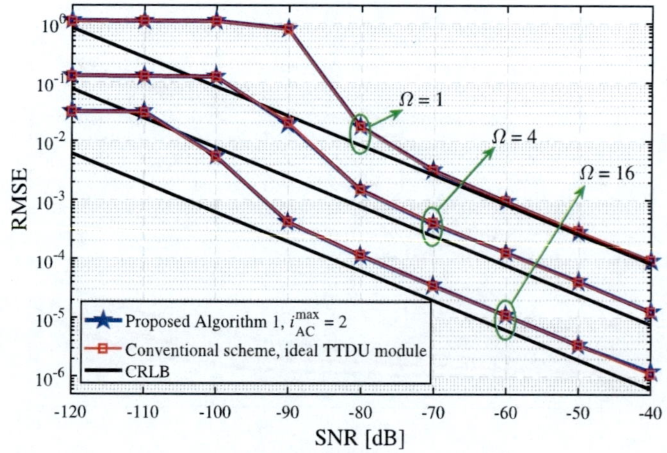

(b) azimuth angle at aircraft: θ^{AC}

Fig. 6.21 The RMSE performance of the accurate angle tracking the pilot-aided channel tracking stage [146].

The RMSE performance of the azimuth and elevation angle estimates in the proposed pilot-aided accurate angle tracking scheme with different sparse spacing $\Omega d = \{1, 4, 16\}d$ is investigated in Fig. 6.21, where the angles $\{\widehat{\theta}_l^{\text{BS}}, \widehat{\varphi}_l^{\text{BS}}\}_{l=1}^{L}$ are estimated at the BS with fixed SNR $= -60$ dB. Since the RMSE performance of the elevation angles φ^{BS} and φ^{AC} is similar to that of the azimuth angles, only θ^{BS} and θ^{AC} are considered in the simulations. As can be seen, the accuracy of angle estimation can be significantly improved by employing sparse arrays, and the RMSE performance gains are consistent with the conclusions in *Remark 2*, i.e., the performance gain of $20 \lg \Omega$ dB with sparse array spacing Ωd.

6.9 Summary

In this chapter, a low-overhead CE and tracking scheme was proposed for the THz UM-MIMO-based aeronautical communication system in SAGIN to effectively solve the triple delay-beam-Doppler squint effects, which was not considered in wireless communication systems below 6 GHz and mmWave massive MIMO systems. The proposed solution consisted of three stages, i.e., initial CE, data-aided channel tracking, and pilot-aided channel tracking. In particular, an initial THz UM-MIMO link can be established according to the rough angle estimates from navigation information, where the delay-beam squint effects on the transceiver can be significantly attenuated through the proposed GTTDU module. By applying the designed subarray selection scheme and the proposed prior-aided iterative angle estimation algorithm, the corresponding azimuth/elevation angles of the low-dimensional equivalent fully-digital array can be accurately estimated. These estimated angles are exploited to achieve high-precision beam alignment and optimize the GTTDU module at the transceivers to further eliminate the delay-beam squint effects. Then, the pilot-aided iterative Doppler shift estimation algorithm was developed to accurately estimate the Doppler shift under the Doppler squint effect, as well as the path delay and channel gain. In the data transmission stage, the proposed DADD-based channel tracking algorithm can effectively track the equivalent channel after beam alignment, ensuring reliable THz communication links. When the data-aided channel tracking failed, the angles can be re-estimated by employing the equivalent fully-digital sparse array in the pilot-aided channel tracking stage, where the angle ambiguity from the sparse array was solved according to the previously estimated angles. Finally, the CRLB of main channel parameters and simulation results demonstrated the effectiveness of the proposed schemes in the THz UM-MIMO-based aeronautical communication.

Chapter 7
Summary and Future Directions

7.1 Summary of This Book

This book aims to study the key physical layer technologies in mmWave/sub-THz ultra-massive MIMO systems that are crucial to achieve various KPIs in future 6G wireless communication networks. The book addresses technologies such as CE and beamforming for mmWave massive MIMO systems, joint AUD and CE in indoor mmWave XL-MIMO systems, and the application of UM-MIMO systems to SAGINs. Dedicated solutions are proposed to improve the transmission efficiency of communication systems, taking into account the specific characteristics of mmWave/sub-THz massive or ultra-massive MIMO in different scenarios. The main objective is to provide a solid theoretical foundation and practical implementation schemes for 6G wireless communication systems and to contribute to related wireless communication standards.

In Chap. 2, we present a closed-loop sparse CE scheme based on multi-dimensional array signal processing for multi-user wideband mmWave FD-MIMO with hybrid beamforming architecture. The scheme consists of a downlink CE stage where the BS broadcasts public signals to all users, and an uplink CE stage dedicated to each user. The downlink makes use of the large transmission power of the BS to broadcast training signals to ensure accurate estimation of AoAs by the UE, while the UE uses the estimated AoAs to design a multi-beam precoding matrix to improve the received SNR at the BS. The combining matrix designed at the BS and the UE can be utilized so that the high-dimensional hybrid MIMO array is equivalently visualized as a low-dimensional fully-digital array. The double sparsity in both the angular and delay domains in mmWave MIMO channels is utilized, and robust array signal processing techniques are applied to obtain the super-resolution estimation of horizontal/vertical AoAs and delays. Simulation results show that the proposed closed-loop channel estimation scheme outperforms its counterparts in terms of accuracy, computational efficiency, and storage requirement.

In Chap. 3, a CS-based CE scheme for mmWave FD-MIMO systems with lens antenna arrays is proposed. First of all, we introduce the theoretical principle of the

Z. Gao et al., *Millimeter-Wave/Sub-Terahertz Ultra-Massive MIMO Transmission Technology*, https://doi.org/10.1007/978-981-99-2388-5_7

lens antenna array and CS. Then, we formulate the CE problem as CS problem using the proposed pilot transmission scheme suitable for the ASN. The redundant dictionary is proposed to represent the sparse channel in a more sparse form, enhancing the robustness of the CS-based CE scheme against the number of effective paths in channels. Based on the CS theory, a pilot optimization scheme aiming at minimizing the total mutual coherence of the sensing matrix is proposed. Simulation results show that the proposed CS-based channel estimation scheme is able to estimate the channel associated with the FD lens antenna array with the simple ASN and a low pilot frequency overhead.

Chapter 4 aims at the hybrid beamforming schemes for mmWave massive MIMO systems under narrowband/broadband channels and fully-/partially-connected structures. To begin with, we introduce a hybrid beamforming scheme suitable for narrowband flat fading channels in a fully-connected mmWave hybrid massive MIMO system. The digital and analog beamforming matrices are designed separately. The MMSE criterion is used to design the digital beamformer and a closed-form solution is obtained through mathematical derivation. The analog beamformer is designed by using an over-sampled codebook, which enhances the degree-of-freedom of the analog beamforming design. Later, the proposed schemes are further extended to broadband frequency-selective channels and partially-connected structures. We present a series of hybrid beamforming algorithms and prove their rationality through mathematics. Finally, we verify the efficiency of the proposed schemes through simulation analysis and performance comparisons with traditional beamforming methods. The proposed schemes offer a good trade-off between system performance, hardware complexity, and power consumption.

In Chap. 5, we present a CS-based joint AUD and CE scheme for the indoor massive IoT access under non-stationary near-field conditions. We first use a designed mmWave XL-MIMO array to facilitate the indoor massive IoT access, which introduces the challenging near-field effect. The mmWave XL-MIMO channel model is established taking into account the near-field spherical wave propagation. The channel model exhibits SNS property because of the ultra-large aperture of the applied XL-MIMO array. The joint AUD and CE problem can be represented as MMV-CS and GMMV-CS problems, given that the channel of each subcarrier has a common support set. Next, based on the common block sparsity structure of mmWave XL-MIMO channels, the MMV-BSOMP and GMMV-BSOMP algorithms dedicated for mmWave XL-MIMO are proposed. The predefined iteration termination threshold can be used to adaptively obtain the support set (number) of active users. Simulation results indicate that the proposed CS-based algorithms provide better AUD and CE performance than other existing schemes.

In Chap. 6, a low-overhead CE and tracking scheme is proposed for the aeronautical communication systems aided by THz UM-MIMO in SAGINs. This scheme effectively addresses the triple delay-beam-Doppler squint effects of THz UM-MIMO channels, which are not considered in existing low-frequency or mmWave communication systems. During the initial CE and pilot-aided channel tracking stage, we propose a prior-assisted iterative angle and Doppler shift estimation algorithm which can estimate the horizontal/vertical angles and Doppler shift components accurately. These estimates are then used to obtain the path delays. In the data transmission stage, the proposed data-aided channel tracking algorithm ensures reliable data transmission by tracking the equivalent channel after beam alignment. The extremely large hybrid array can be equivalently visualized as a low-dimensional fully-digital sparse array using the designed sub-array selection scheme, resulting in the reduced signal dimension and more accurate angle estimation. The angle ambiguity caused by the sparse array is effectively handled by the previously estimated angles. The CRLBs of the key channel parameters are derived and simulation results verify the effectiveness of the proposed scheme.

7.2 Future Directions

Although this book has discussed the specific characteristics of various communication scenarios in mmWave/sub-THz ultra-massive MIMO systems and has proposed some key physical layer solutions, there are still many challenging problems that need to be solved in the future, which are summarized as follows.

1. Closed-loop CE scheme for mmWave FD-MIMO systems.

- How can we design an effective uplink multi-user scheduling scheme when the number of users is large?
- Considering the fast-varying mMIMO channels between aerial BS and ground users, it is necessary to consider the effect of Doppler shifts and estimate the Doppler parameters corresponding to each path.

2. MmWave lens antenna array systems.

- For multi-user wideband scenarios, how can we design a CS-based channel estimation scheme that is aware of the delay-angular double squint effects introduced by large bandwidth?
- In practice, there is still a large gap between the actual array and the theoretical derivation. To avoid the performance loss caused by the mismatch, the dictionary learning algorithm can be considered to further improve the performance of CS-based CE.

3. Hybrid beamforming design for mmWave massive MIMO systems.

- It is interesting and crucial to further extend the massive MIMO systems to ultra-massive MIMO systems by considering the beam squints, beam split, and the near-field spherical wave propagation, where the hybrid beamforming needs to be redesigned to adapt to such scenarios.
- To further reduce the computational complexity of hybrid beamforming, advanced AI algorithms can be considered, and the end-to-end training strategy can be adopted to achieve better performance at a lower complexity.

4. Massive IoT access for indoor mmWave XL-MIMO systems.

- More accurate models of near-field channels are needed, and one possible modeling approach is to use AI (deep learning) to set up complex near-field environments.
- Advanced AUD and CE algorithms, such as Bayesian message-passing algorithms, deep learning, etc., can be adopted.
- To solve the problem of intermittent LoS path in mmWave communications, reconfigurable intelligent surface (RIS) with extremely large aperture (i.e., XL-RIS) can be introduced to enhance the coverage by reflecting and/or refracting signals. The corresponding channel model and transmission strategy need further investigation.
- The radio stripe is also a promising form of antenna arrays, and it can also be used to facilitate indoor mmWave XL-MIMO systems.

5. THz UM-MIMO aeronautical communications in SAGINs.

- Hardware design is essential for THz UM-MIMO communications, such as the THz UM-MIMO systems based on Rotman lenses, and the advanced digital signal processing modules that support extremely high data rates of the order of Tbps.
- More accurate modeling of THz UM-MIMO channels, and the low complexity channel tracking and signal transmission methods under the large bandwidth and high dynamic environments are needed. Advanced modulation and coding schemes should also be integrated into the systems.
- It is necessary to consider the network optimization under various constraints, such as the location of the aerial BS, energy consumption, and inter-BS interference.

Appendix A
Proof of Equation (2.7)

Sampling the delay-domain continuous $\boldsymbol{H}_q(\tau)$ in (2.3) with the sampling period T_s yields

$$\begin{aligned}\boldsymbol{H}_q(nT_s) &= \beta_q \sum_{l=1}^{L_q} \boldsymbol{H}_{q,l}\, p\left(\tau-\tau_{q,l}\right) \circledast \sum_{n=-\infty}^{\infty} \delta\left(\tau-nT_s\right) \\ &= \beta_q \sum_{n=-\infty}^{\infty} \sum_{l=1}^{L_q} \boldsymbol{H}_{q,l}\, p\left(nT_s-\tau_{q,l}\right),\end{aligned} \tag{A.1}$$

where $\circledast$ and $\delta(\cdot)$ represent the linear convolution operation and Dirac delta function, respectively. The Fourier transform of $\boldsymbol{H}_q(nT_s)$ is then given by

$$\boldsymbol{H}_q(f) = \frac{\beta_q}{T_s} \sum_{l=1}^{L_q} \sum_{n=-\infty}^{\infty} \boldsymbol{H}_{q,l} P(f) e^{-\mathrm{j}2\pi f \tau_{q,l}} \delta\left(f-nf_s\right), \tag{A.2}$$

where $P(f)$ is the Fourier transform of $p(\tau)$. Obviously, $\boldsymbol{H}_q(f)$ exhibits periodicity with period f_s. Thus, $\boldsymbol{H}_q(f)$ within a period of $f \in [-f_s/2,\ f_s/2]$ can be expressed as

$$\boldsymbol{H}_q(f) = \frac{\beta_q}{T_s} \sum_{l=1}^{L_q} P(f) \boldsymbol{H}_{q,l} e^{-\mathrm{j}2\pi f \tau_{q,l}} \approx \frac{\beta_q}{T_s} \sum_{l=1}^{L_q} C \boldsymbol{H}_{q,l} e^{-\mathrm{j}2\pi f \tau_{q,l}}. \tag{A.3}$$

The approximation in (A.3) is valid because the PSF $p(\tau)$ is designed to realize the ideal passband filter characteristics of $P(f) = C$ for $f \in [-f_s/2,\ f_s/2]$ and $P(f) \approx 0$ for $f \notin [-f_s/2,\ f_s/2]$. For convenience, we consider $C = T_s$. Therefore, the frequency-domain channel matrix $\boldsymbol{H}_q[k]$ at the kth subcarrier, where $0 \leq k \leq K-1$, can be written as

$$\boldsymbol{H}_q[k] = \boldsymbol{H}_q\left(\frac{kf_s}{K}\right) = \beta_q \sum_{l=1}^{L_q} \boldsymbol{H}_{q,l} e^{-\mathrm{j}\frac{2\pi k f_s \tau_{q,l}}{K}}. \tag{A.4}$$

Z. Gao et al., *Millimeter-Wave/Sub-Terahertz Ultra-Massive MIMO Transmission Technology*, https://doi.org/10.1007/978-981-99-2388-5

Appendix B
The Proof of Lemma 3.2.1

For a point $(0, y, z)$ on the EM lens, $y \in \left[-D^{\mathrm{h}}/2, D^{\mathrm{h}}/2\right]$, $z \in [-D^{\mathrm{v}}/2, D^{\mathrm{v}}/2]$, let $\Phi(y, z)$ be the phase shift provided by the lens to the incident EM signal. In order to guarantee the normally-incident EM signal to be focused (constructively superposed) at the focal point $(F, 0, 0)$, $\Phi(y, z)$ should satisfy

$$\Phi(y, z) = -k_0\sqrt{F^2 + y^2 + z^2}, \tag{B.1}$$

where $k_0 = 2\pi/\lambda$ is the wavenumber of the EM signal with the wavelength λ. Then, the phase shift (including the phase shift brought by the EM lens and the free space propagation) from $(0, y, z)$ to the m-th antenna element located at $p_m = (F\cos\theta_m\cos\phi_m, F\cos\theta_m\sin\phi_m, F\sin\theta_m)$ can be represented as

$$\begin{aligned}\psi_m(y, z) &= \Phi(y, z) + k_0 d_m(y, z)\\ &= -k_0\sqrt{F^2 + y^2 + z^2}\\ &\quad + \sqrt{F^2 + y^2 + z^2 - 2yF\cos\theta_m\sin\phi_m - 2zF\sin\theta_m}\\ &\approx -k_0 y\cos\theta_m\sin\phi_m - k_0 z\sin\theta_m,\end{aligned} \tag{B.2}$$

where $d_m(y, z)$ is the distance between the point $(0, y, z)$ and p_m. Note that we apply the first order Taylor approximation under $F \gg D^{\mathrm{h}}$ and $F \gg D^{\mathrm{h}}$ to obtain (B.2).

Consider that the arrived signal with the azimuth angle θ and elevation angle ϕ at the point $(0, y, z)$ is $s(y, z)$, then we have $s(y, z) = x_0 e^{-jk_0(y\cos\theta\sin\phi + z\sin\theta)}$, where x_0 is the signal at the reference point (i.e., the center of the EM lens). Accordingly, the received signal at the m-th antenna element can be expressed as

Z. Gao et al., *Millimeter-Wave/Sub-Terahertz Ultra-Massive MIMO Transmission Technology*, https://doi.org/10.1007/978-981-99-2388-5

$$
\begin{aligned}
r_m(\theta,\phi) &= C \int_{-D^{\mathrm{v}}/2}^{D^{\mathrm{v}}/2} \int_{-D^{\mathrm{h}}/2}^{D^{\mathrm{h}}/2} s(y,z)\, e^{-j\psi_m(y,z)} \mathrm{d}y\mathrm{d}z \\
&\approx C x_0 \int_{-D^{\mathrm{v}}/2}^{D^{\mathrm{v}}/2} \int_{-D^{\mathrm{h}}/2}^{D^{\mathrm{h}}/2} e^{-jk_0(y\cos\theta\sin\phi + z\sin\theta)} e^{j(k_0 y\cos\theta_m\sin\phi_m + k_0 z\sin\theta_m)} \mathrm{d}y\mathrm{d}z \\
&= C x_0 \int_{-D^{\mathrm{v}}/2}^{D^{\mathrm{v}}/2} e^{-jk_0 z(\sin\theta - \sin\theta_m)} \mathrm{d}y \int_{-D^{\mathrm{h}}/2}^{D^{\mathrm{h}}/2} e^{-jk_0 y(\cos\theta\sin\phi - \cos\theta_m\sin\phi_m)} \mathrm{d}y \\
&= C x_0 \mathrm{sinc}\left[\tilde{D}^{\mathrm{v}}(\sin\theta_m - \sin\theta)\right] \mathrm{sinc}\left[\tilde{D}^{\mathrm{h}}(\cos\theta_m\sin\phi_m - \cos\theta\sin\phi)\right],
\end{aligned}
\tag{B.3}
$$

where C is a constant accounting for the received power. By normalizing the received signal $r_m(\theta,\phi)$, we obtain the array response $a_m(\theta,\phi) = r_m(\theta,\phi) / (C x_0)$ as shown in (3.1). This completes the proof of Lemma 3.2.1.

Appendix C
The Proof of Theorem 3.3.2

We first introduce the following two simple lemmas without proof. The interested readers are referred to [176] for more details.

Lemma C.1 *For a Hermitian matrix* $\boldsymbol{X}$ *(i.e.,* $\boldsymbol{X} = \boldsymbol{X}^{\mathrm{H}}$*), if* $[\mathbf{X}]_{i,i} > \sum_{j \neq i} \left|[\mathbf{X}]_{i,j}\right|, \forall i$*,* $\boldsymbol{X}$ *is positive definite.*

Lemma C.2 *Given a matrix* $\boldsymbol{X}$*, its Gram matrix* $\boldsymbol{X}^{\mathrm{H}}\boldsymbol{X}$ *is positive definite, if and only if* $\boldsymbol{X}$ *is of full column rank.*

For a sensing matrix $\boldsymbol{A}$ and its Gram matrix $\boldsymbol{G} = \boldsymbol{A}^{\mathrm{H}}\boldsymbol{A}$, we define $\mu_1(\boldsymbol{G})$ as the smallest integer m such that there exist m off-diagonal elements from the single row of $\boldsymbol{G}$ and the sum of their magnitudes ≥ 1. In other words, when we choose $\mu_1(\boldsymbol{G}) - 1$ off-diagonal elements from any single row of $\boldsymbol{G}$, the sum of their magnitudes is bound to be < 1. We collect the column indices of $\mu_1(\boldsymbol{G}) - 1$ off-diagonal elements from the same row of $\boldsymbol{G}$ as the ordered set $\mathcal{M}$, and construct $\boldsymbol{G}_{\mathrm{sub}} = \boldsymbol{G}_{\{\mathcal{M},\mathcal{M}\}}$ as a sub-matrix of $\boldsymbol{G}$. Note that all the diagonal elements of $\boldsymbol{G}_{\mathrm{sub}}$ are 1, so according to Lemma C.1, $\boldsymbol{G}_{\mathrm{sub}}$ is positive definite. Then, according to Lemma C.2, the sub-matrix $\boldsymbol{A}_{\mathrm{sub}} = \boldsymbol{A}_{\{:,\mathcal{M}\}}$ (obviously $\boldsymbol{G}_{\mathrm{sub}} = \boldsymbol{A}_{\mathrm{sub}}^{\mathrm{H}}\boldsymbol{A}_{\mathrm{sub}}$) is of full column rank. Since we can choose any m off-diagonal elements from any single row of $\boldsymbol{G}$ to form $\mathcal{M}$, we can draw the conclusion that any $\mu_1(\boldsymbol{G}) - 1$ columns from $\boldsymbol{A}$ are linearly independent. Given the definition of the Spark, we have Spark $(\boldsymbol{A}) > \mu_1(\boldsymbol{G}) - 1$, that is

$$\mathrm{Spark}(\boldsymbol{A}) \geq \mu_1(\boldsymbol{G}) \tag{C.1}$$

Furthermore, recall the mutual coherence $\mu(\boldsymbol{A})$ defined in Definition 3.3.2. Since the magnitudes of off-diagonal elements in $\boldsymbol{G}$ are bounded by $\mu(\boldsymbol{A})$, we should choose at least $\lceil 1/\mu(\boldsymbol{A}) \rceil$ off-diagonal elements to make the sum of their magnitudes ≥ 1. Therefore $\mu_1(\boldsymbol{G}) \geq \lceil 1/\mu(\boldsymbol{A}) \rceil$, and together with (C.1) this yields (3.8).

Z. Gao et al., *Millimeter-Wave/Sub-Terahertz Ultra-Massive MIMO Transmission Technology*, https://doi.org/10.1007/978-981-99-2388-5

Appendix D
The Proof Theorem 3.3.3

Denote the rank and the eigenvalues of $\boldsymbol{G} = \boldsymbol{A}^{\mathrm{H}}\boldsymbol{A}$ as $R \leq M$ and $\{\lambda_i\}_{i=1}^{G}$, respectively. Obviously among $\{\lambda_i\}_{i=1}^{G}$ there are only R non-zero values. Without loss of generality, let the non-zero eigenvalues be $\{\lambda_i\}_{i=1}^{R}$. Based on Cauchy-Schwarz inequality, we have

$$\mathrm{tr}^2\left(\boldsymbol{G}\right) = \left(\sum_{i=1}^{R}\lambda_i\right)^2 \leq r\sum_{i=1}^{R}\lambda_i^2 \leq M\sum_{i=1}^{G}\lambda_i^2 \tag{D.1}$$

Note that the diagonal elements of $\boldsymbol{G}$ are all 1, and thus

$$\|\boldsymbol{G}\|_F^2 = \sum_{i=1}^{G}\sum_{j=1}^{G}\left|[\boldsymbol{G}]_{i,j}\right|^2 = G + \sum_{i\neq j}\left|[\boldsymbol{G}]_{i,j}\right|^2 \tag{D.2}$$

Also note that $\|\boldsymbol{G}\|_F^2 = \sum\limits_{i=1}^{G}\lambda_i^2$, so

$$G + \sum_{i\neq j}\left|[\boldsymbol{G}]_{i,j}\right|^2 \geq \frac{\mathrm{tr}^2\left(\boldsymbol{G}\right)}{M} = \frac{G^2}{M} \tag{D.3}$$

that is

$$\sum_{i\neq j}\left|[\boldsymbol{G}]_{i,j}\right|^2 \geq \frac{G\left(G-M\right)}{M} \tag{D.4}$$

On the other hand, it is a self-evident conclusion that the average of a set of non-negative numbers cannot be greater than the largest number in the set. Apply this conclusion to the $G\left(G-1\right)$ off-diagonal elements in $\boldsymbol{G}$, we have

Z. Gao et al., *Millimeter-Wave/Sub-Terahertz Ultra-Massive MIMO Transmission Technology*, https://doi.org/10.1007/978-981-99-2388-5

$$\mu^2(\boldsymbol{A}) = \max_{i \neq j} \left|[\boldsymbol{G}]_{i,j}\right|^2 \geq \frac{1}{G(G-1)} \sum_{i \neq j} \left|[\boldsymbol{G}]_{i,j}\right|^2$$

$$\geq \frac{1}{G(G-1)} \cdot \frac{G(G-M)}{M} = \frac{G-M}{M(G-1)} \tag{D.5}$$

and then (3.9) is proved.

Appendix E
Derivation of Equation (6.4)

First, the frequency response of $\{\bar{\boldsymbol{H}}_{\mathrm{DL},l}^{(t)}(\tau)\}_{n_{\mathrm{AC}},n_{\mathrm{BS}}}$ in Eq. (6.3) can be obtained by taking the Fourier transform with respect to τ, i.e.,

$$\{\bar{\boldsymbol{H}}_{\mathrm{DL},l}^{(t)}(f_c)\}_{n_{\mathrm{AC}},n_{\mathrm{BS}}} = \sqrt{G_l}\alpha_l e^{\mathrm{j}2\pi\psi_l t} e^{-\mathrm{j}2\pi f_c \tau_l} e^{-\mathrm{j}2\pi f_c \tau_l^{[n_{\mathrm{AC}}]}} e^{-\mathrm{j}2\pi f_c \tau_l^{[n_{\mathrm{BS}}]}}, \tag{E.1}$$

where the large-scale fading coefficient G_l can be modeled as $G_l = \lambda_c^2/(4\pi D_l)^2$ according to the Friis formula [129], and D_l denotes the communication distance between the aircraft and the l-th BS. Considering the ultra-large system bandwidth f_s, the carrier frequency is denoted as $f_c = f_z + f$, where f $(-f_s/2 \le f \le f_s/2)$ denotes the baseband frequency and f_z denotes the central carrier frequency (the corresponding wavelength λ_z). After down-converting Eq. (E.1), the $(n_{\mathrm{AC}}, n_{\mathrm{BS}})$-th element of the downlink space-frequency domain baseband channel matrix $\boldsymbol{H}_{\mathrm{DL},l}^{(t)}(f)$ can be obtained as [154, 158, 177]

$$\begin{aligned}&\{\boldsymbol{H}_{\mathrm{DL},l}^{(t)}(f)\}_{n_{\mathrm{AC}},n_{\mathrm{BS}}}\\ &= \sqrt{G_l}\alpha_l e^{\mathrm{j}2\pi\psi_l t} e^{-\mathrm{j}2\pi f \tau_l} e^{\mathrm{j}\frac{2d}{\lambda_c}\left((n_{\mathrm{AC}}^{\mathrm{h}}-1)\mu_l^{\mathrm{AC}}+(n_{\mathrm{AC}}^{\mathrm{v}}-1)\nu_l^{\mathrm{AC}}\right)} e^{-\mathrm{j}\frac{2d}{\lambda_c}\left((n_{\mathrm{BS}}^{\mathrm{h}}-1)\mu_l^{\mathrm{BS}}+(n_{\mathrm{BS}}^{\mathrm{v}}-1)\nu_l^{\mathrm{BS}}\right)}.\end{aligned} \tag{E.2}$$

Due to the ultra-large bandwidth of the THz UM-MIMO system, the carrier frequency and wavelength associated with each subcarrier are quite different. Hence, the frequency-dependent Doppler shift of the k-th subcarrier can be defined as $\psi_{l,k} = \psi_{z,l} + \frac{v_l}{c}\left(\frac{k-1}{K} - \frac{1}{2}\right) f_s$, where $\psi_{z,l} = \underline{v}_l/\lambda_z$. Set the antenna spacing $d = \lambda_z/2$, the baseband frequency response in Eq. (E.2) can be further expressed as the space-frequency domain channel coefficient at the k-th subcarrier, i.e.,

$$\begin{aligned}\{\boldsymbol{H}_{\mathrm{DL},l}^{(t)}[k]\}_{n_{\mathrm{AC}},n_{\mathrm{BS}}} =& \sqrt{G_l}\alpha_l e^{\mathrm{j}2\pi\psi_{l,k} t} e^{-\mathrm{j}2\pi\left(\frac{k-1}{K}-\frac{1}{2}\right) f_s \tau_l} \left\{\boldsymbol{a}_{\mathrm{AC}}(\mu_l^{\mathrm{AC}}, \nu_l^{\mathrm{AC}}, k)\right\}_{n_{\mathrm{AC}}}\\ &\times \left\{\boldsymbol{a}_{\mathrm{BS}}^*(\mu_l^{\mathrm{BS}}, \nu_l^{\mathrm{BS}}, k)\right\}_{n_{\mathrm{BS}}},\end{aligned} \tag{E.3}$$

Z. Gao et al., *Millimeter-Wave/Sub-Terahertz Ultra-Massive MIMO Transmission Technology*, https://doi.org/10.1007/978-981-99-2388-5

where $\boldsymbol{a}_{\mathrm{AC}}(\mu_l^{\mathrm{AC}}, \nu_l^{\mathrm{AC}}, k) \in \mathbb{C}^{N_{\mathrm{AC}}}$ and $\boldsymbol{a}_{\mathrm{BS}}(\mu_l^{\mathrm{BS}}, \nu_l^{\mathrm{BS}}, k) \in \mathbb{C}^{N_{\mathrm{BS}}}$ denote the array response vector corresponding to the k-th subcarrier at the aircraft and the l-th BS, respectively. To be more specific,

$$\left\{\boldsymbol{a}_{\mathrm{AC}}(\mu_l^{\mathrm{AC}}, \nu_l^{\mathrm{AC}}, k)\right\}_{n_{\mathrm{AC}}} = e^{\mathrm{j}\left((n_{\mathrm{AC}}^{\mathrm{h}}-1)\mu_l^{\mathrm{AC}}+(n_{\mathrm{AC}}^{\mathrm{v}}-1)\nu_l^{\mathrm{AC}}\right)} e^{\mathrm{j}\left(\frac{k-1}{K}-\frac{1}{2}\right)\frac{f_s}{f_z}\left((n_{\mathrm{AC}}^{\mathrm{h}}-1)\mu_l^{\mathrm{AC}}+(n_{\mathrm{AC}}^{\mathrm{v}}-1)\nu_l^{\mathrm{AC}}\right)}, \tag{E.4}$$

$$\left\{\boldsymbol{a}_{\mathrm{BS}}(\mu_l^{\mathrm{BS}}, \nu_l^{\mathrm{BS}}, k)\right\}_{n_{\mathrm{BS}}} = e^{\mathrm{j}\left((n_{\mathrm{BS}}^{\mathrm{h}}-1)\mu_l^{\mathrm{BS}}+(n_{\mathrm{BS}}^{\mathrm{v}}-1)\nu_l^{\mathrm{BS}}\right)} e^{\mathrm{j}\left(\frac{k-1}{K}-\frac{1}{2}\right)\frac{f_s}{f_z}\left((n_{\mathrm{BS}}^{\mathrm{h}}-1)\mu_l^{\mathrm{BS}}+(n_{\mathrm{BS}}^{\mathrm{v}}-1)\nu_l^{\mathrm{BS}}\right)}. \tag{E.5}$$

Considering the N_{AC} antennas of the THz UM-MIMO array at the aircraft and the N_{BS} antennas at the l-th BS, the complete downlink space-frequency domain channel matrix at the k-th subcarrier of the n-th OFDM symbol, i.e., $\boldsymbol{H}_{\mathrm{DL},l}^{[n]}[k]$ in Eq. (6.2), can be obtained as

$$\boldsymbol{H}_{\mathrm{DL},l}^{[n]}[k] = \sqrt{G_l}\alpha_l e^{\mathrm{j}2\pi\psi_{l,k}(n-1)T_{\mathrm{sym}}} e^{-\mathrm{j}2\pi\left(\frac{k-1}{K}-\frac{1}{2}\right)f_s\tau_l} \boldsymbol{A}_{\mathrm{DL},l}[k], \tag{E.6}$$

where the downlink array response matrix $\boldsymbol{A}_{\mathrm{DL},l}[k] \in \mathbb{C}^{N_{\mathrm{AC}}\times N_{\mathrm{BS}}}$ associated with the array response vector of the aircraft and the BS can be expressed as follows

$$\begin{aligned}
\boldsymbol{A}_{\mathrm{DL},l}[k] &= \boldsymbol{a}_{\mathrm{AC}}(\mu_l^{\mathrm{AC}}, \nu_l^{\mathrm{AC}}, k)\boldsymbol{a}_{\mathrm{BS}}^{\mathrm{H}}(\mu_l^{\mathrm{BS}}, \nu_l^{\mathrm{BS}}, k) \\
&= \left(\boldsymbol{a}_{\mathrm{AC}}(\mu_l^{\mathrm{AC}}, \nu_l^{\mathrm{AC}}) \circ \bar{\boldsymbol{a}}_{\mathrm{AC}}(\mu_l^{\mathrm{AC}}, \nu_l^{\mathrm{AC}}, k)\right)\left(\boldsymbol{a}_{\mathrm{BS}}(\mu_l^{\mathrm{BS}}, \nu_l^{\mathrm{BS}}) \circ \bar{\boldsymbol{a}}_{\mathrm{BS}}(\mu_l^{\mathrm{BS}}, \nu_l^{\mathrm{BS}}, k)\right)^{\mathrm{H}} \\
&\overset{(a)}{=} \left(\boldsymbol{a}_{\mathrm{AC}}(\mu_l^{\mathrm{AC}}, \nu_l^{\mathrm{AC}})\boldsymbol{a}_{\mathrm{BS}}^{\mathrm{H}}(\mu_l^{\mathrm{BS}}, \nu_l^{\mathrm{BS}})\right) \circ \left(\bar{\boldsymbol{a}}_{\mathrm{AC}}(\mu_l^{\mathrm{AC}}, \nu_l^{\mathrm{AC}}, k)\bar{\boldsymbol{a}}_{\mathrm{BS}}^{\mathrm{H}}(\mu_l^{\mathrm{BS}}, \nu_l^{\mathrm{BS}}, k)\right).
\end{aligned} \tag{E.7}$$

Here, equation (a) holds due to the identity $(\boldsymbol{a} \circ \boldsymbol{b})(\boldsymbol{c} \circ \boldsymbol{d})^{\mathrm{H}} = (\boldsymbol{a}\boldsymbol{c}^{\mathrm{H}}) \circ (\boldsymbol{b}\boldsymbol{d}^{\mathrm{H}})$ [178].

Appendix F
Proof of Lemma (6.1)

After the compensation on the transmission delay of the antenna at the transceiver with ideal TTDU module, the $(n_{\rm AC}, n_{\rm BS})$-th element of the channel matrix $\bar{\boldsymbol{H}}_{{\rm DL},l}^{(t)}(\tau)$ in Eq. (6.3) can be written as follows

$$\{\widetilde{\bar{\boldsymbol{H}}}_{{\rm DL},l}^{(t)}(\tau)\}_{n_{\rm AC},n_{\rm BS}} = \delta(\tau - \widetilde{\tau}_l^{[n_{\rm AC}]}) \circledast \{\bar{\boldsymbol{H}}_{{\rm DL},l}^{(t)}(\tau)\}_{n_{\rm AC},n_{\rm BS}} \circledast \delta(\tau - \widetilde{\tau}_l^{[n_{\rm BS}]}), \tag{F.1}$$

where $\widetilde{\tau}_l^{[n_{\rm AC}]}$ and $\widetilde{\tau}_l^{[n_{\rm BS}]}$ denote the compensated transmission delay from the ideal TTDU module at the aircraft and the BS, respectively, i.e.,

$$\widetilde{\tau}_l^{[n_{\rm AC}]} = \left((n_{\rm AC}^{\rm h} - 1)d\sin(\widetilde{\theta}_l^{\rm AC})\cos(\widetilde{\varphi}_l^{\rm AC}) + (n_{\rm AC}^{\rm v} - 1)d\sin(\widetilde{\varphi}_l^{\rm AC})\right)/c, \tag{F.2}$$

$$\widetilde{\tau}_l^{[n_{\rm BS}]} = -\left((n_{\rm BS}^{\rm h} - 1)d\sin(\widetilde{\theta}_l^{\rm BS})\cos(\widetilde{\varphi}_l^{\rm BS}) + (n_{\rm BS}^{\rm v} - 1)d\sin(\widetilde{\varphi}_l^{\rm BS})\right)/c. \tag{F.3}$$

By taking the Fourier transform of Eq. (F.1), the baseband frequency response of $\{\widetilde{\bar{\boldsymbol{H}}}_{{\rm DL},l}^{(t)}(\tau)\}_{n_{\rm AC},n_{\rm BS}}$ can be obtained after the down-conversion, i.e.,

$$\{\widetilde{\boldsymbol{H}}_{{\rm DL},l}^{(t)}(f)\}_{n_{\rm AC},n_{\rm BS}} = e^{-{\rm j}2\pi f\widetilde{\tau}_l^{[n_{\rm AC}]}}\{\boldsymbol{H}_{{\rm DL},l}^{(t)}(f)\}_{n_{\rm AC},n_{\rm BS}} e^{-{\rm j}2\pi f\widetilde{\tau}_l^{[n_{\rm BS}]}}. \tag{F.4}$$

Moreover, the space-frequency domain channel coefficient at the k-th subcarrier can be written as follows

$$\{\widetilde{\boldsymbol{H}}_{{\rm DL},l}^{(t)}[k]\}_{n_{\rm AC},n_{\rm BS}} = \left\{\bar{\boldsymbol{a}}_{\rm AC}^{*}(\widetilde{\mu}_l^{\rm AC}, \widetilde{\nu}_l^{\rm AC}, k)\right\}_{n_{\rm AC}} \{\boldsymbol{H}_{{\rm DL},l}^{(t)}[k]\}_{n_{\rm AC},n_{\rm BS}} \left\{\bar{\boldsymbol{a}}_{\rm BS}(\widetilde{\mu}_l^{\rm BS}, \widetilde{\nu}_l^{\rm BS}, k)\right\}_{n_{\rm BS}}. \tag{F.5}$$

Considering the $N_{\rm AC}$ antennas of the THz UM-MIMO array at the aircraft and the $N_{\rm BS}$ antennas at the l-th BS, the compensated downlink space-frequency domain channel matrix at the k-th subcarrier of the n-th OFDM symbol, i.e., $\widetilde{\boldsymbol{H}}_{{\rm DL},l}^{[n]}[k]$ in Eq. (6.12), can be obtained as

$$\widetilde{\boldsymbol{H}}_{{\rm DL},l}^{[n]}[k] = \sqrt{G_l}\alpha_l e^{{\rm j}2\pi\psi_{l,k}(n-1)T_{\rm sym}} e^{-{\rm j}2\pi\left(\frac{k-1}{K}-\frac{1}{2}\right)f_s\tau_l}\widetilde{\boldsymbol{A}}_{{\rm DL},l}[k], \tag{F.6}$$

Z. Gao et al., *Millimeter-Wave/Sub-Terahertz Ultra-Massive MIMO Transmission Technology*, https://doi.org/10.1007/978-981-99-2388-5

where the compensated downlink array response matrix $\widetilde{\boldsymbol{A}}_{\mathrm{DL},l}[k]$ can be expressed as follows

$$\begin{aligned}\widetilde{\boldsymbol{A}}_{\mathrm{DL},l}[k] &= \left(\boldsymbol{a}_{\mathrm{AC}}(\mu_l^{\mathrm{AC}}, \nu_l^{\mathrm{AC}}) \circ \bar{\boldsymbol{a}}_{\mathrm{AC}}(\mu_l^{\mathrm{AC}}, \nu_l^{\mathrm{AC}}, k) \circ \bar{\boldsymbol{a}}_{\mathrm{AC}}^*(\widetilde{\mu}_l^{\mathrm{AC}}, \widetilde{\nu}_l^{\mathrm{AC}}, k)\right) \\ &\quad \times \left(\boldsymbol{a}_{\mathrm{BS}}(\mu_l^{\mathrm{BS}}, \nu_l^{\mathrm{BS}}) \circ \bar{\boldsymbol{a}}_{\mathrm{BS}}(\mu_l^{\mathrm{BS}}, \nu_l^{\mathrm{BS}}, k) \circ \bar{\boldsymbol{a}}_{\mathrm{BS}}^*(\widetilde{\mu}_l^{\mathrm{BS}}, \widetilde{\nu}_l^{\mathrm{BS}}, k)\right)^{\mathrm{H}} \\ &= \boldsymbol{A}_{\mathrm{DL},l}[k] \circ \left(\bar{\boldsymbol{a}}_{\mathrm{AC}}^*(\widetilde{\mu}_l^{\mathrm{AC}}, \widetilde{\nu}_l^{\mathrm{AC}}, k)\bar{\boldsymbol{a}}_{\mathrm{BS}}^{\mathrm{T}}(\widetilde{\mu}_l^{\mathrm{BS}}, \widetilde{\nu}_l^{\mathrm{BS}}, k)\right). \end{aligned} \tag{F.7}$$

References

1. You, X., Wang, C.X., Huang, J., Gao, X., Zhang, Z., Wang, M., Huang, Y., Zhang, C., Jiang, Y., Wang, J., et al.: Towards 6G wireless communication networks: Vision, enabling technologies, and new paradigm shifts. Science China Information Sciences **64**(1), 1–74 (2021)
2. Dang, S., Amin, O., Shihada, B., Alouini, M.S.: What should 6g be? Nature Electronics **3**(1), 20–29 (2020)
3. Heath, R.W., Gonzalez-Prelcic, N., Rangan, S., Roh, W., Sayeed, A.M.: An overview of signal processing techniques for millimeter wave MIMO systems. IEEE Journal of Selected Topics in Signal Processing **10**(3), 436–453 (2016)
4. Akyildiz, I.F., Jornet, J.M., Han, C.: Terahertz band: Next frontier for wireless communications. Physical Communication **12**, 16–32 (2014)
5. Akyildiz, I.F., Jornet, J.M.: Realizing ultra-massive MIMO (1024×1024) communication in the (0.06–10) terahertz band. Nano Communication Networks **8**, 46–54 (2016)
6. El Ayach, O., Rajagopal, S., Abu-Surra, S., Pi, Z., Heath, R.W.: Spatially sparse precoding in millimeter wave MIMO systems. IEEE Transactions on Wireless Communications **13**(3), 1499–1513 (2014)
7. Saeed, A., Gurbuz, O., Akkas, M.A.: Terahertz communications at various atmospheric altitudes. Physical Communication **41**, 101113 (2020)
8. Chen, X., Ng, D.W.K., Yu, W., Larsson, E.G., Al-Dhahir, N., Schober, R.: Massive access for 5G and beyond. IEEE Journal on Selected Areas in Communications **39**(3), 615–637 (2020)
9. Yuan, Y., Wang, S., Wu, Y., Poor, H.V., Ding, Z., You, X., Hanzo, L.: NOMA for next-generation massive IoT: Performance potential and technology directions. IEEE Communications Magazine **59**(7), 115–121 (2021)
10. Gong, S., Lu, X., Hoang, D.T., Niyato, D., Shu, L., Kim, D.I., Liang, Y.C.: Toward smart wireless communications via intelligent reflecting surfaces: A contemporary survey. IEEE Communications Surveys & Tutorials **22**(4), 2283–2314 (2020)
11. Wei, L., Hu, R.Q., Qian, Y., Wu, G.: Key elements to enable millimeter wave communications for 5G wireless systems. IEEE Wireless Communications **21**(6), 136–143 (2014)
12. Rappaport, T.S., Xing, Y., MacCartney, G.R., Molisch, A.F., Mellios, E., Zhang, J.: Overview of millimeter wave communications for fifth-generation (5G) wireless networks—With a focus on propagation models. IEEE Transactions on Antennas and Propagation **65**(12), 6213–6230 (2017)
13. Rusek, F., Persson, D., Lau, B.K., Larsson, E.G., Marzetta, T.L., Edfors, O., Tufvesson, F.: Scaling up MIMO: Opportunities and challenges with very large arrays. IEEE Signal Processing Magazine **30**(1), 40–60 (2012)

Z. Gao et al., *Millimeter-Wave/Sub-Terahertz Ultra-Massive MIMO Transmission Technology*, https://doi.org/10.1007/978-981-99-2388-5

14. Akyildiz, I.F., Han, C., Nie, S.: Combating the distance problem in the millimeter wave and terahertz frequency bands. IEEE Communications Magazine **56**(6), 102–108 (2018)
15. Hillger, P., Grzyb, J., Jain, R., Pfeiffer, U.R.: Terahertz imaging and sensing applications with silicon-based technologies. IEEE Transactions on Terahertz Science and Technology **9**(1), 1–19 (2018)
16. Gao, Z., Dai, L., Mi, D., Wang, Z., Imran, M.A., Shakir, M.Z.: MmWave massive-MIMO-based wireless backhaul for the 5G ultra-dense network. IEEE Wireless communications **22**(5), 13–21 (2015)
17. IEEE standard for high data rate wireless multi-media networks–amendment 2: 100 Gb/s wireless switched point-to-point physical layer (2017)
18. Yang, P., Xiao, Y., Xiao, M., Li, S.: 6G wireless communications: Vision and potential techniques. IEEE Network **33**(4), 70–75 (2019)
19. Akyildiz, I.F., Kak, A., Nie, S.: 6G and beyond: The future of wireless communications systems. IEEE Access **8**, 133995–134030 (2020)
20. Sesia, S., Toufik, I., Baker, M.: LTE-the UMTS long term evolution: from theory to practice. John Wiley & Sons (2011)
21. Marzetta, T.L.: Noncooperative cellular wireless with unlimited numbers of base station antennas. IEEE Transactions on Wireless Communications **9**(11), 3590–3600 (2010)
22. Huang, Y., Zhang, J., Xiao, M.: Constant envelope hybrid precoding for directional millimeter-wave communications. IEEE Journal on Selected Areas in Communications **36**(4), 845–859 (2018)
23. Larsson, E.G., Edfors, O., Tufvesson, F., Marzetta, T.L.: Massive MIMO for next generation wireless systems. IEEE Communications Magazine **52**(2), 186–195 (2014)
24. Björnson, E., Larsson, E.G., Marzetta, T.L.: Massive MIMO: Ten myths and one critical question. IEEE Communications Magazine **54**(2), 114–123 (2016)
25. Shepard, C., Yu, H., Anand, N., Li, L.E., Marzetta, T., Yang, R., Zhong, L.: Practical many-antenna base stations. In: ACM Int Conf. Mobile Computing and Networking (MobiCom), Istanbul, Turkey, p. 52 (2012)
26. Manteuffel, D., Martens, R.: Compact multimode multielement antenna for indoor UWB massive MIMO. IEEE Transactions on Antennas and Propagation **64**(7), 2689–2697 (2016)
27. Zeng, Y., Zhang, R.: Cost-effective millimeter-wave communications with lens antenna array. IEEE Wireless Communications **24**(4), 81–87 (2017)
28. Liu, A., Lau, V.K.: Impact of CSI knowledge on the codebook-based hybrid beamforming in massive MIMO. IEEE Transactions on Signal Processing **64**(24), 6545–6556 (2016)
29. Arun, V., Balakrishnan, H.: RFocus: Beamforming using thousands of passive antennas. In: 17th USENIX symposium on networked systems design and implementation (NSDI 20), pp. 1047–1061 (2020)
30. Cui, M., Wu, Z., Chen, Y., Xu, F., Dai, L.: Demo: Low-power communications based on RIS and AI for 6G. In: Proceedings of IEEE International Conference on Communications, Demo Session (2022)
31. De Carvalho, E., Ali, A., Amiri, A., Angjelichinoski, M., Heath, R.W.: Non-stationarities in extra-large-scale massive MIMO. IEEE Wireless Communications **27**(4), 74–80 (2020)
32. Marinello, J.C., Abrão, T., Amiri, A., De Carvalho, E., Popovski, P.: Antenna selection for improving energy efficiency in XL-MIMO systems. IEEE Transactions on Vehicular Technology **69**(11), 13305–13318 (2020)
33. Wireless LAN medium access control (MAC) and physical layer (PHY) specifications. Amendment 3: Enhancements for very high throughput in the 60 GHz band (2012)
34. Kokshoorn, M., Chen, H., Wang, P., Li, Y., Vucetic, B.: Millimeter wave MIMO channel estimation using overlapped beam patterns and rate adaptation. IEEE Transactions on Signal Processing **65**(3), 601–616 (2016)
35. Xiao, Z., Xia, P., Xia, X.G.: Codebook design for millimeter-wave channel estimation with hybrid precoding structure. IEEE Transactions on Wireless Communications **16**(1), 141–153 (2016)

36. Xiao, Z., Xia, P., Xia, X.G.: Channel estimation and hybrid precoding for millimeter-wave MIMO systems: A low-complexity overall solution. IEEE Access **5**, 16100–16110 (2017)
37. Lee, J., Gil, G.T., Lee, Y.H.: Channel estimation via orthogonal matching pursuit for hybrid MIMO systems in millimeter wave communications. IEEE Transactions on Communications **64**(6), 2370–2386 (2016)
38. Wu, Y., Gu, Y., Wang, Z.: Channel estimation for mmWave MIMO with transmitter hardware impairments. IEEE Communications Letters **22**(2), 320–323 (2017)
39. Zhou, Z., Fang, J., Yang, L., Li, H., Chen, Z., Li, S.: Channel estimation for millimeter-wave multiuser MIMO systems via PARAFAC decomposition. IEEE Transactions on Wireless Communications **15**(11), 7501–7516 (2016)
40. Gao, Z., Hu, C., Dai, L., Wang, Z.: Channel estimation for millimeter-wave massive MIMO with hybrid precoding over frequency-selective fading channels. IEEE Communications Letters **20**(6), 1259–1262 (2016)
41. Dong, Y., Chen, C., Yi, N., Lu, G., Jin, Y.: Channel estimation using low-resolution PSs for wideband mmWave systems. In: 2017 IEEE 85th Vehicular Technology Conference (VTC Spring), pp. 1–5. IEEE (2017)
42. Venugopal, K., Alkhateeb, A., Prelcic, N.G., Heath, R.W.: Channel estimation for hybrid architecture-based wideband millimeter wave systems. IEEE Journal on Selected Areas in Communications **35**(9), 1996–2009 (2017)
43. Rodríguez-Fernández, J., González-Prelcic, N., Venugopal, K., Heath, R.W.: Frequency-domain compressive channel estimation for frequency-selective hybrid millimeter wave MIMO systems. IEEE Transactions on Wireless Communications **17**(5), 2946–2960 (2018)
44. Ma, X., Yang, F., Liu, S., Song, J., Han, Z.: Design and optimization on training sequence for mmWave communications: A new approach for sparse channel estimation in massive MIMO. IEEE Journal on Selected Areas in Communications **35**(7), 1486–1497 (2017)
45. Zhou, Z., Fang, J., Yang, L., Li, H., Chen, Z., Blum, R.S.: Low-rank tensor decomposition-aided channel estimation for millimeter wave MIMO-OFDM systems. IEEE Journal on Selected Areas in Communications **35**(7), 1524–1538 (2017)
46. Yang, L., Zeng, Y., Zhang, R.: Channel estimation for millimeter-wave mimo communications with lens antenna arrays. IEEE Transactions on Vehicular Technology **67**(4), 3239–3251 (2017)
47. Gao, X., Dai, L., Han, S., Chih-Lin, I., Wang, X.: Reliable beamspace channel estimation for millimeter-wave massive mimo systems with lens antenna array. IEEE Transactions on Wireless Communications **16**(9), 6010–6021 (2017)
48. Ma, W., Qi, C.: Channel estimation for 3-d lens millimeter wave massive mimo system. IEEE Communications Letters **21**(9), 2045–2048 (2017)
49. He, H., Wen, C.K., Jin, S., Li, G.Y.: Deep learning-based channel estimation for beamspace mmwave massive mimo systems. IEEE Wireless Communications Letters **7**(5), 852–855 (2018)
50. Yang, J., Wen, C.K., Jin, S., Gao, F.: Beamspace channel estimation in mmwave systems via cosparse image reconstruction technique. IEEE Transactions on Communications **66**(10), 4767–4782 (2018)
51. Liang, L., Xu, W., Dong, X.: Low-complexity hybrid precoding in massive multiuser mimo systems. IEEE Wireless Communications Letters **3**(6), 653–656 (2014)
52. Alkhateeb, A., Leus, G., Heath, R.W.: Limited feedback hybrid precoding for multi-user millimeter wave systems. IEEE transactions on wireless communications **14**(11), 6481–6494 (2015)
53. Sohrabi, F., Yu, W.: Hybrid digital and analog beamforming design for large-scale antenna arrays. IEEE Journal of Selected Topics in Signal Processing **10**(3), 501–513 (2016)
54. Ni, W., Dong, X.: Hybrid block diagonalization for massive multiuser mimo systems. IEEE transactions on communications **64**(1), 201–211 (2015)
55. Venugopal, K., González-Prelcic, N., Heath, R.W.: Optimal frequency-flat precoding for frequency-selective millimeter wave channels. IEEE Transactions on Wireless Communications **18**(11), 5098–5112 (2019)

56. Alkhateeb, A., Heath, R.W.: Frequency selective hybrid precoding for limited feedback millimeter wave systems. IEEE Transactions on Communications **64**(5), 1801–1818 (2016)
57. González-Coma, J.P., González-Prelcic, N., Castedo, L., Heath, R.W.: Frequency selective multiuser hybrid precoding for mmwave systems with imperfect channel knowledge. In: 2016 50th Asilomar Conference on Signals, Systems and Computers, pp. 291–295. IEEE (2016)
58. González-Coma, J.P., Rodriguez-Fernandez, J., González-Prelcic, N., Castedo, L., Heath, R.W.: Channel estimation and hybrid precoding for frequency selective multiuser mmwave mimo systems. IEEE Journal of Selected Topics in Signal Processing **12**(2), 353–367 (2018)
59. Rodriguez-Fernández, J., Gonzálcz-Prelcic, N.: Low-complexity multiuser hybrid precoding and combining for frequency selective millimeter wave systems. In: 2018 IEEE 19th International Workshop on Signal Processing Advances in Wireless Communications (SPAWC), pp. 1–5. IEEE (2018)
60. Gao, X., Dai, L., Han, S., Chih-Lin, I., Heath, R.W.: Energy-efficient hybrid analog and digital precoding for mmwave mimo systems with large antenna arrays. IEEE Journal on Selected Areas in Communications **34**(4), 998–1009 (2016)
61. Yu, X., Shen, J.C., Zhang, J., Letaief, K.B.: Alternating minimization algorithms for hybrid precoding in millimeter wave mimo systems. IEEE Journal of Selected Topics in Signal Processing **10**(3), 485–500 (2016)
62. Park, S., Alkhateeb, A., Heath, R.W.: Dynamic subarrays for hybrid precoding in wideband mmwave mimo systems. IEEE Transactions on Wireless Communications **16**(5), 2907–2920 (2017)
63. Jiang, J.C., Wang, H.M.: Massive random access with sporadic short packets: Joint active user detection and channel estimation via sequential message passing. IEEE Transactions on Wireless Communications **20**(7), 4541–4555 (2021)
64. Jiang, S., Yuan, X., Wang, X., Xu, C., Yu, W.: Joint user identification, channel estimation, and signal detection for grant-free NOMA. IEEE Transactions on Wireless Communications **19**(10), 6960–6976 (2020)
65. Mei, Y., Gao, Z., Wu, Y., Chen, W., Zhang, J., Ng, D.W.K., Di Renzo, M.: Compressive sensing based joint activity and data detection for grant-free massive IoT access. IEEE Transactions on Wireless Communications **21**(3), 1851–1869 (2022)
66. Liu, L., Yu, W.: Massive connectivity with massive MIMO-Part I: Device activity detection and channel estimation. IEEE Transactions on Signal Processing **66**(11), 2933–2946 (2018)
67. Li, T., Wu, Y., Zheng, M., Zhang, W., Xing, C., An, J., Xia, X.G., Xiao, C.: Joint device detection, channel estimation, and data decoding with collision resolution for MIMO massive unsourced random access. IEEE Journal on Selected Areas in Communications **40**(5), 1535–1555 (2022)
68. Jiang, J.C., Wang, H.M.: Grouping-based joint active user detection and channel estimation with massive MIMO. IEEE Transactions on Wireless Communications **21**(4), 2305–2319 (2022)
69. Ke, M., Gao, Z., Wu, Y., Gao, X., Schober, R.: Compressive sensing-based adaptive active user detection and channel estimation: Massive access meets massive MIMO. IEEE Transactions on Signal Processing **68**, 764–779 (2020)
70. Chen, W., Xiao, H., Sun, L., Ai, B.: Joint activity detection and channel estimation in massive MIMO systems with angular domain enhancement. IEEE Transactions on Wireless Communications **21**(5), 2999–3011 (2022)
71. Zhang, Y., Guo, Q., Wang, Z., Xi, J., Wu, N.: Block sparse bayesian learning based joint user activity detection and channel estimation for grant-free NOMA systems. IEEE Transactions on Vehicular Technology **67**(10), 9631–9640 (2018)
72. Wang, Y., Zhu, X., Lim, E.G., Wei, Z., Jiang, Y.: Grant-free communications with adaptive period for IIoT: Sparsity and correlation based joint channel estimation and signal detection. IEEE Internet of Things Journal **9**(6), 4624–4638 (2022)
73. Du, Y., Cheng, C., Dong, B., Chen, Z., Wang, X., Fang, J., Li, S.: Block-sparsity-based multiuser detection for uplink grant-free NOMA. IEEE Transactions on Wireless Communications **17**(12), 7894–7909 (2018)

74. Wu, L., Sun, P., Wang, Z., Yang, Y.: Joint user activity identification and channel estimation for grant-free NOMA: A spatial–temporal structure-enhanced approach. IEEE Internet of Things Journal **8**(15), 12339–12349 (2021)
75. Shao, X., Chen, X., Jia, R.: A dimension reduction-based joint activity detection and channel estimation algorithm for massive access. IEEE Transactions on Signal Processing **68**, 420–435 (2019)
76. Djelouat, H., Leinonen, M., Ribeiro, L., Juntti, M.: Joint user identification and channel estimation via exploiting spatial channel covariance in mMTC. IEEE Wireless Communications Letters **10**(4), 887–891 (2021)
77. Qiang, Y., Shao, X., Chen, X.: A model-driven deep learning algorithm for joint activity detection and channel estimation. IEEE Communications Letters **24**(11), 2508–2512 (2020)
78. Mao, Z., Liu, X., Peng, M., Chen, Z., Wei, G.: Joint channel estimation and active-user detection for massive access in Internet of Things—A deep learning approach. IEEE Internet of Things Journal **9**(4), 2870–2881 (2022)
79. Ahn, Y., Kim, W., Shim, B.: Active user detection and channel estimation for massive machine-type communication: Deep learning approach. IEEE Internet of Things Journal **9**(14), 11904–11917 (2022)
80. Yang, X., Cao, F., Matthaiou, M., Jin, S.: On the uplink transmission of extra-large scale massive MIMO systems. IEEE Transactions on Vehicular Technology **69**(12), 15229–15243 (2020)
81. González-Coma, J.P., López-Martínez, F.J., Castedo, L.: Low-complexity distance-based scheduling for multi-user XL-MIMO systems. IEEE Wireless Communications Letters **10**(11), 2407–2411 (2021)
82. Hou, S., Wang, Y., Zeng, T., Wu, S.: Sparse channel estimation for spatial non-stationary massive MIMO channels. IEEE Communications Letters **24**(3), 681–684 (2019)
83. Han, Y., Li, M., Jin, S., Wen, C.K., Ma, X.: Deep learning-based FDD non-stationary massive MIMO downlink channel reconstruction. IEEE Journal on Selected Areas in Communications **38**(9), 1980–1993 (2020)
84. Chen, J., Zhang, P., Ma, N., Xu, X.: Hierarchical-block sparse Bayesian learning for spatial non-stationary massive MIMO channel estimation. IEEE Wirless Communications Letters **11**(5), 888–892 (2022)
85. Wei, X., Dai, L.: Channel estimation for extremely large-scale massive MIMO: Far-field, near-field, or hybrid-field? IEEE Communications Letters **26**(1), 177–181 (2022)
86. Cui, M., Dai, L.: Channel estimation for extremely large-scale MIMO: Far-field or near-field? IEEE Transactions on Communications **70**(4), 2663–2677 (2022)
87. Gong, Z., Jiang, F., Li, C.: Angle domain channel tracking with large antenna array for high mobility V2I millimeter wave communications. IEEE Journal of Selected Topics in Signal Processing **13**(5), 1077–1089 (2019)
88. Ma, J., Zhang, S., Li, H., Gao, F., Jin, S.: Sparse Bayesian learning for the time-varying massive MIMO channels: Acquisition and tracking. IEEE Transactions on Communications **67**(3), 1925–1938 (2018)
89. Li, M., Zhang, S., Zhao, N., Zhang, W., Wang, X.: Time-varying massive MIMO channel estimation: Capturing, reconstruction, and restoration. IEEE Transactions on Communications **67**(11), 7558–7572 (2019)
90. Qin, Q., Gui, L., Cheng, P., Gong, B.: Time-varying channel estimation for millimeter wave multiuser MIMO systems. IEEE Transactions on Vehicular Technology **67**(10), 9435–9448 (2018)
91. Cheng, L., Yue, G., Xiong, X., Liang, Y., Li, S.: Tensor decomposition-aided time-varying channel estimation for millimeter wave MIMO systems. IEEE Wireless Communications Letters **8**(4), 1216–1219 (2019)
92. Gao, X., Dai, L., Zhang, Y., Xie, T., Dai, X., Wang, Z.: Fast channel tracking for terahertz beamspace massive MIMO systems. IEEE Transactions on Vehicular Technology **66**(7), 5689–5696 (2016)

93. Peng, B., Kürner, T.: Three-dimensional angle of arrival estimation in dynamic indoor terahertz channels using a forward–backward algorithm. IEEE Transactions on Vehicular Technology **66**(5), 3798–3811 (2016)
94. Xiao, Z., He, T., Xia, P., Xia, X.G.: Hierarchical codebook design for beamforming training in millimeter-wave communication. IEEE Transactions on Wireless Communications **15**(5), 3380–3392 (2016)
95. Zhang, J., Huang, Y., Shi, Q., Wang, J., Yang, L.: Codebook design for beam alignment in millimeter wave communication systems. IEEE Transactions on Communications **65**(11), 4980–4995 (2017)
96. Tsai, Y., Zheng, L., Wang, X.: Millimeter-wave beamformed full-dimensional MIMO channel estimation based on atomic norm minimization. IEEE Transactions on Communications **66**(12), 6150–6163 (2018)
97. Hu, C., Dai, L., Mir, T., Gao, Z., Fang, J.: Super-resolution channel estimation for mmWave massive MIMO with hybrid precoding. IEEE Transactions on Vehicular Technology **67**(9), 8954–8958 (2018)
98. Yu, P., Li, W., Zhou, F., Feng, L., Yin, M., Guo, S., Gao, Z., Qiu, X.: Capacity enhancement for 5G networks using mmWave aerial base stations: Self-organizing architecture and approach. IEEE Wireless Communications **25**(4), 58–64 (2018)
99. Zoltowski, M.D., Haardt, M., Mathews, C.P.: Closed-form 2-D angle estimation with rectangular arrays in element space or beamspace via unitary ESPRIT. IEEE Transactions on Signal Processing **44**(2), 316–328 (1996)
100. Van der Veen, A.J., Vanderveen, M.C., Paulraj, A.: Joint angle and delay estimation using shift-invariance techniques. IEEE Transactions on Signal Processing **46**(2), 405–418 (1998)
101. Liao, A., Gao, Z., Wang, H., Chen, S., Alouini, M.S., Yin, H.: Closed-loop sparse channel estimation for wideband millimeter-wave full-dimensional MIMO systems. IEEE Transactions on Communications **67**(12), 8329–8345 (2019)
102. Zhao, L., Ng, D.W.K., Yuan, J.: Multi-user precoding and channel estimation for hybrid millimeter wave systems. IEEE Journal on Selected Areas in Communications **35**(7), 1576–1590 (2017)
103. Gao, Z., Dai, L., Han, S., Chih-Lin, I., Wang, Z., Hanzo, L.: Compressive sensing techniques for next-generation wireless communications. IEEE Wireless Communications **25**(3), 144–153 (2018)
104. Xiao, Z., Xia, P., Xia, X.G.: Enabling UAV cellular with millimeter-wave communication: Potentials and approaches. IEEE Communications Magazine **54**(5), 66–73 (2016)
105. Mao, J., Gao, Z., Wu, Y., Alouini, M.S.: Over-sampling codebook-based hybrid minimum sum-mean-square-error precoding for millimeter-wave 3D-MIMO. IEEE Wireless Communications Letters **7**(6), 938–941 (2018)
106. Haardt, M., Nossek, J.A.: Simultaneous Schur decomposition of several nonsymmetric matrices to achieve automatic pairing in multidimensional harmonic retrieval problems. IEEE Transactions on Signal Processing **46**(1), 161–169 (1998)
107. Vanderveen, M.C., Van der Veen, A.J., Paulraj, A.: Estimation of multipath parameters in wireless communications. IEEE Transactions on Signal Processing **46**(3), 682–690 (1998)
108. Donoho, D.L.: De-noising by soft-thresholding. IEEE Transactions on Information Theory **41**(3), 613–627 (1995)
109. Golub, G.H., Van Loan, C.F.: Matrix Computations. The Johns Hopkins Univ. Press (2013)
110. Liao, A., Gao, Z., Wu, Y., Wang, H., Alouini, M.S.: 2D unitary ESPRIT based super-resolution channel estimation for millimeter-wave massive MIMO with hybrid precoding. IEEE Access **5**, 24747–24757 (2017)
111. Sun, Y., Gao, Z., Wang, H., Shim, B., Gui, G., Mao, G., Adachi, F.: Principal component analysis-based broadband hybrid precoding for millimeter-wave massive MIMO systems. IEEE Transactions on Wireless Communications **19**(10), 6331–6346 (2020)
112. Donoho, D.L., Elad, M.: Optimally sparse representation in general (nonorthogonal) dictionaries via l_1 minimization. Proceedings of the National Academy of Sciences **100**(5), 2197–2202 (2003)

113. Candès, E.J., Tao, T.: Decoding by linear programming. IEEE Transactions on Information Theory **51**(12), 4203–4215 (2005)
114. Donoho, D.L.: Compressed sensing. IEEE Transactions on Information Theory **52**(4), 1289–1306 (2006)
115. Candès, E.J., Romberg, J., Tao, T.: Robust uncertainty principles: Exact signal reconstruction from highly incomplete frequency information. IEEE Transactions on Information Theory **52**(2), 489–509 (2006)
116. Candès, E.J., Wakin, M.B.: An introduction to compressive sampling. IEEE Signal Processing Magazine **25**(2), 21–30 (2008)
117. Welch, L.: Lower bounds on the maximum cross correlation of signals (corresp.). IEEE Transactions on Information theory **20**(3), 397–399 (1974)
118. Xia, P., Zhou, S., Giannakis, G.B.: Achieving the welch bound with difference sets. IEEE Transactions on Information Theory **51**(5), 1900–1907 (2005)
119. Elad, M.: Optimized projections for compressed sensing. IEEE Transactions on Signal Processing **55**(12), 5695–5702 (2007)
120. Zeng, Y., Yang, L., Zhang, R.: Multi-user millimeter wave mimo with full-dimensional lens antenna array. IEEE Transactions on Wireless Communications **17**(4), 2800–2814 (2018)
121. Wan, Z., Gao, Z., Shim, B., Yang, K., Mao, G., Alouini, M.S.: Compressive sensing based channel estimation for millimeter-wave full-dimensional mimo with lens-array. IEEE Transactions on Vehicular Technology **69**(2), 2337–2342 (2019)
122. Zelnik-Manor, L., Rosenblum, K., Eldar, Y.C.: Sensing matrix optimization for block-sparse decoding. IEEE Transactions on Signal Processing **59**(9), 4300–4312 (2011)
123. Zhang, X.D.: Matrix analysis and applications. Cambridge University Press (2017)
124. Choi, J., Ding, J., Le, N.P., Ding, Z.: Grant-free random access in machine-type communication: Approaches and challenges. IEEE Wireless Communications **29**(1), 151–158 (2022)
125. Ma, Y., Ma, G., Wang, N., Zhong, Z., Ai, B.: OTFS-TSMA for massive Internet of Things in high-speed railway. IEEE Transactions on Wireless Communications **21**(1), 519–531 (2022)
126. Jiao, J., Wu, S., Lu, R., Zhang, Q.: Massive access in space-based Internet of Things: Challenges, opportunities, and future directions. IEEE Wireless Communications **28**(5), 118–125 (2021)
127. Ke, M., Gao, Z., Huang, Y., Ding, G., Ng, D.W.K., Wu, Q., Zhang, J.: An edge computing paradigm for massive IoT connectivity over high-altitude platform networks. IEEE Wireless Communications **28**(5), 102–109 (2021)
128. Du, Y., Dong, B., Chen, Z., Wang, X., Liu, Z., Gao, P., Li, S.: Efficient multi-user detection for uplink grant-free NOMA: Prior-information aided adaptive compressive sensing perspective. IEEE Journal on Selected Areas in Communications **35**(12), 2812–2828 (2017)
129. Stutzman, W.L., Thiele, G.A.: Antenna theory and design. John Wiley & Sons (2012)
130. Han, Y., Jin, S., Wen, C.K., Ma, X.: Channel estimation for extremely large-scale massive MIMO systems. IEEE Wireless Communications Letters **9**(5), 633–637 (2020)
131. Iimori, H., Takahashi, T., Ishibashi, K., De Abreu, G.T.F., González, G.D., Gonsa, O.: Joint activity and channel estimation for extra-large MIMO systems. IEEE Transactions on Wireless Communications **21**(9), 7253–7270 (2022)
132. López, O.L., Kumar, D., Souza, R.D., Popovski, P., Tölli, A., Latva-aho, M.: Massive MIMO with radio stripes for indoor wireless energy transfer. IEEE Transactions on Wireless Communications **21**(9): 7088–7104(2022)
133. Shaik, Z.H., Björnson, E., Larsson, E.G.: MMSE-optimal sequential processing for cell-free massive MIMO with radio stripes. IEEE Transactions on Communications **69**(11), 7775–7789 (2021)
134. Tropp, J.A., Gilbert, A.C., Strauss, M.J.: Algorithms for simultaneous sparse approximation. Part I: Greedy pursuit. Signal processing **86**(3), 572–588 (2006)
135. Gao, Z., Dai, L., Wang, Z., Chen, S.: Spatially common sparsity based adaptive channel estimation and feedback for FDD massive MIMO. IEEE Transactions on Signal Processing **63**(23), 6169–6183 (2015)

136. Tropp, J.A., Wakin, M.B., Duarte, M.F., Baron, D., Baraniuk, R.G.: Random filters for compressive sampling and reconstruction. In: 2006 IEEE International Conference on Acoustics Speech and Signal Processing Proceedings, vol. 3, pp. 872–875. IEEE (2006)
137. Ke, M., Gao, Z., Wu, Y., Gao, X., Wong, K.K.: Massive access in cell-free massive MIMO-based Internet of Things: Cloud computing and edge computing paradigms. IEEE Journal on Selected Areas in Communications **39**(3), 756–772 (2020)
138. Han, C., Bicen, A.O., Akyildiz, I.F.: Multi-ray channel modeling and wideband characterization for wireless communications in the terahertz band. IEEE Transactions on Wireless Communications **14**(5), 2402–2412 (2014)
139. Sarieddeen, H., Alouini, M.S., Al-Naffouri, T.Y.: Terahertz-band ultra-massive spatial modulation MIMO. IEEE Journal on Selected Areas in Communications **37**(9), 2040–2052 (2019)
140. Akyildiz, I.F., Jornet, J.M., Nie, S.: A new CubeSat design with reconfigurable multi-band radios for dynamic spectrum satellite communication networks. Ad Hoc Networks **86**, 166–178 (2019)
141. Han, C., Chen, Y.: Propagation modeling for wireless communications in the terahertz band. IEEE Communications Magazine **56**(6), 96–101 (2018)
142. Han, C., Bicen, A.O., Akyildiz, I.F.: Multi-wideband waveform design for distance-adaptive wireless communications in the terahertz band. IEEE Transactions on Signal Processing **64**(4), 910–922 (2015)
143. Han, C., Akyildiz, I.F.: Distance-aware bandwidth-adaptive resource allocation for wireless systems in the terahertz band. IEEE Transactions on Terahertz Science and Technology **6**(4), 541–553 (2016)
144. He, X., Xu, X.: Physics-based prediction of atmospheric transfer characteristics at terahertz frequencies. IEEE Transactions on Antennas and Propagation **67**(4), 2136–2141 (2019)
145. Elayan, H., Amin, O., Shihada, B., Shubair, R.M., Alouini, M.S.: Terahertz band: The last piece of RF spectrum puzzle for communication systems. IEEE Open Journal of the Communications Society **1**, 1–32 (2019)
146. Liao, A., Gao, Z., Wang, D., Wang, H., Yin, H., Ng, D.W.K., Alouini, M.S.: Terahertz ultra-massive MIMO-based aeronautical communications in space-air-ground integrated networks. IEEE Journal on Selected Areas in Communications **39**(6), 1741–1767 (2021)
147. Liu, J., Shi, Y., Fadlullah, Z.M., Kato, N.: Space-air-ground integrated network: A survey. IEEE Communications Surveys & Tutorials **20**(4), 2714–2741 (2018)
148. Huang, X., Zhang, J.A., Liu, R.P., Guo, Y.J., Hanzo, L.: Airplane-aided integrated networking for 6G wireless: Will it work? IEEE Vehicular Technology Magazine **14**(3), 84–91 (2019)
149. Zhang, J., Chen, S., Maunder, R.G., Zhang, R., Hanzo, L.: Regularized zero-forcing precoding-aided adaptive coding and modulation for large-scale antenna array-based air-to-air communications. IEEE Journal on Selected Areas in Communications **36**(9), 2087–2103 (2018)
150. Zhang, J., Chen, T., Zhong, S., Wang, J., Zhang, W., Zuo, X., Maunder, R.G., Hanzo, L.: Aeronautical Ad Hoc networking for the Internet-above-the-clouds. Proceedings of the IEEE **107**(5), 868–911 (2019)
151. Chen, Z., Ma, X., Zhang, B., Zhang, Y., Niu, Z., Kuang, N., Chen, W., Li, L., Li, S.: A survey on terahertz communications. China Communications **16**(2), 1–35 (2019)
152. Ogbe, D., Love, D.J., Rebholz, M., Bidigare, T.P.: Efficient channel estimation for aerial wireless communications. IEEE Transactions on Aerospace and Electronic Systems **55**(6), 2774–2785 (2019)
153. Wang, B., Gao, F., Jin, S., Lin, H., Li, G.Y., Sun, S., Rappaport, T.S.: Spatial-wideband effect in massive MIMO with application in mmWave systems. IEEE Communications Magazine **56**(12), 134–141 (2018)
154. Wang, B., Gao, F., Jin, S., Lin, H., Li, G.Y.: Spatial-and frequency-wideband effects in millimeter-wave massive MIMO systems. IEEE Transactions on Signal Processing **66**(13), 3393–3406 (2018)
155. Krozer, V., Löffler, T., Dall, J., Kusk, A., Eichhorn, F., Olsson, R.K., Buron, J.D., Jepsen, P.U., Zhurbenko, V., Jensen, T.: Terahertz imaging systems with aperture synthesis techniques. IEEE Transactions on Microwave Theory and Techniques **58**(7), 2027–2039 (2010)

156. Chen, Y., Xiong, Y., Chen, D., Jiang, T., Ng, S.X., Hanzo, L.: Hybrid precoding for wideband millimeter wave MIMO systems in the face of beam squint. IEEE Transactions on Wireless Communications **20**(3), 1847–1860 (2020)
157. Jian, M., Gao, F., Tian, Z., Jin, S., Ma, S.: Angle-domain aided UL/DL channel estimation for wideband mmWave massive MIMO systems with beam squint. IEEE Transactions on Wireless Communications **18**(7), 3515–3527 (2019)
158. Wang, B., Jian, M., Gao, F., Li, G.Y., Lin, H.: Beam squint and channel estimation for wideband mmWave massive MIMO-OFDM systems. IEEE Transactions on Signal Processing **67**(23), 5893–5908 (2019)
159. Wang, M., Gao, F., Shlezinger, N., Flanagan, M.F., Eldar, Y.C.: A block sparsity based estimator for mmWave massive MIMO channels with beam squint. IEEE Transactions on Signal Processing **68**, 49–64 (2019)
160. Han, C., Akyildiz, I.F.: Three-dimensional end-to-end modeling and analysis for graphene-enabled terahertz band communications. IEEE Transactions on Vehicular Technology **66**(7), 5626–5634 (2016)
161. Jornet, J.M., Akyildiz, I.F.: Channel modeling and capacity analysis for electromagnetic wireless nanonetworks in the terahertz band. IEEE Transactions on Wireless Communications **10**(10), 3211–3221 (2011)
162. Yuan, H., Yang, N., Yang, K., Han, C., An, J.: Hybrid beamforming for terahertz multi-carrier systems over frequency selective fading. IEEE Transactions on Communications **68**(10), 6186–6199 (2020)
163. Hashemi, H., Chu, T.S., Roderick, J.: Integrated true-time-delay-based ultra-wideband array processing. IEEE Communications Magazine **46**(9), 162–172 (2008)
164. Lin, C., Li, G.Y., Wang, L.: Subarray-based coordinated beamforming training for mmWave and sub-THz communications. IEEE Journal on Selected Areas in Communications **35**(9), 2115–2126 (2017)
165. Han, C., Jornet, J.M., Akyildiz, I.: Ultra-massive MIMO channel modeling for graphene-enabled terahertz-band communications. In: 2018 IEEE 87th Vehicular Technology Conference (VTC Spring), pp. 1–5. IEEE (2018)
166. Yan, L., Han, C., Yuan, J.: A dynamic array-of-subarrays architecture and hybrid precoding algorithms for terahertz wireless communications. IEEE Journal on Selected Areas in Communications **38**(9), 2041–2056 (2020)
167. Kodheli, O., Lagunas, E., Maturo, N., Sharma, S.K., Shankar, B., Montoya, J.F.M., Duncan, J.C.M., Spano, D., Chatzinotas, S., Kisseleff, S., et al.: Satellite communications in the new space era: A survey and future challenges. IEEE Communications Surveys & Tutorials **23**(1), 70–109 (2020)
168. Rotman, R., Tur, M., Yaron, L.: True time delay in phased arrays. Proceedings of the IEEE **104**(3), 504–518 (2016)
169. Gao, Y., Khaliel, M., Zheng, F., Kaiser, T.: Rotman lens based hybrid analog–digital beamforming in massive MIMO systems: Array architectures, beam selection algorithms and experiments. IEEE Transactions on Vehicular Technology **66**(10), 9134–9148 (2017)
170. Wang, X., Akbarzadeh, A., Zou, L., Caloz, C.: Flexible-resolution, arbitrary-input, and tunable Rotman lens spectrum decomposer. IEEE Transactions on Antennas and Propagation **66**(8), 3936–3947 (2018)
171. Roy, R., Kailath, T.: ESPRIT-estimation of signal parameters via rotational invariance techniques. IEEE Transactions on Acoustics, speech, and signal processing **37**(7), 984–995 (1989)
172. Tse, D., Viswanath, P.: Fundamentals of wireless communication. Cambridge university press (2005)
173. Chuang, S.F., Wu, W.R., Liu, Y.T.: High-resolution AoA estimation for hybrid antenna arrays. IEEE Transactions on Antennas and Propagation **63**(7), 2955–2968 (2015)
174. Kay, S.M.: Fundamentals of statistical signal processing: Estimation theory. Prentice-Hall, Inc. (1993)
175. Stoica, P., Nehorai, A.: MUSIC, maximum likelihood, and Cramer-Rao bound. IEEE Transactions on Acoustics, speech, and signal processing **37**(5), 720–741 (1989)

176. Horn, R.A., Johnson, C.R.: Matrix analysis. Cambridge University Press (2012)
177. Lin, X., Wu, S., Kuang, L., Ni, Z., Meng, X., Jiang, C.: Estimation of sparse massive MIMO-OFDM channels with approximately common support. IEEE Communications Letters **21**(5), 1179–1182 (2017)
178. Liu, S., Trenkler, G., et al.: Hadamard, Khatri-Rao, Kronecker and other matrix products. International Journal of Information and Systems Sciences **4**(1), 160–177 (2008)